GEOMETRIC PARTIAL DIFFERENTIAL EQUATIONS AND IMAGE ANALYSIS

This book provides an introduction to the use of geometric partial differential equations in image processing and computer vision. This research area brings a number of new concepts into the field, providing a very fundamental and formal approach to image processing. State-of-the-art practical results in a large number of real problems are achieved with the techniques described in this book. Applications covered include image segmentation, shape analysis, image enhancement, and tracking.

This book will be a useful resource for researchers and practitioners. It is intended to provide information for people investigating new solutions to image processing problems as well as for people searching for existing advanced solutions.

Guillermo Sapiro is a Professor of Electrical and Computer Engineering at the University of Minnesota, where he works on differential geometry and geometric partial differential equations, both in theory and applications in computer vision, image analysis, and computer graphics.

GEOMETRIC PARTIAL DIFFERENTIAL EQUATIONS AND IMAGE ANALYSIS

GUILLERMO SAPIRO
University of Minnesota

CAMBRIDGE UNIVERSITY PRESS
Cambridge, New York, Melbourne, Madrid, Cape Town, Singapore, São Paulo, Delhi

Cambridge University Press
32 Avenue of the Americas, New York, NY 10013-2473, USA

www.cambridge.org
Information on this title: www.cambridge.org/9780521685078

First published 2001
Reprinted 2002
First paperback edition 2006

A catalog record for this publication is available from the British Library

Library of Congress Cataloging in Publication data
Sapiro, Guillermo, 1966 –
Geometric partial differential equations and image analysis / Guillermo Sapiro.
p. cm
ISBN 0-521-79075-1
1. Image analysis. 2. Differential equations, Partial. 3. Geometry, Differential. I. Title.
TA1637 .S26 2000
621.36'7 - dc21 00-040354

ISBN 978-0-521-79075-8 hardback
ISBN 978-0-521-68507-8 paperback

Transferred to digital printing 2009

Cover Figure: The original data for the image appearing on the cover was obtained from
http://www-graphics.stanford.edu/software/volpack/ and
then processed using the algorithms in this book.

To Eitan, Dalia, and our little one on the way...

They make the end of my working journey and return home something to look forward to from the moment the day starts....

Contents

List of Figures

*These figures are available in color at http://www.ece.umn.edu/users/guille/book.html

Preface

This book is an introduction to the use of geometric partial differential equations (PDEs) in image processing and computer vision. This relatively new research area brings a number of new concepts into the field, providing, among other things, a very fundamental and formal approach to image processing. State-of-the-art practical results in problems such as image segmentation, stereo, image enhancement, distance computations, and object tracking have been obtained with algorithms based on PDE's formulations.

This book begins with an introduction to classical mathematical concepts needed for understanding both the subsequent chapters and the current published literature in the area. This introduction includes basic differential geometry, PDE theory, calculus of variations, and numerical analysis. Next we develop the PDE approach, starting with curves and surfaces deforming with intrinsic velocities, passing through surfaces moving with image-based velocities, and escalating all the way to the use of PDEs for families of images. A large number of applications are presented, including image segmentation, shape analysis, image enhancement, stereo, and tracking. The book also includes some basic connections among PDEs themselves as well as connections to other more classical approaches to image processing.

This book will be a useful resource for researchers and practitioners. It is intended to provide information for people investigating new solutions to image processing problems as well as for people searching for existent advanced solutions. One of the main goals of this book is to provide a resource for a graduate course in the topic of PDEs in image processing. Exercises are provided in each chapter to help with this.

Acknowledgments

Writing the acknowledgments part of this book is a great pleasure for me. Not only because it means that I am getting close to the end of the writing process (and although writing a book is a very enjoyable experience, finishing it is a great satisfaction!), but also because it gives me the opportunity to thank a number of people who have been very influential to my career. Although a book is always covered with subjective opinions, even a book in applied mathematics and engineering like this, in the acknowledgments part, the author is officially allowed to be completely subjective. This gives to the writing of this part of the book an additional very relaxed pleasure.

This books comes after approximately 10 years of research in image processing and computer vision. My interest in image processing and computer and biological vision began when I was an undergraduate assistant at the Signal and Image Processing Lab, Department of Electrical Engineering, Technion, Haifa, Israel. This lab, as well as the Electrical Engineering Department at the Technion, provides the perfect research environment, and without any doubt, this was extremely influential in my career. I always miss that place! I was deeply influenced as well by the basic undergraduate course given by Y. Y. Zeevi on computational vision; this opened my eyes (and brain!) to this exciting area. Later, I became a graduate student in the same department. I was very lucky to be accepted by David Malah as one of his M.Sc. students. David is a TEACHER and EDUCATOR (with capital letters). David's lessons go way beyond image processing; he basically teaches his students all the basics of research (Research 101 we could call it). His students are the ones that spend the most time working for their Master's degree, but after those difficult 2 years, you realize that all of it was worth the effort. I could not continue in a better direction, and I was very lucky again when Allen Tannenbaum accepted me as his Ph.D. student.

This gave me the chance to move more toward mathematical aspects, my all-time academic love, and to learn a whole new area of science. But this is not as important as the fact that being his student gave me the opportunity to get to know a great person. Allen and I became colleagues and great friends. He was my long-distance Ph.D. advisor, and between ski, planes, Rosh HaShana in Cambridge, steaks in Montreal, all-you-can-eat restaurants in Minneapolis, and zoos in San Diego, we managed to have a lot of fun and also to do some research. We continue to have a lot of fun and sometimes to do some research as well. Both David and Allen go way beyond the basic job description of being graduate advisors, and without their continuous help and teaching, I would not be writing this book and most probably I would be doing something completely different. I learned a great deal from you, and for this I thank you both.

After getting my education at the Technion, I spent one very enjoyable year at MIT. I thank Sanjoy Mitter for inviting me and the members of the Laboratory for Information and Decisions Systems for their hospitality. From MIT I moved to Hewlett-Packard Labs in Palo Alto, California, where some of the research reported in this book was performed. I thank members of the labs for the three great years I spent there.

Getting to know great people and researchers and having the chance to work with them is without any doubt the most satisfactory part of my job. During my still short academic life, I had the great pleasure to have long collaborations with and to learn from Freddy Bruckstein, Vicent Caselles, Peter Giblin, Ron Kimmel, Jean-Michael Morel, Peter Olver, Stan Osher, and Dario Ringach. We are not just colleagues, but friends, and this makes our work much more enjoyable. Thanks for all that I learned from you. I also enjoyed shorter, but very productive, collaborations with Michael Black, David Heeger, Patrick Teo, and Brian Wandell. Parts of this book are based on material developed with these collaborators (and with David and Allen), and a big thanks goes to them.

My graduate and post-doctoral students Alberto Bartesaghi, Santiago Betelu, Marcelo Bertalmio, Do Hyun Chung, Facundo Memoli, Hong Shan Neoh, and Bei Tang, "adopted" graduate students Andres Martinez Sole, Alvaro Pardo, and Luis Vazquez, as well as students from my courses in advanced image processing are, in part, the reason for this book. I wanted them and future students to have a reference book, and I hope this is it. My graduate students have also provided a number of the images used in this book, and our joint research is reported in this book as well.

Many other people provided figures for this book, and they are explicitly acknowledged when those figures appear. They helped me to make this book more illustrative, and this is a necessary condition when writing a book in image analysis.

I started this book while enjoying a quarter off teaching at the University of Minnesota, my current institution. I thank Mos Kaveh and the people who took over my teaching responsibilities for this break and the Electrical and Computer Engineering department for supporting me in this task.

Stan Osher, Peter Olver, Ron Kimmel, Vicent Caselles, Marcelo Bertalmio, and Dario Ringach read parts of this book and provided a lot of corrections and ideas. I thank them enormously for this volunteer job.

While writing this book, I was supported by grants from the U.S. Office of Naval Research and the U.S. National Science Foundation. The support of these two agencies allowed me to spend hours, days, and months writing, and it is greatfully acknowledged. I especially thank my Program Managers John Cozzens (National Science Foundation), Dick Lau (U.S. Office of Naval Research), Wen Masters (U.S. Office of Naval Research), and Carey Schwartz (U.S. Office of Naval Research) for supporting and encouraging me from the early days of my career.

My editor, Dr. Alan Harvey, helped a lot with my first experience writing a book.

The long hours I spent writing this book I could have spent with my family. So, without their acceptance of this project I could not write the book. This is why the book is dedicated to them.

Collaborations

As mentioned above, portions of this book describe results obtained in collaboration with colleagues and friends, particularly the following:

- The work on affine curve evolution was done with Allen Tannenbaum, and Peter Olver joined us when it was generalized to other groups and dimensions.
- The work on mathematical morphology is the result of collaboration with Ronny Kimmel, Doron Shaked, and Alfred Bruckstein.
- The initial work on geodesic active contours is in collaboration with Vicent Caselles and Ronny Kimmel.
- The work on morphing contours and the work on tracking regions on level sets is in collaboration with Marcelo Bertalmio and Gregory Randall.

- The work on edge tracing is in collaboration with Luis Vazquez and Do Hyun Chung.
- The work on affine invariant detection is in collaboration with Allen Tannenbaum and Peter Olver.
- The work on robust diffusion is in collaboration with Michael Black, David Heeger, and Dave Marimont.
- The work on the introduction of prior knowledge into anisotropic diffusion started with Brian Wandell and Patrick Teo and was followed up with Steven Haker, Allen Tannenbaum, and Alvaro Pardo.
- The work on vector-valued diffusion is in part with Dario Ringach and in part with Vicent Caselles and Do Hyun Chung.
- The work on image inpainting is in collaboration with Marcelo Bertalmio, Vicent Caselles, and Coloma Ballester.
- The original work on diffusion on general manifolds by means of harmonic maps is in collaboration with Vicent Caselles and Bei Tang. It continued with Alvaro Pardo, and extensions follow with Stan Osher, Marcelo Bertalmio, Facundo Memoli, and Li-Tien Cheng.
- The work on contrast enhancement started with Vicent Caselles and then continued with Vicent Caselles, Jose Luis Lisani, and Jean Michael Morel.
- Many additional images were produced with the software package developed in collaboration with Do Hyun Chung.
- Images from their own work were provided by V. Caselles, T. Chan, R. Deriche, R. Kimmel, J. M. Morel, S. Osher, N. Paragios, L. Vese, C. K. Wong, and H. Zhao. Additional images were provided by my graduate students.

Introduction

The use of partial differential equations (PDEs) and curvature-driven flows in image analysis has become an interest-raising research topic in the past few years. The basic idea is to deform a given curve, surface, or image with a PDE, and obtain the desired result as the solution of this PDE. Sometimes, as in the case of color images, a system of coupled PDEs is used. The art behind this technique is in the design and analysis of these PDEs.

Partial differential equations can be obtained from variational problems. Assume a variational approach to an image processing problem formulated as

$$\arg\{\mathrm{Min}_I \mathcal{U}(u)\},$$

where $\mathcal{U}$ is a given energy computed over the image (or surface) I. Let $\mathcal{F}(\Phi)$ denote the Euler derivative (first variation) of $\mathcal{U}$. Because under general assumptions a necessary condition for I to be a minimizer of $\mathcal{U}$ is that $\mathcal{F}(I) = 0$, the (local) minima may be computed by means of the steady-state solution of the equation

$$\frac{\partial I}{\partial t} = \mathcal{F}(I),$$

where t is an artificial time-marching parameter. PDEs obtained in this way have already been used for quite some time in computer vision and image processing, and the literature is large. The most classical example is the Dirichlet integral,

$$\mathcal{U}(I) = \int |\nabla I|^2(x)\,\mathrm{d}x,$$

which is associated with the linear heat equation

$$\frac{\partial I}{\partial t}(t, x) = \Delta I(x).$$

More recently, extensive research is being done on the direct derivation of evolution equations that are not necessarily obtained from the energy approaches. Both types of PDEs are studied in this book.

Clearly, when introducing a new approach to a given research area, one must justify its possible advantages. Using partial differential equations and curve/surface flows in image analysis leads to modeling images in a continuous domain. This simplifies the formalism, which becomes grid-independent and isotropic. The understanding of discrete local nonlinear filters is facilitated when one lets the grid mesh tend to zero and, thanks to an asymptotic expansion, one rewrites the discrete filter as a partial differential operator.

Conversely, when the image is represented as a continuous signal, PDEs can be seen as the iteration of local filters with an infinitesimal neighborhood. This interpretation of PDEs allows one to unify and classify a number of the known iterated filters as well as to derive new ones. Actually, we can classify all the PDEs that satisfy several stability requirements for image processing such as locality and causality [5].

Another important advantage of the PDE approach is the possibility of achieving high speed, accuracy, and stability with the help of the extensive available research on numerical analysis. Of course, when considering PDEs for image processing and numerical implementations, we are dealing with derivatives of nonsmooth signals, and the right framework must be defined. The theory of *viscosity solutions* provides a framework for rigorously using a partial differential formalism, in spite of the fact that the image may be not smooth enough to give a classical sense to derivatives involved in the PDE. Last, but not least, this area has a unique level of formal analysis, giving the possibility of providing not only successful algorithms but also useful theoretical results such as existence and uniqueness of solutions.

Ideas on the use of PDEs in image processing go back at least to Gabor [146] and, a bit more recently, to Jain [196]. However, the field really took off thanks to the independent works of Koenderink [218] and Witkin [413]. These researchers rigorously introduced the notion of *scale space*, that is, the representation of images simultaneously at multiple scales. Their seminal contribution is, to a large extent, the basis of most of the research in PDEs for image processing. In their work, the multiscale image representation is obtained by Gaussian filtering. This is equivalent to deforming the original image by means of the classical heat equation, obtaining in this way an isotropic diffusion flow. In the late 1980s, Hummel [191] noted that the heat flow is not the only parabolic PDE that can be used to create a scale space, and indeed he argued that an evolution equation that satisfies the maximum principle will define a scale space as well (all these concepts will be described in this book). Maximum principle appears to be a natural mathematical translation of *causality*. Koenderink once again made a major contribution into the PDE arena (this time probably involuntarily, as the consequences were not clear at all in his original formulation), when he suggested adding a thresholding operation to the process of Gaussian filtering. As later suggested by Osher and his colleagues and as proved by a number of groups, this leads to a geometric PDE, actually, one of the most famous ones: curvature motion.

The work of Perona and Malik [310] on anisotropic diffusion has been one of the most influential papers in the area. They proposed replacing Gaussian smoothing, equivalent to isotropic diffusion by means of the heat flow, with a selective diffusion that preserves edges. Their work opened a number of theoretical and practical questions that continue to occupy the PDE image processing community, e.g., Refs. [6 and 324]. In the same framework, the seminal works of Osher and Rudin on shock filters [293] and Rudin et al. [331] on total variation decreasing methods explicitly stated the importance and the need for understanding PDEs for image processing applications. At approximately the same time, Price et al. published a very interesting paper on the use of Turing's reaction-diffusion theory for a number of image processing problems [319]. Reaction-diffusion equations were also suggested to create artificial textures [394, 414]. In Ref. [5] the authors showed that a number of basic axioms lead to basic and fundamental PDEs.

Many of the PDEs used in image processing and computer vision are based on moving curves and surfaces with curvature-based velocities. In this area, the level-set numerical method developed by Osher and Sethian [294] was very influential and crucial. Early developments on this idea were provided by Ohta et al. [274], and their equations were first suggested for shape analysis in computer vision in Ref. [204]. The basic idea is to represent the deforming curve, surface, or image as the level set of a higher dimensional hypersurface. This technique not only provides more accurate numerical implementations but also solves topological issues that were previously very difficult to treat. The representation of objects as level sets (zero sets) is of course not completely new to the computer vision and image processing communities, as it is one of the fundamental techniques in mathematical morphology [356]. Considering the image itself as a collection of its level sets and not just as the level set of a higher dimensional function is a key concept in the PDE community [5].

Other works, such as the segmentation approach of Mumford and Shah [265] and the snakes of Kass et al. [198] have been very influential in the PDE community as well.

It should be noted that a number of the above approaches rely quite heavily on a large number of mathematical advances in differential geometry for curve evolution [162] and in viscosity solution theory for curvature motion (see, e.g., Evans and Spruck [126].)

Of course, the frameworks of PDEs and geometry-driven diffusion have been applied to many problems in image processing and computer vision since the seminal works mentioned above. Examples include continuous mathematical morphology, invariant shape analysis, shape from shading, segmentation, tracking, object detection, optical flow, stereo, image denoising, image sharpening, contrast enhancement, and image quantization. Many of these contributions are discussed in this book.

This book provides the basic mathematical background necessary for understanding the literature in PDEs applied to image analysis. Fundamental topics such as differential geometry, PDEs, calculus of variations, and numerical analysis are covered. Then the basic concepts and applications of surface evolution theory and PDEs are presented.

It is technically impossible to cover in a single book all the great literature in the area, especially when the area is still very active. This book is based on the author's own experience and view of the field. I apologize in advance to those researchers whose outstanding contributions are not covered in this book (maybe there is a reason for a sequel!). I expect that the reader of this book will be prepared to continue reading the abundant literature related to PDEs. Important sources of literature are the excellent collection of papers in the book edited by Romeny [324], the book by Guichard and Morel [168], which contains an outstanding description of the topic from the point of view of iterated infinitesimal filters, Sethian's book on level sets [361], which are covered in a very readable and comprehensive form, Osher's long-expected book (hopefully ready soon; until then see the review paper in Ref. [290]), Lindeberg's book, a classic in scale-space theory [242], Weickert's book on anisotropic diffusion in image processing [404], Kimmel's lecture notes [207], Toga's book on brain warping [389], which includes a number of PDE-based algorithms for this, the special issue (March 1998) of the *IEEE Transactions on Image Processing* (March 1998), the special issues of the *Journal of Visual Communication and Image Representation* (April 2000 and 2001, to appear), a series of Special Sessions at a number of IEEE International Conferences on Image Processing (ICIP 95, 96, 97), and the Proceedings of the Scale Space Workshop (June 1997 and September 1999). Finally, additional information for this book can be found in my home page (including the pages with color figures) at http://www.ece.umn.edu/users/guille/book.html.

Enjoy!

CHAPTER ONE

Basic Mathematical Background

The goal of this chapter is twofold: first, to provide the basic mathematical background needed to read the rest of this book, and second, to give the reader the basic background and motivation to learn more about the topics covered in this chapter by use of, for example, the referenced books and papers. This background is necessary to better prepare the reader to work in the area of partial differential equations (PDEs) applied to image processing and computer vision. Topics covered include differential geometry, PDEs, variational formulations, and numerical analysis. Extensive treatment on these topics can be found in the following books, which are considered essential for the shelves of everybody involved in this topic:

1. Guggengheimer's book on differential geometry [166]. This is one of the few simple-to-read books that covers affine differential geometry, Cartan moving frames, and basic Lie group theory. A very enjoyable book.
2. Spivak's "encyclopedia" on differential geometry [374]. Reading any of the comprehensive five volumes is a great pleasure. The first volume provides the basic mathematical background, and the second volume contains most of the basic differential geometry needed for the work described in this book. The very intuitive way Spivak writes makes this book a great source for learning the topic.
3. DoCarmo's book on differential geometry [56]. This is a very formal presentation of the topic, and one of the classics in the area.
4. Blaschke's book on affine differential geometry [39]. This is the source of basic affine differential geometry. A few other books have been published, but this is still very useful, and may be the most useful of all. Unfortunately, it is in German. A translation of some of the parts of the book appears in Ref. [53]. Be aware that this translation

contains a large number of errors. I suggest you check with the original every time you want to be sure about a certain formula.

5. Cartan's book on moving frames [61]. What can I say, this is a must. It is comprehensive coverage of the moving-frames theory, including the projective case, which is not covered in this book (projective differential geometry can be found in Ref. [412]). If you want to own this book, ask any French mathematician and he or she will point you to a place in Paris where you can buy it (and all the rest of the classical French literature). And if you want to learn about the recent developments in this theory, read the recent papers by Fels and Olver [140, 141].
6. Olver's books on Lie theory and differential invariants [281, 283]. A comprehensive coverage of the topic by one of the leaders in the field.
7. Many books have been written on PDEs. Basic concepts can be found in almost any book on applied mathematics. I strongly recommend the relatively new book by Evans [125] and the classic book by Protter and Weinberger for the maximum principle [321].
8. For numerical analysis, an almost infinite number of books have been published. Of special interest for the topics of this book are the books by Sod [371] and LeVeque [240]. As mentioned in the Introduction, Sethian's book [361] is also an excellent source of the basic numerical analysis needed to implement many of the equations discussed in this book. We are all expecting Osher's book as well, so keep an eye open, and, until then, check his papers at the website given in Ref. [290].
9. For applied mathematics in general and calculus of variations in particular, I strongly suggest looking at the classics, the books of Strang [375] and Courant and Hilbert [104].

1.1. Planar Differential Geometry

To start the mathematical background, basic concepts on planar differential geometry are presented. A planar curve, which can be considered as the boundary of a planar object, is given by a map from the real line into the real plane. More formally, a map $\mathcal{C}(p): [a, b] \in \mathbb{R} \to \mathbb{R}^2$ defines a planar curve, where p parameterizes the curve. For each value $p_0 \in [a, b]$, we obtain a point $\mathcal{C}(p_0) = [x(p_0), y(p_0)]$ on the plane.

If $\mathcal{C}(a) = \mathcal{C}(b)$, the curve is said to be a closed curve. If there exists at least one pair of parameters $p_0 \neq p_1$, $p_0, p_1 \in (a, b)$, such that $\mathcal{C}(p_0) = \mathcal{C}(p_1)$, then the curve has a self-intersection. Otherwise, the curve is said to be a simple curve. Throughout this section, we will assume that the curve is at least two times differentiable.

Up to now, the parameter p has been arbitrary. Basically, p tells us the velocity at which the curve travels. This velocity if given by the tangent vector

$$\frac{\partial C}{\partial p}.$$

We now search for a very particular parameterization, denoted as Euclidean arc length(s), such that this vector is always a unit vector,

$$\left\| \frac{\partial C}{\partial p} \right\| = 1,$$

where $\|\cdot\| = \langle \cdot, \cdot \rangle^{1/2}$ is the classical Euclidean length. In terms of the arc length, the curve is not defined anymore on the interval $[a, b]$ but on some interval $[0, L]$, where L is the (Euclidean) length of the curve. The arc length is unique (up to a constant), and is obtained by means of the relation

$$\frac{dC}{ds} = \frac{dC}{dp}\frac{dp}{ds},$$

which leads to

$$\frac{ds}{dp} = \left[\left(\frac{dx}{dp}\right)^2 + \left(\frac{dy}{dp}\right)^2 \right]^{1/2}.$$

We should note that throughout this book we consider only rectifiable curves. These are curves with a finite length between every two points. This is also equivalent to saying that the functions $x(p)$ and $y(p)$ have bounded variation.

From the definition of arc length, the (Euclidean) length of a curve between two points $C(p_0)$ and $C(p_1)$ is then given by

$$L(p_0, p_1) = \int_{p_0}^{p_1} \left[\left(\frac{dx}{dp}\right)^2 + \left(\frac{dy}{dp}\right)^2 \right]^{1/2} dp = \int_{s(p_0)}^{s(p_1)} ds.$$

Euclidean Curvature

The Euclidean arc length is one of the two basic concepts in planar differential geometry. The second one is that of curvature, which we proceed to define now.

The condition for the arc-length parameterization means that the inner product of the tangent $\mathcal{C}_s$ with itself is a constant, equal to one (throughout this book, subscripts indicate derivatives):

$$\langle \mathcal{C}_s, \mathcal{C}_s \rangle = 1.$$

Computing derivatives, we obtain

$$\langle \mathcal{C}_s, \mathcal{C}_{ss} \rangle = 0.$$

The first and the second derivatives, according to arc length, are then vectors perpendicular to each other. Ignoring for a moment the sign, we can define the Euclidean curvature κ as the absolute value of the normal vector $\mathcal{C}_{ss}$:

$$\kappa := \|\mathcal{C}_{ss}\|. \tag{1.1}$$

If $\vec{\mathcal{T}}$ and $\vec{\mathcal{N}}$ stand for the unit Euclidean tangent and the unit Euclidean normal, respectively ($\vec{\mathcal{T}} \perp \vec{\mathcal{N}}$), then (now κ has the sign back)

$$\frac{\mathrm{d}\mathcal{C}}{\mathrm{d}s} = \vec{\mathcal{T}},$$

$$\frac{\mathrm{d}^2\mathcal{C}}{\mathrm{d}s^2} = \kappa\vec{\mathcal{N}},$$

and from this we obtain the Frenet equations:

$$\frac{\mathrm{d}\vec{\mathcal{T}}}{\mathrm{d}s} = \kappa\vec{\mathcal{N}},$$

$$\frac{\mathrm{d}\vec{\mathcal{N}}}{\mathrm{d}s} = -\kappa\vec{\mathcal{T}}.$$

Many other definitions of curvature, all leading of course to the same concept, exist, and all of them can be derived from each other. For example, if θ stands for the angle between $\vec{\mathcal{T}}$ and the x axis, then

$$\kappa = \frac{\mathrm{d}\theta}{\mathrm{d}s}.$$

This is easy to show:

$$\frac{\mathrm{d}\vec{\mathcal{T}}}{\mathrm{d}s} = \frac{\mathrm{d}(\cos\theta, \sin\theta)}{\mathrm{d}s}$$

$$= \frac{\mathrm{d}\theta}{\mathrm{d}s}(-\sin\theta, \cos\theta) = \frac{\mathrm{d}\theta}{\mathrm{d}s}\vec{\mathcal{N}},$$

and the result follows from the Frenet equations.

The curvature $\kappa(s)$ at a given point $\mathcal{C}(s)$ is also the inverse of the radius of the disk that best fits the curve at $\mathcal{C}(s)$. Best fit means that the disk is tangent to the curve at $\mathcal{C}(s)$ (and therefore its center is on the normal direction). This is called the osculating circle.

A curve is not always given by an explicit representation of the form $\mathcal{C}(p)$. In many cases, as we will see later in this book, a curve is given in implicit form as the level set of a two-dimensional (2D) function $u(x, y) : \mathbb{R}^2 \to \mathbb{R}$, that is,

$$\mathcal{C} \equiv \{(x, y) : u(x, y) = 0\}.$$

It is important then to be able to compute the curvature of $\mathcal{C}$ given in this form. It is possible to show that

$$\kappa = \frac{u_{xx}u_y^2 - 2u_x u_y u_{xy} + u_{yy}u_x^2}{\left(u_x^2 + u_y^2\right)^{3/2}}.$$

Basically, this result can easily be obtained from the following simple facts:

1. The unit normal $\vec{\mathcal{N}}$ is perpendicular to the level sets, and

$$\vec{\mathcal{N}} = +(-)\frac{\nabla u}{\|\nabla u\|}, \quad (1.2)$$

   where the sign depends on the direction selected for $\vec{\mathcal{N}}$. This follows from the definition of the gradient vector

$$\nabla u := \frac{\partial u}{\partial x}\vec{x} + \frac{\partial u}{\partial y}\vec{y}.$$

   Of course, the tangent $\vec{\mathcal{T}}$ to the curve $\mathcal{C}$ is also tangent to the level sets.
2. If $\vec{\mathcal{N}} = (n_1, n_2)$, then $\kappa = \mathrm{d}n_1/\mathrm{d}x + \mathrm{d}n_2/\mathrm{d}y$.

Curve Representation by Means of Curvature

A curve is uniquely represented, up to a rotation and a translation, by the function $\kappa(s)$, that is, by its curvature as a function of the arc length. This is a very important property, which means that the curvature is invariant to Euclidean motions. In other words, two curves obtained from each other by a rotation and a translation have exactly the same curvature function $\kappa(s)$. Moreover, a curve $\mathcal{C} = (x, y)$ can be reconstructed from the curvature by the following equations:

$$x = x_0 + \cos\alpha \int_{\theta_0}^{\theta} \frac{\cos(\theta - \theta_0)}{\kappa(\theta)} \mathrm{d}\theta,$$

$$y = y_0 + \cos\beta \int_{\theta_0}^{\theta} \frac{\sin(\theta - \theta_0)}{\kappa(\theta)} \mathrm{d}\theta,$$

where the constants x_0, y_0, α, β, and θ_0 represent the fact that the reconstruction is unique up to a rotation and a translation.

Some Global Properties and the Evolute

A number of basic global facts related to the Euclidean curvature are now presented:

1. There are only two curves with constant curvature: straight lines (zero curvature) and circles (curvature equal to the inverse of the radius). The only closed curve with constant curvature is then the circle.
2. Vertices are the points at which the first derivative of the curvature vanishes. Every closed curve has at least four of these points (four-vertex theorem).
3. The total curvature of a closed curve is a multiple of 2π (exactly 2π in the case of a simple curve).
4. Isoperimetric inequality: Among all closed single curves of length (perimeter) L, the circle of radius $L/2\pi$ defines the one with the largest area.

As pointed out when defining the osculating circle, the curvature is the inverse of the radius of the osculating circle. The centers of these circles are called centers of curvature, and their loci define the Euclidean evolute of the curve:

$$\mathcal{E}_{\mathcal{C}}(s) := \mathcal{C}(s) + \frac{1}{\kappa(s)} \vec{\mathcal{N}}(s). \tag{1.3}$$

The basic geometric properties of the evolute, such as tangent, normal, arc length, and curvature, can be directly computed from those of the curve. The fact that the evolute of a closed curve is not a smooth curve it is of particular interest, as it is easy to show that the evolute has a cusp for every vertex of the curve.

1.2. Affine Differential Geometry

All the concepts presented in Section 1.1 are just Euclidean invariant, that is, invariant to rotations and translations. We now extend the concepts to the affine group. For the projective group see for example [226].

A general affine transformation in the plane ($\mathbb{R}^2$) is defined as

$$\tilde{\mathcal{X}} = A\mathcal{X} + B, \tag{1.4}$$

where $\mathcal{X} \in \mathbb{R}^2$ is a vector, $A \in \mathrm{GL}_2^+(\mathbb{R})$ (the group of invertible real 2×2 matrices with positive determinant) is the affine matrix, and $B \in \mathbf{R}^2$ is a translation vector. It is easy to show that transformations of the type of Eq. (1.4) form a real algebraic group $\mathcal{A}$, called the group of proper affine motions. We also consider the case in which we restrict $A \in \mathrm{SL}_2(\mathbb{R})$ (i.e., the determinant of A is 1), in which case Eq. (1.4) gives us the group of special affine motions, $\mathcal{A}_{\mathrm{sp}}$.

Before proceeding, let us briefly recall the notion of invariant. (For more detailed and rigorous discussions, see Refs. [51, 111, and 166] and Section 1.7 on Lie groups later in this chapter.) A quantity Q is called an invariant of a Lie group G if whenever Q transforms into $\tilde{Q}$ by any transformation G, we obtain $\tilde{Q} = \Psi Q$, where Ψ is a function of the transformation alone. If $\Psi = 1$ for all transformations in $\mathcal{G}$, Q is called an absolute invariant [111]. What we call invariant here is sometimes referred to in the literature as relative invariant. (We discuss more on Lie groups in Section 1.7.)

In the case of Euclidean motions (A in Eq. (1.4) being a rotation matrix), we have already seen that the Euclidean curvature κ of a given plane curve, as defined in Section 1.1, is a differential invariant of the transformation. In the case of general affine transformations, in order to keep the invariance property, a new definition of curvature is necessary. In this section, this affine curvature is presented [51, 53, 166, 374]. See also Refs. [39 and 53] for general properties of affine differential geometry.

Let $\mathcal{C}(p): S^1 \to \mathbb{R}^2$ be a simple curve with curve parameter p (where S^1 denotes the unit circle). We assume throughout this section that all of our mappings are sufficiently smooth, so that all the relevant derivatives may be defined. A reparameterization of $\mathcal{C}(p)$ to a new parameter s can be performed such that

$$[\mathcal{C}_s, \mathcal{C}_{ss}] = 1, \tag{1.5}$$

where $[\mathcal{X}, \mathcal{Y}]$ stands for the determinant of the 2×2 matrix whose columns are given by the vectors $\mathcal{X},\ \mathcal{Y} \in \mathbb{R}^2$. This is also the area of the parallelogram defined by the vectors. This relation is invariant under special affine transformations, and the parameter s is called the affine arc length. (As commonly done in the literature, we use s for both the Euclidean and the affine arc lengths, and the meaning will be clear from the context.) Setting

$$g(p) := [\mathcal{C}_p, \mathcal{C}_{pp}]^{1/3}, \tag{1.6}$$

we find that the parameter s is explicitly given by

$$s(p) = \int_0^p g(\xi)\mathrm{d}\xi. \tag{1.7}$$

This is easily obtained by means of the relation

$$1 = [\mathcal{C}_s, \mathcal{C}_{ss}] = \left[\mathcal{C}_p \frac{\mathrm{d}p}{\mathrm{d}s}, \mathcal{C}_{pp}\left(\frac{\mathrm{d}p}{\mathrm{d}s}\right)^2 + \mathcal{C}_p \frac{\mathrm{d}^2 p}{\mathrm{d}s^2}\right].$$

Note that in the above standard definitions, we have assumed (of course) that g (the affine metric) is different from zero at each point of the curve, i.e., the curve has no inflection points. This assumption will be made throughout this section unless explicitly stated otherwise. In particular, the convex curves we consider will be strictly convex, i.e., will have strictly positive (Euclidean) curvature. Fortunately, inflection points, that is, points with $\kappa = 0$, are affine invariant. Therefore limiting ourself to convex curves is not a major limitation for most image processing and computer vision problems.

It is easy to see that the following relations hold:

$$\mathrm{d}s = g\mathrm{d}p, \tag{1.8}$$

$$\vec{\mathbf{T}} := \mathcal{C}_s = \mathcal{C}_p \frac{\mathrm{d}p}{\mathrm{d}s}, \tag{1.9}$$

$$\vec{\mathbf{N}} := \mathcal{C}_{ss} = \mathcal{C}_{pp}\left(\frac{\mathrm{d}p}{\mathrm{d}s}\right)^2 + \mathcal{C}_p \frac{\mathrm{d}^2 p}{\mathrm{d}s^2}. \tag{1.10}$$

$\vec{\mathbf{T}}$ is called the affine tangent vector and $\vec{\mathbf{N}}$ is the affine normal vector. These formulas help to derive the relations between the Euclidean and the affine arc lengths, tangents, and normals. For example, considering v to be the Euclidean arc length, we have that

$$\mathrm{d}s = \kappa^{1/3}\mathrm{d}v, \tag{1.11}$$

where $\mathrm{d}s$ is still the affine arc length and $\mathrm{d}v$ is the Euclidean arc length:

$$\vec{\mathbf{T}} = \kappa^{-1/3}\vec{\mathcal{T}},$$

$$\vec{\mathbf{N}} = \kappa^{1/3}\vec{\mathcal{N}} + f(\kappa, \kappa_p)\vec{\mathcal{T}},$$

where f is a function of the first and second derivatives of the Euclidean curvature.

Affine Curvature

We now follow the same procedure as in the Euclidean case in order to obtain the affine curvature. By differentiating Eq. (1.5) we obtain

$$[\mathcal{C}_s, \mathcal{C}_{sss}] = 0. \tag{1.12}$$

Hence the two vectors $\mathcal{C}_s$ and $\mathcal{C}_{sss}$ are linearly dependent, and so there exists μ such that

$$\mathcal{C}_{sss} + \mu \mathcal{C}_s = 0. \tag{1.13}$$

Equation (1.13) implies (just compare the corresponding areas and recall that $[\mathcal{C}_s, \mathcal{C}_{ss}] = 1$) that

$$\mu = [\mathcal{C}_{ss}, \mathcal{C}_{sss}], \tag{1.14}$$

and μ is called the affine curvature. The affine curvature is the simplest nontrivial differential affine invariant of the curve $\mathcal{C}$ [53]. Note that μ can also be computed as

$$\mu = [\mathcal{C}_{ssss}, \mathcal{C}_s]. \tag{1.15}$$

For the exact expression of μ as a function of the original parameter p, see Ref. [53].

As pointed out in Section 1.1, in the Euclidean case constant Euclidean curvature κ is obtained for only circular arcs and straight lines. Further, the Euclidean osculating figure of a curve $\mathcal{C}(p)$ at a given point is always the circle with radius $1/\kappa$ whose center lies on the normal at the given point. In the affine case, the conics (parabola, ellipse, and hyperbola) are the only curves with constant affine curvature μ ($\mu = 0$, $\mu > 0$, and $\mu < 0$, respectively). Therefore the ellipse is the only closed curve with constant affine curvature. The affine osculating conic of a curve C at a noninflexion point is a parabola, ellipse, or hyperbola, depending on whether the affine curvature μ is zero, positive, or negative, respectively. This conic has a triple-point contact with the curve $\mathcal{C}$ at that point (same point, tangent, and second derivative, or Euclidean curvature).

Affine Invariants

Assuming the group of special affine motions, we can easily prove the absolute invariance of some of the concepts introduced above when $\tilde{\mathcal{C}}$ is obtained from $\mathcal{C}$ by means of an affine transformation, that is, the affine arc length, tangent, normal, and curvature, as well as the area, are absolute invariants for the group of special affine motions. In general, for $A \in \mathrm{GL}_2^+(\mathbb{R})$, we obtain

$$\begin{aligned} d\tilde{s} &= [A]^{1/3} ds, \\ \tilde{\mathcal{C}}_{\tilde{s}} &= A[A]^{-1/3}\mathcal{C}_s, \\ \tilde{\mathcal{C}}_{\tilde{s}\tilde{s}} &= A[A]^{-2/3}\mathcal{C}_{ss}, \\ \tilde{\mu} &= [A]^{-2/3}\mu, \\ \text{area}(\tilde{\mathcal{C}}) &= [A]\text{area}(\mathcal{C}). \end{aligned}$$

Thus the affine properties remain invariant (relative) but not absolute invariants. For an extended analysis about curvature like invariants, see Ref. [51].

Global Affine Differential Geometry

As in the Euclidean case, we now give a number of global properties regarding affine differential geometry:

1. There are at least six points with $\mu_s = 0$ (affine vertices) in a closed convex curve.
2. Define the affine perimeter of a closed curve as

$$L := \oint g dp = \int ds.$$

   Then, from all closed convex curves with a constant area, the ellipse, and only the ellipse, attains the greatest affine perimeter. In other words, for an oval (strictly convex closed curve) the following relation holds:

$$8\pi^2 \text{area} - L^3 \geq 0,$$

   and equality holds for only the ellipse.
3. For closed convex sets (ovals), the following affine isoperimetric inequality holds:

$$2\oint \mu ds \leq \frac{L^2}{\text{area}}. \tag{1.16}$$

   See [246,247] for other inequalities and [147] for a related Euclidean result.
4. In the important case of the ellipse, the relation between the affine curvature and the area is given by

$$\mu = \left(\frac{\pi}{\text{area}}\right)^{2/3},$$

   where area $= \pi r_1 r_2$ is the area, and r_1 and r_2 are the ellipse radii.

5. The greatest osculating ellipse of an elliptically curved oval ($\mu > 0$) contains the oval entirely, whereas the smallest osculating ellipse of the oval is contained entirely within it.

Affine Invariant Distance

A key missing component of affine differential geometry so far it is the definition of affine invariant distance. Fortunately this concept is well know in the affine geometry literature [39, 195, 272]. The affine distance is now defined and some of its basic properties are presented.

Definition 1.1. *Let $\mathcal{X}$ be a generic point on the plane and $\mathcal{C}(s)$ be a strictly convex planar curve parameterized by the affine arc length. The affine distance between $\mathcal{X}$ and a point $\mathcal{C}(s)$ on the curve is given by*

$$d(\mathcal{X}, s) := [\mathcal{X} - \mathcal{C}(s), \mathcal{C}_s(s)]. \tag{1.17}$$

To be consistent with the Euclidean case and the geometric interpretation of the affine arclength, $d(\mathcal{X}, s)$ should be defined as the $1/3$ power of the determinant above. Because this does not imply any conceptual difference, we keep the definition above to avoid introducing further notation. In Ref. [195] the above function $d(\mathcal{X}, s)$ is called the affine distance-cubed function.

From the definition of $d(\mathcal{X}, s)$, we note that the affine distance between an arbitrary point $\mathcal{X}$ and the curve point $\mathcal{C}(s)$ is given by the area of the parallelogram defined by the vector $\mathcal{X} - \mathcal{C}(s)$ and the affine tangent at $\mathcal{C}(s)$ (Fig. 1.1). Unfortunately, we cannot define an affine invariant distance between two points; we need at least three points or a point and a line. This is because area, not length, is the basic affine invariant.

Some of the basic properties of $d(\mathcal{X}, s)$ are now presented. When the affine concepts are replaced with Euclidean ones, the same properties hold

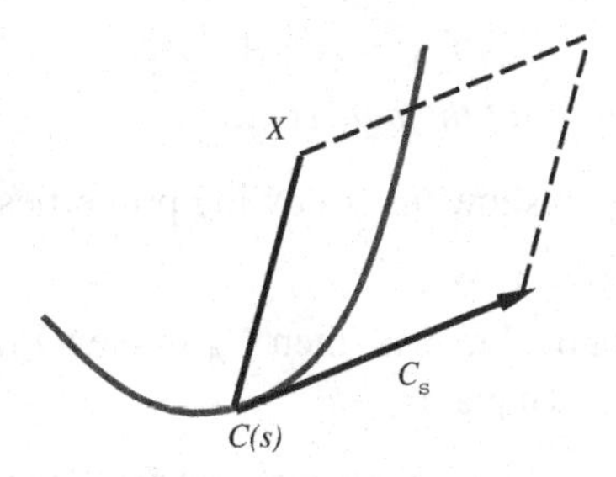

Fig. 1.1. The affine distance between an arbitrary point $\mathcal{X}$ on the plane and the curve point $\mathcal{C}(s)$ is given by the area of the parallelogram defined by the vector $\mathcal{X} - \mathcal{C}(s)$ and the affine tangent at $\mathcal{C}(s)$, $\mathcal{C}_s$.

for the classical (squared) Euclidean invariant distance $\langle \mathcal{X} - \mathcal{C}, \mathcal{X} - \mathcal{C} \rangle$. This shows that the definition is consistent.

Lemma 1.1. *The affine distance satisfies the following properties:*

1. *$d(\mathcal{X}, s)$ is an extremum (i.e., $d_s = 0$) if and only if $\mathcal{X} - \mathcal{C}(s)$ is parallel to $\mathcal{C}_{ss}$, i.e., $\mathcal{X}$ lies on the affine normal to the curve at $\mathcal{C}(s)$.*
2. *For nonparabolic points (where the affine curvature is nonzero), $d_s(\mathcal{X}, s) = d_{ss}(\mathcal{X}, s) = 0$ if and only if $\mathcal{X}$ is on the affine evolute given by the curve $\mathcal{C} + \frac{1}{\mu}\mathcal{C}_{ss}$.*
3. *$\mathcal{C}$ is a conic and $\mathcal{X}$ is its center if and only if $d(\mathcal{X}, s)$ is constant.*
4. *$\mathcal{X} - \mathcal{C}(s) = \alpha \mathcal{C}_s(s) - d(\mathcal{X}, s)\mathcal{C}_{ss}(s)$ for some real number α.*

Regarding the third property, note that the center is defined as the unique point equidistant from all the points on the curve. In both the Euclidean and the affine cases the center is also the intersection of all the normals.

The fourth property basically tells us what happens when the point $\mathcal{X}$ is not on the affine normal. The distance is basically the affine projection of the point onto the affine normal.

1.3. Cartan Moving Frames

The basic Euclidean and affine invariants, that is, arc length and curvature, can also be obtained from Cartan moving frames. The basic concepts are presented now. Details can be found in Refs. [61 and 166] as well as in some very recent contributions by Fels and Olver [140, 141]. This is a very elegant and complete technique that, thanks to these new results, is completely algorithmic as well.

Definition 1.2. *The Cartan matrix C_A of a differentiable nonsingular square matrix $A(p)$ is given by*

$$C_A = A' \cdot A^{-1},$$

where A' is the derivative of the matrix A.

From this it is easy to show the following properties:

1. $C_{AB} = C_A + AC_B A^{-1}$.
2. If A is an orthogonal matrix, then C_A is skew symmetric, that is, the transpose of C_A is equal to $-C_A$.

Note that if M is a nonsingular constant matrix, then $C_A = C_{AM}$. This will be critical in computing the basic differential invariants for the Euclidean and affine groups.

Euclidean Case

Let $\{\vec{x}, \vec{y}\}$ indicate a fixed coordinate system on the plane (both are unit vectors). A curve $\mathcal{C}(s)$ is given as $[x(s), y(s)]\{\vec{x}, \vec{y}\}$. The unit tangent $\vec{\mathcal{T}}$ is then $(x', y')\{\vec{x}, \vec{y}\}$ and the unit normal $\vec{\mathcal{N}}$ is $(-y', x')\{\vec{x}, \vec{y}\}$, obtained from the tangent by a $\pi/2$ rotation. The pair $\{\vec{\mathcal{T}}, \vec{\mathcal{N}}\}$ is the (Euclidean) moving frame. We can consider this frame as a moving coordinate system, making the curve point $\mathcal{C}(s)$ always at the origin and the first coordinate axes tangent to the curve in the direction of increasing arc length. We can also relate the moving frame to the origin O of the fixed frame (the moving frame, given as a function of s, uniquely determines the curve up to a translation). The moving frame is then obtained from the fixed one by a rotation $A(s)$ about the angle θ between $\vec{x}$ and $\vec{\mathcal{T}}$:

$$\{\vec{\mathcal{T}}, \vec{\mathcal{N}}\} = A(s)\{\vec{x}, \vec{y}\},$$

where the rotation matrix $A(s)$, called the frame matrix, is given by

$$A(s) = \begin{bmatrix} \cos\theta & \sin\theta \\ -\sin\theta & \cos\theta \end{bmatrix} = \begin{bmatrix} x' & y' \\ -y' & x' \end{bmatrix}.$$

In a similar way as the tangent (or moving frame) defines the curve up to a translation, so does the frame matrix. If we change the fixed frame by a rotation B,

$$\{\vec{x}, \vec{y}\} = B\{\vec{i}, \vec{j}\},$$

the frame matrix is changed to $A(s)B$:

$$\{\vec{\mathcal{T}}, \vec{\mathcal{N}}\} = A(s)B\{\vec{i}, \vec{j}\}.$$

The variation of the moving frame is given by

$$\frac{\mathrm{d}}{\mathrm{d}s}\{\vec{\mathcal{T}}, \vec{\mathcal{N}}\} = C_A\{\vec{\mathcal{T}}, \vec{\mathcal{N}}\},$$

providing the Frenet equation of planar (Euclidean) differential geometry. Because $A(s)$ is orthonormal, C_A is skew symmetric and then of the form

$$C_A(s) = \begin{bmatrix} 0 & \kappa(s) \\ -\kappa(s) & 0 \end{bmatrix},$$

and the Frenet equation can explicitly be computed. From the properties of the Cartan matrix we also know that C_A is invariant to the rotation B, and therefore $\kappa(s)$ is Euclidean invariant. The explicit computation of C_A from

$A(s)$ above gives

$$\kappa(s) = \frac{d\theta}{ds},$$

which is exactly one of the definitions for the Euclidean curvature given before.

Affine Case

In analogy to the Euclidean case, we are searching for a moving frame $\{\vec{\mathbf{T}}, \vec{\mathbf{N}}\}$ such that

$$\{\vec{\mathbf{T}}, \vec{\mathbf{N}}\} = A(p)\{\vec{x}, \vec{y}\},$$

where this time $A(p)$ is not a rotation matrix but a nonsingular matrix with a unit determinant.

If we consider the general frame $\{C', C''\}$ (this is a frame as long as the two vectors are linearly independent, meaning that the curve has no inflections), then the frame matrix $B(p)$ given by

$$\{C', C''\} = B(p)\{\vec{x}, \vec{y}\}$$

does not necessarily have a unit determinant. If we normalize these vectors,

$$\vec{\mathbf{T}} := \frac{C'}{[C', C'']^{1/2}},$$

$$\vec{\mathbf{N}} := \frac{C''}{[C', C'']^{1/2}},$$

then the matrix

$$A(t) = \frac{1}{[C', C'']^{1/2}} \begin{bmatrix} x' & y' \\ x'' & y'' \end{bmatrix}$$

appearing in $\{\vec{\mathbf{T}}, \vec{\mathbf{N}}\} = A(p)\{\vec{x}, \vec{y}\}$ has a unit determinant. The next step is to compute the Cartan matrix, which is

$$C_A = \begin{bmatrix} \frac{-[C', C''']}{2[C', C'']} & 1 \\ \frac{-[C'', C''']}{[C', C'']} & \frac{[C', C''']}{[C', C'']} \end{bmatrix}.$$

The idea is now to try to simplify, as much as possible, C_A, which is affine invariant. For example, the diagonal elements become zero if

$$[C', C'''] = [C', C'']' = 0,$$

which means that the affine arc length is given by the condition

$$[\mathcal{C}', \mathcal{C}''] = 1,$$

as previously shown in Section 1.2 by other methods. The remaining nonzero coefficient of C_A becomes $[\mathcal{C}'', \mathcal{C}''']$, which is the affine curvature, the second fundamental affine differential invariant.

To recap for these simple cases: The moving-frame method is based on finding an appropriate transformation, belonging to the group, from a fixed frame to a curve-attached one and then computing the differential invariants from the Cartan matrix corresponding to this transformation.

1.4. Space Curves

To complete the picture of differential geometry for curves, the basic Euclidean (and affine) concepts for curves in three-dimensional (3D) space are now presented.

Let us start with the Euclidean case and assume that we are given a three times differentiable space curve $\mathcal{C}(p) = [x(p), y(p), z(p)] : \mathbb{R} \to \mathbb{R}^3$. As in the planar case, the basic idea is to search for an Euclidean invariant arc length s given by

$$s(p) = \int_{p_0}^{p} \|\mathcal{C}_v\| \mathrm{d}v,$$

where, as in Section 1.1, as $\|\cdot\|$ stands for the Euclidean length of a vector. In other words, we search for a unit velocity parameterization

$$\|\mathcal{C}_s\| = 1.$$

Therefore the unit tangent vector is given by

$$\vec{\mathcal{T}} = \frac{\mathrm{d}\mathcal{C}}{\mathrm{d}s}.$$

We can proceed as in the planar case, following the Cartan moving-frame method, and look for a local coordinate system obtained from the fixed Cartesian system by a 3D rotation and translation, that is, given the fixed Cartesian system $\{\vec{x}, \vec{y}, \vec{z}\}$, we search for a moving (local) coordinate system $\{\vec{\mathcal{T}}, \vec{\mathcal{N}}, \vec{\mathcal{B}}\}(s)$ such that

$$\{\vec{\mathcal{T}}, \vec{\mathcal{N}}, \vec{\mathcal{B}}\}(s) = A(s)\{\vec{x}, \vec{y}, \vec{z}\},$$

where $A(s)$ is a rotation matrix. Then

$$\{\vec{\mathcal{T}}, \vec{\mathcal{N}}, \vec{\mathcal{B}}\}'(s) = C_A(s)\{\vec{\mathcal{T}}, \vec{\mathcal{N}}, \vec{\mathcal{B}}\}(s),$$

and we obtain a skew-symmetric Cartan matrix $C_A(s)$.

By analyzing the corresponding Cartan matrix, we obtain the basic invariant and the Frenet equations. Because this is straightforward from what we have learned from the planar case, we just give the results and leave the derivations as an exercise.

The tangent $\vec{\mathcal{T}}$ is the first vector of the local coordinate system. The second one, the principal normal, is given by

$$\vec{\mathcal{N}} = \frac{\vec{\mathcal{T}'}}{\|\vec{\mathcal{T}'}\|}.$$

The third vector $\vec{\mathcal{B}}$, the binormal, is just the unique vector perpendicular to both the tangent and the principal normal, such that the cube spanned by the three vectors has a volume equal to 1. With this construction, the Cartan matrix becomes

$$C = \begin{bmatrix} 0 & \kappa(s) & 0 \\ -\kappa(s) & 0 & t(s) \\ 0 & -t(s) & 0 \end{bmatrix}.$$

Here $\kappa(s)$ is the curvature and $t(s)$ is the tension, and they uniquely define the curve up to a rotation and translation in three dimensions. The Frenet equations are then given by

$$\mathcal{C}_s = \vec{\mathcal{T}},$$

$$\mathcal{C}_{ss} = \kappa\vec{\mathcal{N}},$$

$$\mathcal{C}_{sss} = -\kappa^2\vec{\mathcal{T}} + \kappa_s\vec{\mathcal{N}} + \kappa t\vec{\mathcal{B}}.$$

The curvature is a second-order invariant, and the tension is a third-order one. The curvature of a 3D curve is always positive, by definition, and the tension is zero if and only if the 3D curve can be embedded in a 3D plane.

As was done for the planar case, this process of Cartan frames can be also used to derive the corresponding affine invariant for space curves. In this case, the affine arc length s will be given by the condition

$$[\mathcal{C}_s, \mathcal{C}_{ss}, \mathcal{C}_{sss}] = 1.$$

The reader is referred to Ref. [166] for the details.

1.5. Three-Dimensional Differential Geometry

We now consider regular surfaces $\mathcal{S}(u, v) = [x(u, v), y(u, v), z(u, v)]: \mathbb{R}^2 \to \mathbb{R}^3$, which are homeomorphisms, for which each one of the coordinates has continuous partial derivatives, and such that the differential map

$$\mathrm{d}\mathcal{S} := \begin{pmatrix} x_u & x_v \\ y_u & y_v \\ z_u & z_v \end{pmatrix}$$

is one to one. Later in this book we will see that we can also consider deformations of surfaces that are not smooth, but classical differential geometry does not include those surfaces. The regularity condition avoids surfaces that have self-intersections.

A particular interesting example of surfaces are surfaces of revolution. We obtain these by rotating a regular planar curve, the generating curve, about an axis in the plane that does not meet the curve. Let us consider (x, z) as the plane containing the curve and the axis z as the rotation axis. Let $x = f(v)$ and $z = g(v)$, $a < v < b$, $f(v) > 0$ be the parameterization for the curve. Considering u as the angle of rotation along the z axis, we obtain

$$\mathcal{S}(u, v) = [f(v) \cos u, f(v) \sin u, g(v)].$$

This defines a regular surface.

The differential map $\mathrm{d}\mathcal{S}$ evaluated at a surface point $P = \mathcal{S}(u_0, v_0)$ defines the tangent plane at the point, which is, of course, independent of the parameterization (u, v). We can consider vectors in this tangent plane (which are vectors in $\mathbb{R}^3$) and define the inner product between them. The inner product of any vector $\vec{w}$ in the tangent plane with itself defines the first fundamental form $I_P(\vec{w})$ of the regular surface $\mathcal{S}$. Let us define

$$\mathcal{S}_u = (x_u, y_u, z_u),$$

$$\mathcal{S}_v = (x_v, y_v, z_v).$$

Any vector in the tangent plane is the tangent to a parameterized curve $\mathcal{C}(p) \in \mathcal{S}[u(p), v(p)]$ going through the point P (p is an arbitrary parameterization). Assuming that $\mathcal{C}(0) = P$, it is easy to show that

$$I_P[\mathcal{C}'(0)] = E(u')^2 + 2Fu'v' + G(v')^2,$$

where

$$E := \langle \mathcal{S}_u, \mathcal{S}_u \rangle_P,$$
$$F := \langle \mathcal{S}_u, \mathcal{S}_v \rangle_P,$$
$$G := \langle \mathcal{S}_v, \mathcal{S}_v \rangle_P.$$

One of the properties of the first fundamental form is that we can compute the Euclidean arc length of $\mathcal{C}$ from it:

$$s = \int_0^p \|\mathcal{C}'\| \mathrm{d}t = \int_0^p \sqrt{I(\mathcal{C}')} = \int_0^p \sqrt{E(u')^2 + 2Fu'v' + G(v')^2}.$$

We also observe that the angle between $\mathcal{S}_u$ and $\mathcal{S}_v$ is

$$\cos\theta = \frac{\langle \mathcal{S}_u, \mathcal{S}_v \rangle}{\|\mathcal{S}_u\| \, \|\mathcal{S}_v\|} = \frac{F}{\sqrt{EG}}.$$

If $F = 0$, then this angle is 90°, and the parameterization (u, v) is called an orthogonal parameterization.

In addition to length, the first fundamental form is related to area as well. It is easy to show that the area of the surface in a region Ω is given by

$$\iint_\Omega \sqrt{EG - F^2} \mathrm{d}u \mathrm{d}v.$$

We have just worked with the tangent plane; let us now describe a few concepts related to the surface normals. The unit normal to the surface is given by

$$\vec{\mathcal{N}} = \frac{\mathcal{S}_u \times \mathcal{S}_v}{\|\mathcal{S}_u \times \mathcal{S}_v\|},$$

where $(\cdot \times \cdot)$ stands for the vector obtained by the exterior product of two vectors. This is basically a map from the surface to the unit sphere S^2 in $\mathbb{R}^3$. This map is the so-called Gauss map of the surface. (Actually, this definition applies to only regular surfaces with an orientation, i.e., that have a differentiable field of unit normals.)

Let us now look at the differential map $\mathrm{d}\vec{\mathcal{N}}_P$ mapping a vector from the tangent plane of $\mathcal{S}$ at P to the tangent plane of the the unit sphere S^2 at $\vec{\mathcal{N}}_P$. This is a linear map, mapping a plane into another parallel plane. Let us have a closer look at this map. For each parameterized curve $\mathcal{C}(p)$ such that $\mathcal{C}(0) = P$, we consider the curve $\vec{\mathcal{N}} \cdot \mathcal{C}(p)$ in the sphere S^2, that is, the normals to $\mathcal{C}(p)$. The tangent vector $\vec{\mathcal{N}}' = \mathrm{d}\vec{\mathcal{N}}[\mathcal{C}'(0)]$ is a vector in the tangent plane at the point P, measuring the rate of change of the normal restricted to the curve $\mathcal{C}$. For curves, this is a number, the curvature. For surfaces it is the linear map. This map gives light to the second fundamental form, which is given by

$$II_P(\vec{v}) := -\langle \mathrm{d}\vec{\mathcal{N}}_p(\vec{v}), \vec{v} \rangle,$$

which is defined on the tangent plane at P.

Having now the definition of the second fundamental form, let us now start looking at curvatures on the surface. Let $\mathcal{C}$ be a curve on $\mathcal{S}$ passing through P, κ be the curvature of $\mathcal{C}$ at P, $\cos\theta = \langle \vec{n}, \vec{\mathcal{N}} \rangle$, where $\vec{n}$ is the normal to the curve, and $\vec{\mathcal{N}}$ be, as above, the normal to the surface (both at the point P). The number $\kappa_n := \kappa \cos\theta$ is called the normal curvature of $\mathcal{C}$ at P. The normal curvature is basically the length of the projection of $\kappa\vec{n}$ over the surface normal. Let us now connect this to the second fundamental form. Because $\langle \vec{\mathcal{N}}, \mathcal{C}' \rangle = 0$,

$$\langle \vec{\mathcal{N}}, \mathcal{C}'' \rangle = -\langle \vec{\mathcal{N}}', \mathcal{C}' \rangle,$$

and from here and the definition of the second fundamental form,

$$II_P = \kappa_n.$$

In other words, the value of the second fundamental form II_P for a unit vector $\vec{v}$ on the tangent plane at P is equal to the normal curvature of a regular curve passing through P and with tangent $\vec{v}$. In addition, all curves lying on the surface and having at a given point the same tangent have the same normal curvature.

From the results above, we can discuss the normal curvature at a given direction. We can then consider all possible directions and obtain the maximal and the minimal normal curvatures κ_1 and κ_2, respectively. These are the principal curvatures of the surface and the corresponding directions are the principal directions. This leads to the definition of the Gaussian curvature

$$\mathbf{K} := \kappa_1 \kappa_2,$$

and the mean curvature

$$\mathbf{H} := \frac{\kappa_1 + \kappa_2}{2}.$$

These curvatures are actually the determinant and the negative half-trace of the differential of the Gauss map, respectively.

The last concept we want to introduce is the concept of geodesic curvature. The geodesic curvature κ_g at a point P of a curve $\mathcal{C} \in \mathcal{S}$ is given by the projection of the vector $\kappa\vec{n}$ onto the tangent plane (this is the absolute value). That is, the absolute value of the geodesic curvature is equal to the length of the tangential component of $\kappa\vec{n}$. From this, it is obvious that

$$\kappa^2 = \kappa_g^2 + \kappa_n^2.$$

A geodesic curve is a curve on the surface with zero geodesic curvature. The minimal path between two points on a surface is always a geodesic curve.

1.6. Discrete Differential Geometry

Shapes in the computer are commonly represented as a collection of points. To compute curvatures in particular and differential invariants in general, we need then to deal with this discrete representation. Many works have addressed this issue. One possible approach is to obtain a continuous curve or surface that fits the set of discrete points, that is, to perform surface fitting. The most popular way is to use B-splines, described in Subsection 1.6.1, to do this. A different approach, deeply investigated by Taubin [382], is to find a function such that the set of points belongs its zero level set. In other words, the problem is reduced to the search for an implicit representation of the set of discrete points. A third approach is to do local approximations, that is, for every point in the set, we locally fit a continuous curve by using the neighboring points and then compute the differential invariants; see Refs. [57 and 58] for an example. The problem becomes more interesting and difficult, of course, if the set of points, especially in three dimensions, is unorganized and the topology of the surface is unknown. The computer graphics literature has addressed this; e.g., Ref. [115]. One of the possibilities is to compute a parameterization corresponding to the set of points and then perform B-spline representation by using this parameterization. For completeness, the basic concepts of planar B-splines are described. We refer the reader to the mentioned references for a deeper investigation of this, still active, research area.

1.6.1. Basic B-Spline Theory

The theory of B-spline approximations for 2D curves is now briefly described. For details and extensions to surfaces, see, for example, Refs. [22, 43, 350, and 351].

Let

$$\mathcal{C}(p) : [a, b] \to \mathbb{R}^2$$

be a planar curve, with Cartesian coordinates $[x(p), y(p)]$. Polynomials are computationally efficient to work with, but it is not always possible to define a satisfactory curve $\mathcal{C}$ when single polynomials are used for x and y. Then, the curve is divided in segments, each one defined by a given polynomial. The segments are joined together to form a piecewise polynomial curve. The joints between the polynomial segments occur at special curve points called knots. The sequence $p_1, p_2, \ldots,$ of knots is required to be nondecreasing. The distance between two consecutive knots can be constant or not. Two successive polynomial segments are joined together at a given knot p_j in such a way that the resulting piecewise polynomial has d continuous derivatives. Of course, the order of the polynomials depends on d. In this way, a basis is obtained, and the curve $\mathcal{C}$ is given by a linear combination of it.

Formally, the curve $\mathcal{C}$ is a B-spline approximation of the series of points $V_i = [x_i, y_i]$, $1 \leq i \leq n$, called control vertices, if it can be written as

$$\mathcal{C}_k(p) = \sum_i^n V_i B_{i,k}, \tag{1.18}$$

where $B_{i,k} = B(\cdot; p_i, p_{i+1}, \ldots, p_{i+k})$ is the ith B-spline basis of the order of k for the knot sequence $[p_1, \ldots, p_{n+k}]$. In particular, $B_{i,k}$ is a piecewise polynomial function of degree $<k$, with breakpoints $p_i, \ldots, p_{i+k}$. To deal with closed curves, the series of points are periodic, that is, the indices are computed modulo $n + 1$.

Several properties can be proven for the basis $B_{i,k}$:

1. $B_{i,k} \geq 0$.
2. $B_{i,k} \equiv 0$ outside the interval $[p_i, p_{i+k}]$. This property shows the locality of the approximation: Changing a given control vertex affects only a corresponding portion of the curve.
3. The basis is normalized for all order k:

$$\sum_i B_{i,k}(p) = 1 \text{ on } [p_k \ldots p_{n+1}].$$

4. The support of the B-spline basis is minimal among all polynomial splines of the order of k. This property shows that this representation is optimal in a certain sense.

The multiplicity of the knots governs the smoothness. If a given number τ occurs r times in the knot sequence $[p_i, \ldots, p_{i+k}]$, then the first $k - r - 1$ derivatives of $B_{j,k}$ are continuous at the breakpoint τ. Therefore, without knot multiplicity, $\mathcal{C}_k \in \mathbf{C}^{k-2}$.

Computation with B-splines is facilitated by use of the following recursive formula [22,43,44]:

$$B_{i,k}(p) = \frac{p - p_i}{p_{i+k-1} - p_i} B_{i,k-1} + \frac{p_{i+k} - p}{p_{i+k} - p_{i+1}} B_{i+1,k-1}, \tag{1.19}$$

together with

$$B_{i,1}(p) = \begin{cases} 1, & p_i \leq p < p_{i+1}, \\ 0, & \text{otherwise.} \end{cases} \tag{1.20}$$

The basis $B_{i,k}$ is in fact the repeated convolution of a Haar function with itself, i.e.,

$$B_{0,k} = (*)^k \chi[0, 1].$$

To represent a given set of point on the plane, we then must find the control points that define a B-spline curve passing through or as close as possible to the set. This is a simple curve-fitting task, which, in the case of a quadratic error norm, reduces to a classical linear least-squares fitting.

1.7. Differential Invariants and Lie Group Theory

We have already presented basic differential invariants for the Euclidean and affine group. We have also shown how to derive those invariant by using Cartan's theory. This was done with special consideration for the group being analyzed. We now want to be more general, that is, to present a number of general concepts that apply to all relevant transformation groups. The theory developed here will be also useful for developing a general formulation for invariant shape deformations in Chap. 2. To do this in a rigorous manner, we first sketch some relevant facts from differential geometry and the theory of Lie groups. The material here is based on the books by Olver [281, 283] to which the interested reader is referred for all the details. See also the classical work of Sophus Lie on the theory of differential invariants [233] as well as Refs. [62, 138, 281, 283, and 374].

Although this section is quite detailed, a certain mathematical background is assumed, i.e., the reader should be familiar with the basic definitions of "manifold" and "smooth function." Accordingly, all the manifolds and mappings we subsequently consider are C^∞. (This type of foundational material may be found in Refs. [166, and 374].) In this section no proofs will be given, and the results will just be stated.

1.7.1. Vector Fields and One-Forms

Because we will be considering the theory of differential invariants, we first review the infinitesimal (differential) structure of manifolds. Accordingly, a tangent vector to a manifold M at a point $x \in M$ is geometrically given as the tangent to a (smooth) curve passing through x. The collection of all such tangent vectors gives the tangent space $TM|_x$ to M at x, which is a vector space of the same dimension m as that of M. In local coordinates, a curve is parameterized by $x = \phi(t) = [\phi^1(t), \ldots, \phi^m(t)]$, and has tangent vector $\mathbf{v} = \xi^1 \frac{\partial}{\partial x^1} + \cdots + \xi^m \frac{\partial}{\partial x^m}$ at $x = \phi(t)$, with components $\xi^i = \frac{d\phi^i}{dt}$ given by the components of the derivative $\phi'(t)$. Here the tangent vectors to the coordinate axes are denoted by $\frac{\partial}{\partial x^i} = \partial_{x^i}$ and form a basis for the tangent space $TM|_x$. If $f : M \to \mathbb{R}$ is any smooth function, then its directional derivative along the curve is

$$\frac{\mathrm{d}}{\mathrm{d}t} f[\phi(t)] = \mathbf{v}(f)[\phi(t)] = \sum_{i=1}^{m} \xi^i[\phi(t)] \frac{\partial f}{\partial x^i}[\phi(t)],$$

which provides one motivation for using derivational notation for tangent vectors. The tangent spaces are patched together to form the tangent bundle $TM = \bigcup_{x \in M} TM|_x$ of the manifold, which is an m-dimensional vector bundle over the m-dimensional manifold M. A vector field $\mathbf{v}$ is a smoothly (or analytically) varying assignment of tangent vector $\mathbf{v}|_x \in TM|_x$. In local coordinates, a vector field has the form

$$\mathbf{v} = \sum_{i=1}^{m} \xi^i(x) \frac{\partial}{\partial x^i},$$

where the coefficients $\xi^i(x)$ are smooth (analytic) functions.

A parameterized curve $\phi : \mathbb{R} \to M$ is called an integral curve of the vector field $\mathbf{v}$ if its tangent vector agrees with the vector field $\mathbf{v}$ at each point; this requires that $x = \phi(t)$ satisfy the first-order system of ordinary differential equations:

$$\frac{\mathrm{d}x^i}{\mathrm{d}t} = \xi^i(t), \qquad 1 \leq i \leq m.$$

Standard existence and uniqueness theorems for systems of ordinary differential equations imply that through each $x \in M$ there passes a unique, maximal integral curve. We use the notation $\exp(t\mathbf{v})x$ to denote the maximal integral curve passing through $x = \exp(0\mathbf{v})x$ at $t = 0$, which may or may not be defined for all t. The family of (locally defined) maps $\exp(t\mathbf{v})$ is called the flow generated by the vector field $\mathbf{v}$ and obeys the usual exponential rules:

$$\begin{aligned}\exp(t\mathbf{v})\exp(s\mathbf{v})x &= \exp[(t+s)\mathbf{v}]x, \quad t, s \in \mathbb{R},\\ \exp(0\mathbf{v})x &= x,\\ \exp(-t\mathbf{v})x &= \exp(t\mathbf{v})^{-1}x,\end{aligned}$$

the equations holding where defined. Conversely, given a flow obeying the latter equalities, we can reconstruct a generating vector field by differentiation:

$$\mathbf{v}\Big|_x = \frac{\mathrm{d}}{\mathrm{d}t}\exp(t\mathbf{v})\Big|_{t=0} x, \qquad x \in M.$$

Applying the vector field $\mathbf{v}$ to a function $f : M \to \mathbb{R}$ determines the infinitesimal change in f under the flow induced by $\mathbf{v}$:

$$\mathbf{v}(f) = \sum_{i=1}^{n} \xi^i(x)\frac{\partial f}{\partial x^i} = \frac{\mathrm{d}}{\mathrm{d}t} f[\exp(t\mathbf{v})x]\Big|_{t=0},$$

so that

$$f[\exp(t\mathbf{v})x] = f(x) + t\mathbf{v}(f)(x) + \frac{1}{2}t^2\mathbf{v}[\mathbf{v}(f)] + \ldots.$$

Next, given a (smooth) mapping $F : M \to N$, we define the differential $\mathrm{d}F : TM|_x \to TN|_{F(x)}$ by

$$[(\mathrm{d}F)(\mathbf{v})](f)[F(x)] := \mathbf{v}(f \circ F)(x),$$

where $f : N \to \mathbb{R}$ is a smooth function and $\mathbf{v}$ is a vector field.

In general, given a point $x \in M$, a one-form at x is a real-valued linear map on the tangent space

$$\omega : TM|_x \to \mathbb{R}.$$

In local coordinates $x = (x^1, \ldots, x^m)$, the differentials $\mathrm{d}x^i$ are characterized by $\mathrm{d}x^i(\partial_{x^j}) = \delta_{ij}$ (the Kronecker delta), where $\partial_{x^1}, \ldots, \partial_{x^m}$ denotes the standard basis of $TM|_x$. Then, locally,

$$\omega = \sum_{i=1}^{m} h_i(x)\mathrm{d}x^i.$$

In particular, for $f : M \to \mathbb{R}$, we get the one-form $\mathrm{d}f$ given by its differential, so that

$$\mathrm{d}f(\mathbf{v}) := \mathbf{v}(f).$$

The vector space of one-forms at x is denoted by $T^*M|_x$ and is called the cotangent space. (It can be regarded as the dual space of $TM|_x$.) As for the tangent bundle, the cotangent spaces can be patched together to form the cotangent bundle T^*M over M.

1.7.2. Lie Groups

In this section, we collect together the basic necessary facts from the theory of Lie groups that will be used in the rest of this section and in Section 2.8. Recall first that a group is a set, together with an associative multiplication operation. The group must also contain an identity element, denoted by e, and each group element g has an inverse g^{-1} satisfying $g \cdot g^{-1} = g^{-1} \cdot g = e$. Historically, it was Galois who made the fundamental observation that the set of symmetries of an object forms a group (this was in his work on the roots of polynomials). However, the groups of Galois were discrete; in this section we study the continuous groups first investigated by Sophus Lie.

Definition 1.3. *A Lie group is a group G that also carries the structure of a smooth manifold so that the operations of group multiplication $(g, h) \mapsto g \cdot h$ and inversion $g \mapsto g^{-1}$ are smooth maps.*

Example. The basic example of a real Lie group is the general linear group $\mathrm{GL}(\mathbb{R}, n)$ consisting of all real invertible $n \times n$ matrices with matrix multiplication as the group operation; it is an n^2-dimensional manifold, the structure arising simply because it is an open subset (namely where the determinant is nonzero) of the space of all $n \times n$ matrices, which is itself isomorphic to $\mathbb{R}^{n^2}$.

A subset $H \subset G$ of a group is a subgroup if and only if it is closed under multiplication and inversion; if G is a Lie group, then a subgroup H is a Lie subgroup if it is also a submanifold. Most Lie groups can be realized as Lie subgroups of $\mathrm{GL}(\mathbb{R}, n)$; these are the so-called matrix Lie groups, and, in this section, we assume that all Lie groups are of this type. We can also define a notion of a local Lie group in the obvious way (see, e.g., Ref. [166]).

Example. We list here some of the key classical groups. The special linear group $\mathrm{SL}(\mathbb{R}, n) = \{A \in \mathrm{GL}(\mathbb{R}, n) : \det A = 1\}$ is the group of volume-preserving linear transformations. The group is connected and has dimension $n^2 - 1$. The orthogonal group $\mathrm{O}(n) = \{A \in \mathrm{GL}(\mathbb{R}, n) : A^T A = I\}$ is the group of norm-preserving linear transformations – rotations and reflections – and has two connected components. The special orthogonal group

$SO(n) = O(n) \cap SL(\mathbb{R}, n)$ consisting of just the rotations is the component containing the identity. This is also called the rotation group.

Transformation Groups. In many cases in vision (and physical) problems, groups are presented to us as a family of transformations acting on a space. In the case of Lie groups, the most natural setting is as groups of transformations acting smoothly on a manifold. More precisely, we have the following:

Definition 1.4. *Let M be a smooth manifold. A group of transformations acting on M is given by a Lie group G and smooth map $\Phi : G \times M \to M$, denoted by $\Phi(g, x) = g \cdot x$, which satisfies*

$$e \cdot x = x, \quad g \cdot (h \cdot x) = (g \cdot h) \cdot x, \quad \text{for all } x \in M, \; g \in G.$$

We can also define in the obvious way the notion of a local Lie group action.

Example. The key example is the usual linear action of the group $GL(\mathbb{R}, n)$ of $n \times n$ matrices acting by matrix multiplication on column vectors $x \in \mathbb{R}^n$. This action includes linear actions (representations) of the subgroups of $GL(\mathbb{R}, n)$ on $\mathbb{R}^n$. Because linear transformations map lines to lines, there is an induced action of $GL(\mathbb{R}, n)$ on the projective space $\mathbf{RP}(\mathbb{R}, n)^{n-1}$. The diagonal matrices λI (I denotes the identity matrix) act trivially, so the action reduces effectively to one of the projective linear groups $PSL(\mathbb{R}, n) = GL(\mathbb{R}, n)/\{\lambda I\}$. If n is odd, $PSL(\mathbb{R}, n) = SL(\mathbb{R}, n)$ can be identified with the special linear group, whereas for even n, because $-I \in SL(\mathbb{R}, n)$ has the same effect as the identity, the projective group is a quotient $PSL(\mathbb{R}, n) = SL(\mathbb{R}, n)/\{\pm I\}$.

In vision, of particular importance is the case of $GL(\mathbb{R}, 2)$, so we discuss this in some detail. The linear action of $GL(\mathbb{R}, 2)$ on $\mathbb{R}^2$ is given by

$$(x, y) \longmapsto (\alpha x + \beta y, \gamma x + \delta y), \qquad A = \begin{bmatrix} \alpha & \beta \\ \gamma & \delta \end{bmatrix} \in GL(\mathbb{R}, 2).$$

As above, we can identify the projective line $\mathbf{RP}(\mathbb{R}, n)^1$ with a circle S^1. If we use the projective coordinate $p = x/y$, the induced action is given by the linear fractional or Möbius transformations:

$$p \longmapsto \frac{\alpha p + \beta}{\gamma p + \delta}, \qquad A = \begin{bmatrix} \alpha & \beta \\ \gamma & \delta \end{bmatrix} \in GL(\mathbb{R}, 2).$$

In this coordinate chart, the x axis $\{(x, 0)\}$ in $\mathbb{R}^2$ is identified with the point $p = \infty$ in $\mathbf{RP}(\mathbb{R}, n)^1$, and the linear fractional transformations have been well defined to include the point at ∞.

Example. Let $\mathbf{v}$ be a vector field on the manifold M. Then the flow $\exp(t\mathbf{v})$ is a (local) action of the one-parameter group $\mathbb{R}$, parameterized by the time t on the manifold M.

Example. In general, if G is a Lie group that acts as a group of transformations on another Lie group H, we define the semidirect product $G \times_s H$ to be the Lie group that, as a manifold, just looks like the Cartesian product $G \times H$, but whose multiplication is given by $(g, h) \cdot (\tilde{g}, \tilde{h}) = [g \cdot \tilde{g}, h \cdot (g \cdot \tilde{h})]$ and hence is different from the Cartesian product Lie group, which has multiplication $(g, h) \cdot (\tilde{g}, \tilde{h}) = (g \cdot \tilde{g}, h \cdot \tilde{h})$.

The (full) affine group A(n) is defined as the group of affine transformations $x \mapsto Ax + a$ in $\mathbb{R}^n$, parameterized by a pair (A, a) consisting of an invertible matrix A and a vector $a \in \mathbb{R}^n$. The group multiplication law is given by $(A, a) \cdot (B, b) = (AB, a + Ab)$, and hence can be identified with the semidirect product $\mathrm{GL}(\mathbb{R}, n) \times_s \mathbb{R}^n$. The affine group can be realized as a subgroup of $\mathrm{GL}(\mathbb{R}, n+1)$ by identifying the pair (A, a) with the $(n+1) \times (n+1)$ matrix

$$\begin{bmatrix} A & a \\ 0 & 1 \end{bmatrix}.$$

Let $\mathrm{GL}_+(\mathbb{R}, n)$ denote the subgroup of $\mathrm{GL}(\mathbb{R}, n)$ with a positive determinant. Then the group of proper affine motions of $\mathbb{R}^n$ is the semidirect product of $\mathrm{GL}_+(\mathbb{R}, n)$ and the translations. Similarly, the special affine group is given by the semidirect product of $\mathrm{SL}(\mathbb{R}, n)$ and $\mathbb{R}^n$.

We may also define the Euclidean group E(n) as the semidirect product of O(n) and translations in $\mathbb{R}^n$, and the group of Euclidean motions as the semidirect product of the rotation group SO(n) and $\mathbb{R}^n$. The similarity group in $\mathbb{R}^n$, Sm(n), is generated by rotations, translations, and isotropic scalings.

In the sequel, we will usually not differentiate between the real affine group and the group of proper affine motions, and the Euclidean group and the group of Euclidean motions.

Example. In what follows, we consider all the above subgroups for $n = 2$, i.e., acting on the the plane $\mathbb{R}^2$. In this case, they are all subgroups of $\mathrm{SL}(\mathbb{R}, 3)$, the so-called group of projective transformations on $\mathbb{R}^2$. More precisely, $\mathrm{SL}(\mathbb{R}, 3)$ acts on $\mathbb{R}^2$ as follows: for $A \in SL(\mathbb{R}, 3)$

$$(\tilde{x}, \tilde{y}) = \left(\frac{a_{11}x + a_{21}y + a_{31}}{a_{13}x + a_{23}y + a_{33}}, \frac{a_{12}x + a_{22}y + a_{32}}{a_{13}x + a_{23}y + a_{33}} \right),$$

where

$$A = [a_{ij}]_{1 \le i, j \le 3}.$$

Representations. Linear actions of Lie groups, that is, representations of the group, play an essential role in applications. Formally, a representation of a group G is defined by a group homomorphism $\rho : G \to \mathrm{GL}(V)$ from G to the space of invertible linear operators on a vector space V. This means that ρ satisfies the properties $\rho(e) = I$, $\rho(g \cdot h) = \rho(g)\rho(h)$, and $\rho(g^{-1}) = \rho(g)^{-1}$.

One important method to turn a nonlinear group action into a linear representation is to look at its induced action on functions on the manifold. Given any action of a Lie group G on a manifold M, there is a naturally induced representation of G on the space $\mathcal{F} = \mathcal{F}(M)$ of real-valued functions $F : M \to \mathbb{R}$, which maps the function F to $\bar{F} := g \cdot F$ defined by

$$\bar{F}(\bar{x}) = F(g^{-1} \cdot \bar{x}),$$

or, equivalently,

$$(g \cdot F)(g \cdot x) = F(x).$$

The introduction of the inverse g^{-1} in this formula ensures that the action of G on $\mathcal{F}$ is a group homomorphism: $g \cdot (h \cdot F) = (g \cdot h) \cdot F$ for all $g, h \in G$, $F \in \mathcal{F}$.

The representation of G on the function space $\mathcal{F}$ will usually decompose into a wide variety of important subrepresentations, e.g., representations on spaces of polynomial functions, representations on spaces of smooth (C^∞) functions, or L^2 functions, etc. In general, representations of a group containing (nontrivial) subrepresentations are called reducible. An irreducible representation, then, is a representation $\rho : G \mapsto \mathrm{GL}(V)$ that contains no (nontrivial) subrepresentations, i.e., there are no subspaces $W \subset V$ that are invariant under the representation $\rho(g)W \subset W$ for all $g \in G$, other than $W = \{0\}$ and $W = V$. The classification of irreducible representations of Lie groups is a major subject of research in this century.

Example. Consider the action of the group $\mathrm{GL}(\mathbb{R}, 2)$ on the space $\mathbb{R}^2$ acting by means of matrix multiplication. This induces a representation on the space of functions

$$\bar{F}(\bar{x}, \bar{y}) = \bar{F}(\alpha x + \beta y, \gamma x + \delta y) = F(x, y),$$

where

$$A = \begin{bmatrix} \alpha & \beta \\ \gamma & \delta \end{bmatrix} \in \mathrm{GL}(\mathbb{R}, 2).$$

Note that if F is a homogeneous polynomial of degree n, $\bar{F}$ is also, so that this representation includes the finite-dimensional irreducible representations $\rho_{n,0}$ of GL($\mathbb{R}$, 2) on $\mathcal{P}^{(n)}$, the space of homogeneous polynomials of degree n. For example, on the space $\mathcal{P}^{(1)}$ of linear polynomials, the coefficients of general linear polynomial $F(x, y) = ax + by$ will transform according to

$$\begin{bmatrix} \alpha & \beta \\ \gamma & \delta \end{bmatrix} \begin{pmatrix} \bar{a} \\ \bar{b} \end{pmatrix} = \begin{pmatrix} a \\ b \end{pmatrix}, \quad A = \begin{bmatrix} \alpha & \beta \\ \gamma & \delta \end{bmatrix} \in \mathrm{GL}(\mathbb{R}, 2),$$

so that the representation $\rho_{1,0}(A) = A^{-T}$ can be identified with the inverse transpose representation.

Orbits and Invariant Functions. Let G be a group of transformations acting on the manifold M. A subset $S \subset M$ is called G invariant if it is unchanged by the group transformations, meaning that $g \cdot x \in S$ whenever $g \in G$ and $x \in S$ (provided that $g \cdot x$ is defined if the action is only local). An orbit of the transformation group is a minimal (nonempty) invariant subset. For a global action, the orbit through a point $x \in M$ is just the set of all images of x under arbitrary group transformations $\mathcal{O}_x = \{g \cdot x : g \in G\}$.

Clearly a subset $S \subset M$ is G invariant if and only if it is the union of orbits. If G is connected, its orbits are connected. The action is called transitive if there is only one orbit, so that (assuming that the group acts globally), for every $x, y \in M$ there exists at least one $g \in G$ such that $g \cdot x = y$.

A group action is called semiregular if its orbits are all submanifolds that have the same dimension. The action is called regular if, in addition, for any $x \in M$ there exist arbitrarily small neighborhoods U of x with the property that each orbit intersects U in a pathwise-connected subset. In particular, each orbit is a regular submanifold, but this condition is not sufficient to guarantee regularity; for instance, the one-parameter group $(r, \theta) \mapsto (e^t(r - 1) + 1, \theta + t)$, written in polar coordinates, is semiregular on $\mathbb{R}^2 \setminus \{0\}$ and has regular orbits, but is not regular on the unit circle.

Example. Consider the 2D torus $T = S^1 \times S^1$ with angular coordinates $(\theta, \varphi), 0 \leq \theta, \varphi < 2\pi$. Let α be a nonzero real number, and consider the one-parameter group action $(\theta, \varphi) \mapsto (\theta + t, \varphi + \alpha t) \bmod 2\pi$, $t \in \mathbb{R}$. If α/π is a rational number, then the orbits of this action are closed curves, diffeomorphic to the circle S^1, and the action is regular. On the other hand, if α/π is an irrational number, then the orbits of this action never close, and, in fact, each orbit is a dense subset of T. Therefore the action in the latter case is semiregular, but not regular.

The quotient space M/G is defined as the space of orbits of the group action, endowed with a topology induced from that of M. As the irrational flow on the torus makes clear, the quotient space can be a very complicated topological space. However, regularity of the group action will ensure that the quotient space is a smooth manifold.

Given a group of transformations acting on a manifold M by a canonical form of an element $x \in M$ we just mean a distinguished, simple representative x_0 of the orbit containing x. Of course, there is not necessarily a uniquely determined canonical form, and some choice, usually based on one's aesthetic sense of simplicity, must be used for such forms.

Now orbits and canonical forms of group actions are characterized by the invariants, which are defined as real-valued functions whose values are unaffected by the group transformations.

Definition 1.5. *An invariant for the transformation group G is a function* $I : M \to \mathbb{R}$ *that satisfies* $I(g \cdot x) = I(x)$ *for all* $g \in G,\ x \in M$.

Proposition 1.1. *Let* $I : M \to \mathbb{R}$ *be a function. Then the following three conditions are equivalent:*

1. *I is a G-invariant function.*
2. *I is constant on the orbits of G.*
3. *The level sets $\{I(x) = c\}$ of I are G-invariant subsets of M.*

For example, in the case of the orthogonal group $O(n)$ acting on $\mathbb{R}^n$, the orbits are spheres $|x| =$ constant, and hence any orthogonal invariant is a function of the radius $I = F(r)$, $r = |x|$. Invariants are essentially classified by their quotient representatives: every invariant of the group action induces a function $\tilde{I} : M/G \to \mathbb{R}$ on the quotient space and conversely. The canonical form x_0 of any element $x \in M$ must have the same invariants: $I(x_0) = I(x)$; this condition is also sufficient if there are enough invariants to distinguish the orbits, i.e., x and y lie in the same orbit if and only if $I(x) = I(y)$ for every invariant I that, according to the next theorem, is the case for regular group actions.

An important problem is the determination of all the invariants of a group of transformations. Note that if $I_1(x), \ldots, I_k(x)$ are invariant functions and $\Phi(y_1, \ldots, y_k)$ is any function, then $\hat{I} = \Phi[I_1(x), \ldots, I_k(x)]$ is also invariant. Therefore, to classify invariants, we need determine only all different functionally independent invariants. Many times globally defined invariants are difficult to find, and so we must be satisfied with the description of locally defined invariants of a group action.

Theorem 1.1. *Let G be a Lie group acting regularly on the m-dimensional manifold M with r-dimensional orbits. Then, locally, near any $x \in M$ there exist exactly $m - r$ functionally independent invariants $I_1, \ldots, I_{m-r}$ with the property that any other invariant can be written as a function of the fundamental invariants: $I = \Phi(I_1, \ldots, I_{m-r})$. Moreover, two points x and y in the coordinate chart lie in the same orbit of G if and only if the invariants all have the same value, $I_\nu(x) = I_\nu(y)$, $\nu = 1, \ldots, m - r$.*

This theorem provides a complete answer to the question of local invariants of group actions. Global and irregular considerations are more delicate; for example, consider the one-parameter isotropy group $(x, y) \mapsto (\lambda x, \lambda y)$, $\lambda \in \mathbb{R}^+$. Locally, away from the origin, x/y or y/x or any function thereof (e.g., $\theta = \tan^{-1}(y/x)$) provides the only invariant. However, if we include the origin, then there are no nonconstant invariants. On the other hand, the scaling group

$$(x, y) \mapsto (\lambda x, \lambda^{-1} y), \quad \lambda \neq 0,$$

has the global invariant xy. In general, if G acts transitively on the manifold M, then the only invariants are constants, which are completely trivial invariants. More generally, if G acts transitively on a dense subset $M_0 \subset M$, then the only continuous invariants are constants. For example, the only continuous invariants of the irrational flow on the torus are the constants, as every orbit is dense in this case. Similarly, the only continuous invariants of the standard action of $\mathrm{GL}(\mathbb{R}, n)$ on $\mathbb{R}^n$ are the constant functions, as the group acts transitively on $\mathbb{R}^2 \setminus \{0\}$. (A discontinuous invariant is provided by the function that is 1 at the origin and 0 elsewhere.)

Lie Algebras. Besides invariant functions, there are other important invariant objects associated with a transformation group, including vector fields, differential forms, differential operators, etc. We begin by considering the case of an invariant vector field, which will, in the particular case of a group acting on itself by right (or left) multiplication, lead to the crucially important concept of a Lie algebra or infinitesimal Lie group. A basic feature of (connected) Lie groups is the ability to work infinitesimally, thereby effectively linearizing complicated invariance criteria.

Definition 1.6. *Let G act on the manifold M. A vector field $\mathbf{v}$ on M is called G invariant if it is unchanged by the action of any group element: $dg(\mathbf{v}|_x) = \mathbf{v}|_{g \cdot x}$ for all $g \in G$, $x \in M$.*

In particular, if we consider the action of G on itself by right multiplication, the space of all invariant vector fields forms the Lie algebra of the group. Given that $g \in G$, let $R_g : h \mapsto h \cdot g$ denote the associated right multiplication map. A vector field $\mathbf{v}$ on G is right invariant if it satisfies $\mathrm{d}R_g(\mathbf{v}) = \mathbf{v}$ for all $g \in G$.

Definition 1.7. *The Lie algebra $\mathcal{G}$ of a Lie group G is the space of all right-invariant vector fields.*

Every right-invariant vector field $\mathbf{v}$ is uniquely determined by its value at the identity e, because $\mathbf{v}|_g = \mathrm{d}R_g(\mathbf{v}|_e)$. Therefore we can identify $\mathcal{G}$ with $TG|_e$, the tangent space to the manifold G at the identity, and hence $\mathcal{G}$ is a finite-dimensional vector space having the same dimension as G.

The Lie algebra associated with a Lie group comes equipped with a natural multiplication, defined by the Lie bracket of vector fields given by

$$[\mathbf{v}, \mathbf{w}](f) := \mathbf{v}[\mathbf{w}(f)] - \mathbf{w}[\mathbf{v}(f)].$$

By the invariance of the Lie bracket under diffeomorphisms, if both $\mathbf{v}$ and $\mathbf{w}$ are right invariant, so is $[\mathbf{v}, \mathbf{w}]$. Note that the bracket satisfies the Jacobi identity

$$[\mathbf{u}, [\mathbf{v}, \mathbf{w}]] + [\mathbf{v}, [\mathbf{w}, \mathbf{u}]] + [\mathbf{w}, [\mathbf{u}, \mathbf{v}]] = 0.$$

The basic properties of the Lie bracket translate into the defining properties of an (abstract) Lie algebra.

Definition 1.8. *A Lie algebra $\mathcal{G}$ is a vector space equipped with a bracket operation $[\,\cdot\,,\,\cdot\,] : \mathcal{G} \times \mathcal{G} \to \mathcal{G}$ that is bilinear, antisymmetric, and satisfies the Jacobi identity.*

Theorem 1.2. *Let G be a connected Lie group with Lie algebra $\mathcal{G}$. Every group element can be written as a product of exponentials: $g = \exp(\mathbf{v}_1)\exp(\mathbf{v}_2)\cdots\exp(\mathbf{v}_k)$, for $\mathbf{v}_1, \ldots, \mathbf{v}_k \in \mathcal{G}$.*

Example. The Lie algebra $\mathcal{GL}_n$ of $\mathrm{GL}(\mathbb{R}, n)$ can be identified with the space of all $n \times n$ matrices. Coordinates on $\mathrm{GL}(\mathbb{R}, n)$ are given by the matrix entries $X = (x_{ij})$. The right-invariant vector field associated with a matrix $A \in \mathcal{GL}_n$ is given by $\mathbf{v}_A = \sum_{i,j,k} a_{ij} x_{jk} \partial_{x^k}$. The exponential map is the usual matrix exponential $\exp(t\mathbf{v}_A) = e^{tA}$. The Lie bracket of two such vector fields is found to be $[\mathbf{v}_A, \mathbf{v}_B] = \mathbf{v}_C$, where $C = BA - AB$. Thus the Lie bracket on $\mathcal{GL}_n$ is identified with the negative of the matrix commutator $[A, B] = AB - BA$.

The formula $\det \exp(tA) = \exp(t \operatorname{tr} A)$ proves that the Lie algebra $\mathcal{SL}_n$ of the unimodular subgroup $\mathrm{SL}(\mathbb{R}, n)$ consists of all matrices with trace 0. The subgroups $O(n)$ and $SO(n)$ have the same Lie algebra, $\mathcal{SO}(n)$, consisting of all skew-symmetric $n \times n$ matrices.

Finally, we want to define the key concept of an invariant one-form. To do this, we first have to define the pullback of a one-form. Let $F : M \to N$ be a smooth mapping of manifolds and let η denote a one-form in $T^*N|_{y=F(x)}$. Then $F^*(\eta) \in T^*M|_x$ is the one-form given by

$$F^*(\eta)(\mathbf{v}) := \eta[\mathrm{d}F(\mathbf{v})],$$

where $\mathbf{v} \in TM|_x$.

Definition 1.9. *Let G act on the manifold M. A one-form ω on M is called* G invariant *if it is unchanged by the pull-back action of any group element*

$$g^*(\omega|_{g\cdot x}) = \omega|_x, \quad \forall g \in G, \quad x \in M.$$

Dual to the right-invariant vector fields forming the Lie algebra of a Lie group are the right-invariant one-forms known as the Maurer–Cartan forms. See Refs. [281 and 374] for details.

The following result follows from the definitions:

Lemma 1.2. *Let G be a transformation group acting on M:*

1. *If I is an invariant function, then* $\mathrm{d}I$ *is an invariant one-form.*
2. *If I is an invariant function and ω an invariant one-form, then $I\omega$ is an invariant one-form.*

Infinitesimal Group Actions. Just as a one-parameter group of transformations is generated as the flow of a vector field, so a general Lie group of transformations G acting on the manifold M will be generated by a set of vector fields on M, known as the infinitesimal generators of the group action, whose flows coincide with the action of the corresponding one-parameter subgroups of G. More precisely, if $\mathbf{v}$ generates the one-parameter subgroup $\{\exp(t\mathbf{v}) : t \in \mathbb{R}\} \subset G$, then we identify $\mathbf{v}$ with the infinitesimal generator $\widehat{\mathbf{v}}$ of the one-parameter group of transformations (or flow) $x \mapsto \exp(t\mathbf{v}) \cdot x$. Note that the infinitesimal generators of the group action are found by differentiation of the various one-parameter subgroups:

$$\widehat{\mathbf{v}}\Big|_x = \frac{\mathrm{d}}{\mathrm{d}t}\exp(t\mathbf{v})\Big|_{t=0} x, \qquad x \in M, \quad \mathbf{v} \in \mathcal{G}. \tag{1.21}$$

If $\Phi_x : G \to M$ is given by $\Phi_x(g) = g \cdot x$ (where defined), so $\widehat{\mathbf{v}}|_x = \mathrm{d}\Phi_x(\mathbf{v}|_e)$, and hence $\mathrm{d}\Phi_x(\mathbf{v}|_g) = \widehat{\mathbf{v}}|_{g\cdot x}$. Therefore resulting vector fields satisfy the same commutation relations as the Lie algebra of G, forming a finite-dimensional Lie algebra of vector fields on the manifold M isomorphic to the Lie algebra of G. Conversely, given a finite-dimensional Lie algebra of vector fields on a manifold M, we can reconstruct a (local) action of the corresponding Lie group by means of the exponentiation process.

Theorem 1.3. *If G is a Lie group acting on a manifold M, then its infinitesimal generators form a Lie algebra of vector fields on M isomorphic to the Lie algebra $\mathcal{G}$ of G. Conversely, any Lie algebra of vector fields on M that is isomorphic to $\mathcal{G}$ will generate a local action of the group G on M.*

Consequently, for a fixed group action, the associated infinitesimal generators will, somewhat imprecisely, be identified with the Lie algebra $\mathcal{G}$ itself, so that we will not distinguish between an element $\mathbf{v} \in \mathcal{G}$ and the associated infinitesimal generator of the action of G, which we also denote as $\mathbf{v}$ from now on.

Given a group action of a Lie group G, the infinitesimal generators also determine the tangent space to, and hence the dimension of, the orbits.

Proposition 1.2. *Let G be a Lie group with Lie algebra $\mathcal{G}$ acting on a manifold M. Then, for each $x \in M$, the tangent space to the orbit through x is the subspace $\mathcal{G}|_x \subset TM|_x$ spanned by the infinitesimal generators $\mathbf{v}|_x$, $\mathbf{v} \in \mathcal{G}$. In particular, the dimension of the orbit equals the dimension of $\mathcal{G}|_x$.*

Infinitesimal Invariance. As alluded to above, the invariants of a connected Lie group of transformations can be effectively computed with purely infinitesimal techniques. Indeed, the practical applications of Lie groups ultimately rely on this basic observation.

Theorem 1.4. *Let G be a connected group of transformations acting on a manifold M. A function $F : M \to \mathbb{R}$ is invariant under G if and only if*

$$\mathbf{v}[F] = 0 \tag{1.22}$$

for all $x \in M$ and every infinitesimal generator $\mathbf{v} \in \mathcal{G}$ of G.

Thus, according to Theorem 1.4, the invariants of a one-parameter group with infinitesimal generator $\mathbf{v} = \sum_i \xi^i(x)\partial_{x^i}$ satisfy the first-order, linear, homogeneous PDE

$$\sum_{i=1}^{m} \xi^i(x) \frac{\partial F}{\partial x^i} = 0. \tag{1.23}$$

The solutions of Eq. (1.23) can be computed by the method of characteristics. We replace the PDE with the characteristic system of ordinary differential equations

$$\frac{\mathrm{d}x^1}{\xi^1(x)} = \frac{\mathrm{d}x^2}{\xi^2(x)} = \cdots = \frac{\mathrm{d}x^m}{\xi^m(x)}. \tag{1.24}$$

The general solution to Eq. (1.24) can be written in the form $I_1(x) = c_1, \ldots, I_{m-1}(x) = c_{m-1}$, where the c_i are the constants of integration. It is not hard to prove that the resulting functions $I_1, \ldots, I_{m-1}$ form a complete set of functionally independent invariants of the one-parameter group generated by $\mathbf{v}$.

Example. We consider the (local) one-parameter group generated by the vector field

$$\mathbf{v} = -y\frac{\partial}{\partial x} + x\frac{\partial}{\partial y} + (1+z^2)\frac{\partial}{\partial z}.$$

The group transformations are

$$(x, y, z) \longmapsto \left(x\cos\varepsilon - y\sin\varepsilon, x\sin\varepsilon + y\cos\varepsilon, \frac{\sin\varepsilon + z\cos\varepsilon}{\cos\varepsilon - z\sin\varepsilon}\right)$$

The characteristic system of Eq. (1.24) for this vector field is

$$\frac{\mathrm{d}x}{-y} = \frac{\mathrm{d}y}{x} = \frac{\mathrm{d}z}{1+z^2}.$$

The first equation reduces to a simple separable ordinary differential equation $(\mathrm{d}y/\mathrm{d}x) = -x/y$, with the general solution $x^2 + y^2 = c_1$ for c_1, a constant of integration; therefore the cylindrical radius $r = \sqrt{x^2+y^2}$ is one invariant. To solve the second characteristic equation, we replace x with $\sqrt{r^2 - y^2}$ and treat r as constant; the solution is $\tan^{-1} z - \sin^{-1}(y/r) = \tan^{-1} z - \tan^{-1}(y/x) = c_2$, where c_2 is a second constant of integration. Therefore $\tan^{-1} z - \tan^{-1}(y/x)$ is a second invariant; we find a more convenient choice by taking the tangent of this invariant, and hence we deduce that $r = \sqrt{x^2+y^2}$ and $w = (xz - y)/(yz + x)$ form a complete system of functionally independent invariants, provided that $yz + x \neq 0$.

Invariant Equations. In addition to the classification of invariant functions of group actions, it is also important to characterize invariant systems of equations. A group G is called a symmetry group of a system of equations

$$F_1(x) = \cdots = F_k(x) = 0, \tag{1.25}$$

defined on an m-dimensional manifold M if it maps solutions to other solutions, i.e., if $x \in M$ satisfies the system and $g \in G$ is any group element such that $g \cdot x$ is defined, then we require that $g \cdot x$ also be a solution to the system. Knowledge of a symmetry group of a system of equations allows us to construct new solutions from old ones, a fact that is particularly useful when we apply these methods to systems of differential equations [281, 283]. Let $\mathcal{S}_\mathcal{F}$ denote the subvariety defined by the functions $\mathcal{F} = \{F_1, \ldots, F_k\}$, meaning the set of all solutions x to system (1.25). (Note that G is a symmetry group of the system if and only if $\mathcal{S}_\mathcal{F}$ is a G–invariant subset.) Recall that the system is regular if the Jacobian matrix $(\frac{\partial F_i}{\partial x^k})$ has constant rank n in a neighborhood of $\mathcal{S}_\mathcal{F}$, which implies (through the implicit function theorem) that the solution set $\mathcal{S}_\mathcal{F}$ is a submanifold of dimension $m - n$. In particular, if the rank is maximal, equaling k, on $\mathcal{S}_\mathcal{F}$, the system is regular.

Proposition 1.3. *Let $F_1(x) = \cdots = F_k(x) = 0$ be a regular system of equations. A connected Lie group G is a symmetry group of the system if and only if*

$$\mathbf{v}[F_\nu(x)] = 0, \qquad \textit{whenever} \quad F_1(x) = \cdots = F_k(x) = 0, \qquad 1 \leq \nu \leq k,$$

for every infinitesimal generator $\mathbf{v} \in \mathcal{G}$ of G.

Example. The equation $x^2 + y^2 = 1$ defines a circle that is rotationally invariant. To check the infinitesimal condition, we apply the generator $\mathbf{v} = -y\partial_x + x\partial_y$ to the defining function $F(x, y) = x^2 + y^2 - 1$. We find that $\mathbf{v}(F) = 0$ everywhere (because F is an invariant). Because $\mathrm{d}F$ is nonzero on the circle, the solution set is rotationally invariant.

1.7.3. Prolongations

In this subsection, we review the theory of jets and prolongations, to formalize the notion of differential invariants.

Point Transformations. We have reviewed linear actions of Lie groups on functions. Although of great importance, such actions are not the most general, and we will have to consider more general nonlinear group actions. Such transformation groups figure prominently in Lie's theory of symmetry groups of differential equations and appear naturally in the geometrically invariant diffusion equations of computer vision that we consider below. The transformation groups will act on the basic space coordinatized by the independent and dependent variables relevant to the system of differential equations under consideration. Because we want to treat differential equations, we must be able to handle the derivatives of the dependent variables on the same footing as the independent and dependent variables themselves. In this subsection, we describe a suitable geometric space for this purpose – the so-called jet space. We then discuss how group transformations are prolonged so that the derivative coordinates are appropriately acted on, and, in the case of infinitesimal generators, we state the fundamental prolongation formula that explicitly determines the prolonged action.

A general system of (partial) differential equations involves p independent variables $x = (x^1, \ldots, x^p)$, which we can view as local coordinates on the space $X \simeq \mathbb{R}^p$, and q dependent variables $u = (u^1, \ldots, u^q)$, coordinates on $U \simeq \mathbb{R}^q$. The total space will be an open subset $M \subset X \times U \simeq \mathbb{R}^{p+q}$.

A solution to the system of differential equations will be described by a smooth function $u = f(x)$. The graph of a function $\Gamma_f = \{[x, f(x)]\}$ is a p-dimensional submanifold of M that is transverse, meaning that it has no vertical tangent directions. A vector field is vertical if it is tangent to the vertical fiber $U_{x_0} \equiv \{x_0\} \times U$, so the transversality condition is $T\Gamma_f|_{z_0} \cap TU_{x_0}|_{z_0} = \{0\}$ for each $z_0 = [x_0, f(x_0)]$ with x_0 in the domain of f. Conversely, the implicit function theorem implies that any p-dimensional submanifold $\Gamma \subset M$ that is transverse at a point $z_0 = (x_0, u_0) \in \Gamma$ locally coincides with the graph of a single-valued smooth function $u = f(x)$.

The most basic type of symmetry we will discuss is provided by a (locally defined) smooth, invertible map on the space of independent and dependent variables:

$$(\bar{x}, \bar{u}) = g \cdot (x, u) = [\varphi(x, u), \psi(x, u)]. \tag{1.26}$$

The general type of transformations defined by Eq. (1.26) are often referred to as point transformations because they act pointwise on the independent and dependent variables. Point transformations act on functions $u = f(x)$ by pointwise transforming their graphs; in other words, if $\Gamma_f = \{[x, f(x)]\}$ denotes the graph of f, then the transformed function $\bar{f} = g \cdot f$ will have the graph

$$\Gamma_{\bar{f}} = \{[\bar{x}, \bar{f}(\bar{x})]\} = g \cdot \Gamma_f = \{g \cdot x[x, f(x)]\}. \tag{1.27}$$

In general, we can assert only that the transformed graph is another p-dimensional submanifold of M, and so the transformed function will not be well defined unless $g \cdot \Gamma_f$ is (at least) transverse to the vertical space at each point. This will be guaranteed if the transformation g is sufficiently close to the identity transformation and the domain of f is compact.

Example. Let

$$g_t \cdot (x, u) = (x \cos t - u \sin t,\ x \sin t + u \cos t)$$

be the one-parameter group of rotations acting on the space $M \simeq \mathbb{R}^2$ consisting of one independent and one dependent variable. Such a rotation transforms a function $u = f(x)$ by rotating its graph; therefore the transformed graph $g_t \cdot \Gamma_f$ will be the graph of a well-defined function only if the rotation angle t is not too large. The equation for the transformed function $\bar{f} = g_t \cdot f$ is given in implicit form,

$$\bar{x} = x \cos t - f(x) \sin t,$$
$$\bar{u} = x \sin t + f(x) \cos t, \tag{1.28}$$

and $\bar{u} = \bar{f}(\bar{x})$ is found when x is eliminated from these two equations. For example, if $u = ax + b$ is affine, then the transformed function is also affine and given explicitly by

$$\bar{u} = \frac{\sin t + a \cos t}{\cos t - a \sin t} \bar{x} + \frac{b}{\cos t - a \sin t},$$

which is defined provided that $\cot t \neq a$, i.e., provided the graph of f has not been rotated to be vertical.

Jets and Prolongations. Because we are interested in symmetries of differential equations, we need to know not only how the group transformations act on the independent and the dependent variables, but also how they act on the derivatives of the dependent variables. In the past century, this was done automatically, without worrying about the precise mathematical foundations of the method; in modern times, geometers have defined the jet space (or bundle) associated with the space of independent and dependent variables, whose coordinates will represent the derivatives of the dependent variables with respect to the independent variables. This gives a rigorous, cleaner, and more geometric interpretation of this theory.

Given a smooth, scalar-valued function $f(x_1, \ldots, x_p)$ of p independent variables, there are

$$p_k = \binom{p + k - 1}{k}$$

different kth-order partial derivatives,

$$\partial_J f(x) = \frac{\partial^k f}{\partial x^{j_1} \partial x^{j_2} \cdots \partial x^{j_k}},$$

indexed by all unordered (symmetric) multi-indices $J = (j_1, \ldots, j_k)$, $1 \leq j_k \leq p$, of the order of $k = \#J$. Therefore, if we have q dependent variables $(u^1, \ldots, u^q)$, we require $q_k = qp_k$ different coordinates u^α_J, $1 \leq \alpha \leq q$, $\#J = k$ to represent all the kth-order derivatives $u^\alpha_J = \partial_J f^\alpha(x)$ of a function

$u = f(x)$. For the space $M = X \times U \simeq \mathbb{R}^p \times \mathbb{R}^q$, the nth jet space $\mathrm{J}^n = \mathrm{J}^n M = X \times U^n$ is the Euclidean space of dimension

$$p + q\binom{p+n}{n},$$

whose coordinates consist of the p independent variables x^i, the q dependent variables u^α, and the derivative coordinates u^α_J, $1 \le \alpha \le q$, of the orders of $1 \le \#J \le n$. The points in the vertical space $U^{(n)}$ are denoted by $u^{(n)}$ and consist of all the dependent variables and their derivatives up to order n; thus a point in J^n has coordinates $[x, u^{(n)}]$.

A smooth function $u = f(x)$ from X to U has nth prolongation $u^{(n)} = \mathrm{pr}^{(n)} f(x)$ (also known as the *n jet*), which is a function from X to $U^{(n)}$, given by evaluation of all the partial derivatives of f up to order n; thus the individual coordinate functions of $\mathrm{pr}^{(n)} f$ are $u^\alpha_J = \partial_J f^\alpha(x)$. Note that the graph of the prolonged function $\mathrm{pr}^{(n)} f$, namely $\Gamma^{(n)}_f = \{[x, \mathrm{pr}^{(n)} f(x)]\}$, will be a p-dimensional submanifold of J^n. At a point $x \in X$, two functions have the same nth-order prolongation and so determine the same point of J^n if and only if they have nth-order contact, meaning that they and their first n derivatives agree at the point. (This is the same as requiring that they have the same nth-order Taylor polynomial at the point x.) Thus a more intrinsic way of defining the jet space J^n is to consider it as the set of equivalence classes of smooth functions by use of the equivalence relation of nth-order contact. If g is (local) point transformation (1.26), then g acts on functions by transforming their graphs and hence also acts on the derivatives of the functions in a natural manner. This allows us to naturally define an induced prolonged transformation $[\bar{x}, \bar{u}^{(n)}] = \mathrm{pr}^{(n)} g \cdot [x, u^{(n)}]$ on the jet space J^n, given directly by the chain rule. More precisely, for any point $[x_0, u_0^{(n)}] = [x_0, \mathrm{pr}^{(n)} f(x_0)] \in \mathrm{J}^n$, the transformed point $[\bar{x}_0, \bar{u}_0^{(n)}] = \mathrm{pr}^{(n)} g \cdot [x_0, u_0^{(n)}] = [\bar{x}_0, \mathrm{pr}^{(n)} \bar{f}(\bar{x}_0)]$ is found by evaluation of the derivatives of the transformed function $\bar{f} = g \cdot f$ at the image point $\bar{x}_0$, defined so that $(\bar{x}_0, \bar{u}_0) = [\bar{x}_0, \bar{f}(\bar{x}_0)] = g \cdot [x_0, f(x_0)]$. This definition assumes that $\bar{f}$ is smooth at $\bar{x}_0$ – otherwise the prolonged transformation is not defined at $[x_0, u_0^{(n)}]$. It is not hard to see that the construction does not depend on the particular function f used to represent the point of J^n; in particular, when the identification of the points in J^n with Taylor polynomials of the order of n is used, it suffices to determine how the point transformations act on polynomials of degree at most n in order to compute their prolongation.

Example. For the one-parameter rotation group considered above, the first prolongation $\mathrm{pr}^{(1)} g_t$ will act on the space coordinatized by (x, u, p) where

p represents the derivative coordinate u_x. Given a point (x_0, u_0, p_0), we choose the linear polynomial $u = f(x) = p_0(x - x_0) + u_0$ to represent it, so $f(x_0) = u_0$, $f'(x_0) = p_0$. The transformed function is given by

$$\bar{f}(\bar{x}) = \frac{\sin t + p_0 \cos t}{\cos t - p_0 \sin t}\bar{x} + \frac{u_0 - p_0 x_0}{\cos t - p_0 \sin t}.$$

Then, by Eqs. (1.28), $\bar{x}_0 = x_0 \cos t - u_0 \sin t$, so $\bar{f}(\bar{x}_0) = \bar{u}_0 = x_0 \sin t + u_0 \cos t$, and $\bar{p}_0 = \bar{f}'(\bar{x}_0) = (\sin t + p_0 \cos t)/(\cos t - p_0 \sin t)$, which is defined provided $p_0 \neq \cot t$. Therefore, with the 0 subscripts dropped, the prolonged group action is

$$\mathrm{pr}^{(1)} g_t \cdot (x, u, p) = \left(x \cos t - u \sin t, x \sin t + u \cos t, \frac{\sin t + p \cos t}{\cos t - p \sin t} \right), \tag{1.29}$$

defined for $p \neq \cot t$. Note that even though the original group action is globally defined, the prolonged group action is only locally given.

Total Derivatives. The chain-rule computations used to compute prolongations are notationally simplified if we introduce the concept of a total derivative. We find the total derivative of a function of x, u and derivatives of u by differentiating the function, treating the u's as functions of the x's.

Formally, let $F[x, u^{(n)}]$ be a function on J^n. Then the total derivative $\mathrm{D}_i F$ of F with respect to x^i is the function on $\mathrm{J}^{(n+1)}$ defined by

$$\mathrm{D}_i\left[x, \mathrm{pr}^{(n+1)} f(x)\right] = \frac{\partial F\left[x, \mathrm{pr}^{(n)} f(x)\right]}{\partial x^i}.$$

For example, in the case of one independent variable x and one dependent variable u, the total derivative D_x with respect to x has the general formula

$$\mathrm{D}_x = \frac{\partial}{\partial x} + u_x \frac{\partial}{\partial u} + u_{xx} \frac{\partial}{\partial u_x} + u_{xxx} \frac{\partial}{\partial u_{xx}} + \cdots.$$

In general, the total derivative with respect to the ith independent variable is the first-order differential operator

$$\mathrm{D}_i = \frac{\partial}{\partial x^i} + \sum_{\alpha=1}^{q} \sum_J u^\alpha_{J,i} \frac{\partial}{\partial u^\alpha_J},$$

where $u^\alpha_{J,i} = \mathrm{D}_i(u^\alpha_J) = u^\alpha_{j_1 \ldots j_k i}$. The latter sum is over all multi-indices J of arbitrary order. Even though D_i involves an infinite summation, when the total derivative is applied to any function $F[x, u^{(n)}]$ defined on the nth jet space, only finitely many terms (namely, those for $\#J \leq n$) are needed.

Higher-order total derivatives are defined in the obvious manner, with $D_J = D_{j_1} \cdot \ldots \cdot D_{j_k}$ for any multi-index $J = (j_1, \ldots, j_k)$, $1 \le j_\nu \le p$.

Prolongation of Vector Fields. Given a vector field $\mathbf{v}$ generating a one-parameter group of transformations $\exp(t\mathbf{v})$ on $M \subset X \times U$, the associated nth order prolonged vector field $\mathrm{pr}^{(n)}\mathbf{v}$ is the vector field on the jet space J^n that is the infinitesimal generator of the prolonged one-parameter group $\mathrm{pr}^{(n)} \exp(t\mathbf{v})$. Thus,

$$\mathrm{pr}^{(n)}\mathbf{v}\Big|_{[x,u^{(n)}]} = \frac{\mathrm{d}}{\mathrm{d}t}\mathrm{pr}^{(n)}[\exp(t\mathbf{v})]\Big|_{t=0} \cdot \left[x, u^{(n)}\right]. \qquad (1.30)$$

The explicit formula for the prolonged vector field is given by the following, very important, prolongation formula (see Ref. [281], Theorem 2.36, for the proof).

Theorem 1.5. *The nth prolongation of the vector field*

$$\mathbf{v} = \sum_{i=1}^{p} \xi^i(x,u)\frac{\partial}{\partial x^i} + \sum_{\alpha=1}^{q} \varphi^\alpha(x,u)\frac{\partial}{\partial u^\alpha}$$

is given explicitly by

$$pr^{(n)}\mathbf{v} = \sum_{i=1}^{p} \xi^i(x,u)\frac{\partial}{\partial x^i} + \sum_{\alpha=1}^{q}\sum_{j=\#J=0}^{n} \varphi_J^\alpha\left[x, u^{(j)}\right]\frac{\partial}{\partial u_J^\alpha}, \qquad (1.31)$$

with coefficients

$$\varphi_J^\alpha = D_J Q^\alpha + \sum_{i=1}^{p} \xi^i\, u_{J,i}^\alpha, \qquad (1.32)$$

where the characteristics of $\mathbf{v}$ *are given by*

$$Q^\alpha\left[x, u^{(1)}\right] := \varphi^\alpha(x,u) - \sum_{i=1}^{p} \xi^i(x,u)\frac{\partial u^\alpha}{\partial x^i}, \quad \alpha = 1, \ldots, q. \quad (1.33)$$

Remark: We can easily prove [281, 283] that a function $u = f(x)$ is invariant under the group generated by $\mathbf{v}$ if and only if it satisfies the characteristic equations

$$Q^\alpha\left[x, \mathrm{pr}^{(1)} f(x)\right] = 0, \quad \alpha = 1, \ldots, q.$$

Example. Suppose we have just one independent and dependent variable. Consider a general vector field $\mathbf{v} = \xi(x,u)\partial_x + \varphi(x,u)\partial_u$ on $M = \mathbb{R}^2$. The characteristic of Eq. (1.33) of $\mathbf{v}$ is the function

$$Q(x, u, u_x) = \varphi(x,u) - \xi(x,u)u_x.$$

From the above Remark, we see that a function $u = f(x)$ is invariant under the one-parameter group generated by $\mathbf{v}$ if and only if it satisfies the ordinary differential equation $\xi(x,u)u_x = \varphi(x,u)$. The second prolongation $\mathbf{v}$ is a vector field

$$\mathrm{pr}^{(2)}\mathbf{v} = \xi(x,u)\frac{\partial}{\partial x} + \varphi(x,u)\frac{\partial}{\partial u} + \varphi^x\big[x,u^{(1)}\big]\frac{\partial}{\partial u_x} + \varphi^{xx}\big[x,u^{(2)}\big]\frac{\partial}{\partial u_{xx}}$$

on J^2, whose coefficients φ^x, φ^{xx} are given by

$$\begin{aligned}\varphi^x &= \mathrm{D}_x Q + \xi u_{xx} = \varphi_x + (\varphi_u - \xi_x)u_x - \xi_u u_x^2,\\ \varphi^{xx} &= \mathrm{D}_x^2 Q + \xi u_{xxx}\\ &= \varphi_{xx} + (2\varphi_{xu} - \xi_{xx})u_x + (\varphi_{uu} - 2\xi_{xu})u_x^2 - \xi_{uu}u_x^3\\ &\quad + (\varphi_u - 2\xi_x)u_{xx} - 3\xi_u u_x u_{xx}.\end{aligned}$$

For example, the second prolongation of the infinitesimal generator $\mathbf{v} = -u\partial_x + x\partial_u$ of the rotation group is given by

$$\mathrm{pr}^{(2)}\mathbf{v} = -u\frac{\partial}{\partial x} + x\frac{\partial}{\partial u} + \left(1 + u_x^2\right)\frac{\partial}{\partial u_x} + 3u_x u_{xx}\frac{\partial}{\partial u_{xx}},$$

where the coefficients are computed as

$$\begin{aligned}\varphi^x &= \mathrm{D}_x Q + u_{xx}\xi = \mathrm{D}_x(x + uu_x) - uu_{xx} = 1 + u_x^2,\ \varphi^{xx}\\ &= \mathrm{D}_x^2 Q + u_{xxx}\xi = \mathrm{D}_x^2(x + uu_x) - uu_{xxx} = 3u_x u_{xx}.\end{aligned}$$

We can then readily recover the group transformations by integrating the system of ordinary differential equations

$$\frac{\mathrm{d}x}{\mathrm{d}t} = -u,\quad \frac{\mathrm{d}u}{\mathrm{d}t} = x,\quad \frac{\mathrm{d}p}{\mathrm{d}t} = 1 + p^2,\quad \frac{\mathrm{d}q}{\mathrm{d}t} = 3pq,$$

where we have used p and q to stand for u_x and u_{xx} to avoid confusing derivatives with jet-space coordinates. We find the second prolongation of the rotation group to be

$$\left[x\cos t - u\sin t,\, x\sin t + u\cos t,\, \frac{\sin t + p\cos t}{\cos t - p\sin t},\, \frac{q}{(\cos t - p\sin t)^3}\right],$$

as could be computed directly.

1.7.4. Differential Invariants

At long last, we can precisely define the notion of differential invariant. Indeed, recall that an invariant of a group G acting on a manifold M is just a function $I : M \to \mathbb{R}$ that is not affected by the group transformations. A differential invariant is an invariant in the standard sense for a prolonged group of transformations acting on the jet space J^n. Just as the ordinary invariants of a group action serve to characterize invariant equations, so differential invariants will completely characterize invariant systems of differential equations for the group as well as invariant variational principles. As such, they form the basic building block of many physical theories, for which we often begin by postulating the invariance of the equations or the variational principle under an underlying symmetry group. In particular, they are essential in understanding the invariant heat-type flows presented below.

Suppose that G is a local Lie group of point transformations acting on an open subset $M \subset X \times U$ of the space of independent and dependent variables, and let $\mathrm{pr}^{(n)}G$ be the nth prolongation of the group action on the nth jet space $\mathrm{J}^n = \mathrm{J}^n M$. A differential invariant is a real-valued function $I : \mathrm{J}^n \to \mathbb{R}$ that satisfies $I\{\mathrm{pr}^{(n)}g \cdot [x, u^{(n)}]\} = I[x, u^{(n)}]$ for all $g \in G$ and all $[x, u^{(n)}] \in \mathrm{J}^n$, where $\mathrm{pr}^{(n)}g \cdot [x, u^{(n)}]$ is defined. Note that I may be only locally defined.

The following gives a characterization of differential invariants:

Proposition 1.4. *A function $I : \mathrm{J}^n \to \mathbb{R}$ is a differential invariant for a connected group G if and only if*

$$pr^{(n)}\mathbf{v}(I) = 0,$$

for all $\mathbf{v} \in \mathcal{G}$ where $\mathcal{G}$ denotes the Lie algebra of G.

A basic problem is to classify the differential invariants of a given group action. Note first that if the prolonged group $\mathrm{pr}^{(n)}G$ acts regularly on (an open subset of) J^n with r_n–dimensional orbits, then, locally, there are

$$p + q^{(n)} - r_n = p + q\begin{pmatrix} p+n \\ n \end{pmatrix} - r_n$$

functionally independent nth-order differential invariants. Furthermore, any lower-order differential invariant $I[x, u^{(k)}]$, $k < n$, is automatically an nth differential invariant and should be included in the preceding count. (Here we are identifying $I : \mathrm{J}^k \to \mathbb{R}$ and its composition $I \circ \pi_k^n$ with the standard projection $\pi_k^n : \mathrm{J}^n \to \mathrm{J}^k$.)

If $\mathcal{O}^{(n)} \subset \mathrm{J}^n$ is an orbit of $\mathrm{pr}^{(n)}G$, then, for any $0 \le k < n$, its projection $\pi_k^n(\mathcal{O}) \subset \mathrm{J}^n$ is an orbit of the kth prolongation $\mathrm{pr}^{(k)}G$. Therefore the generic orbit dimension r_n of $\mathrm{pr}^{(n)}G$ is a nondecreasing function of n bounded by r, the dimension of G itself. This implies that the orbit dimension eventually stabilizes, $r_n = r^*$ for all $n \ge n_0$. We call r^* the stable orbit dimension, we call the minimal order n_0 for which $r_{n_0} = r^*$ the order of stabilization of the group.

Now a transformation group G acts effectively on a space M if

$$g \cdot x = h \cdot x, \quad \forall x \in M,$$

if and only if $g = h$. Define the global isotropy group

$$G_M := \{g : g \cdot x = x \ \forall x \in M\}.$$

Then G acts effectively if and only if G_M is trivial. Moreover, G acts locally effectively if the global isotropy group G_M is a discrete subgroup of G, in which case G/G_M has the same dimension and the same Lie algebra as G. We can now state the following remarkable result [297]:

Theorem 1.6. *The transformation group G acts locally effectively if and only if its dimension is the same as its stable orbit dimension, so that*

$$r_n = r^* = \dim G$$

for all n sufficiently large.

There are a number of important results on the stabilization dimensions, maximal orbit dimensions, and their relationship to invariants; see Refs. [282 and 283]. We will make do with the following theorem, which is very useful for counting the number of independent differential invariants of large order:

Theorem 1.7. *Suppose that, for each $k \geq n$, the (generic) orbits of $\mathrm{pr}^{(n)}G$ have the same dimension $r_k = r_n$. Then for every $k > n$ there are precisely $q_k = q\binom{p+k-1}{k}$ independent kth-order differential invariants that are not given by lower-order differential invariants.*

Next we note that the basic method for constructing a complete system of differential invariants of a given transformation group is to use invariant differential operators [281–283]. A differential operator is said to be G invariant if it maps differential invariants to higher-order differential invariants and thus, by iteration, produces hierarchies of differential invariants of arbitrarily large order. For sufficiently high orders, we can guarantee the existence of sufficiently many such invariant operators in order to completely generate all the higher-order independent differential invariants of the group by successively differentiating lower-order differential invariants. Hence a complete description of all the differential invariants is obtained by a set of low-order fundamental differential invariants along with the requisite invariant differential operators.

In our case (one independent variable), the following theorem is fundamental:

Theorem 1.8. *Suppose that G is a group of point transformations acting on a space M having one independent variable and q dependent variables. Then there exist (locally) a G-invariant one-form* $\mathrm{d}r = g\mathrm{d}x$ *of lowest order and q fundamental, independent differential invariants* $J_1, \ldots, J_q$ *such that every differential invariant can be written as a function of these differential invariants and their derivatives* $\mathcal{D}^m J_\nu$, *where*

$$\mathcal{D} := \frac{\mathrm{d}}{\mathrm{d}r} = \frac{1}{g}\frac{\mathrm{d}}{\mathrm{d}x},$$

is the invariant differential operator associated with $\mathrm{d}r$. *The parameter r gives an invariant parametrization of the curve and is called arc length.*

Remark: A version of Theorem 1.8 is true more generally. See Refs. [282 and 283].

With this, we have completed our sketch of the theory of differential invariants. Once again, the reader is referred to the texts of Refs. [281–283] for a full modern treatment of the subject, including methods for constructing and counting differential invariants. The reader is also referred to Refs. [140 and 141] for the most modern treatment on differential invariants.

1.8. Basic Concepts of Partial Differential Equations

PDEs are commonly divided in three classes, elliptic, parabolic, and hyperbolic. The classical examples for each one of these groups are, respectively,

$$\frac{\partial^2 u}{\partial x^2} + \frac{\partial^2 u}{\partial y^2} = f(x, y), \tag{1.34}$$

$$\frac{\partial u}{\partial t} = \frac{\partial^2 u}{\partial x^2} + \frac{\partial^2 u}{\partial y^2}, \tag{1.35}$$

$$\frac{\partial^2 u}{\partial t^2} = \frac{\partial^2 u}{\partial x^2}, \tag{1.36}$$

where, in the first two cases, we considered 2D functions whereas a one-dimensional one is considered for the hyperbolic example. The elliptic example is the Poisson equation, which for the particular case of $f(x, y) = 0$ becomes the Laplace equation. The parabolic equation is the classical linear heat flow or isotropic diffusion, and the hyperbolic example is the wave equation. The parabolic and the hyperbolic equations define initial value or Cauchy problems: Information is given on u at a certain time u_0 and the equations describe how $u(u(x, y, t)$ or $u(x, t))$ evolves in time. The elliptic equation defines a single static function $u(x, y)$ that satisfies the Laplace equation together with some conditions in the domain's boundaries. This is then a boundary-value problem. In an initial boundary problem we normally solve for a given region of interest. Therefore boundary conditions are given as well. Examples are Dirichlet conditions, in which the values at the boundary of the region of interest are given for all t; Newman conditions, in which the values of the normal gradients are given at the boundary; and outgoing-wave boundary conditions. The same type of boundary conditions can be given for elliptic equations, although more elaborate conditions are not uncommon.

The classification given by these simple examples can of course be generalized. Consider the following operator for an n-dimensional function:

$$L := \sum_{i,j=1}^{n} a_{ij}(x_1, x_2, \ldots, x_n) \frac{\partial^2}{\partial x_i \partial x_j}, \tag{1.37}$$

with $a_{ij} = a_{ji}$. This operator is elliptic at a given point if there exists a positive quantity μ such that at this point

$$\sum_{i,j=1}^{n} a_{ij} \xi_i \xi_j \geq \mu \sum_{i=1}^{n} \xi_i^2$$

for all n-tuples of real numbers $(\xi_1, \xi_2, \ldots, \xi_n)$. The operator is elliptic in a domain D, if it is elliptic in all the points in the domain, and it is uniformly elliptic in D if $\mu \geq \mu_0$ in the whole domain for some positive μ_0. (This elliptic operator can be transformed in a Laplacian by a linear transformation of the coordinates.) The elliptic operator L can be further generalized to obtain

$$\mathcal{L} := L + \sum_{i=1}^{n} b_i \frac{\partial}{\partial x_i} + h. \tag{1.38}$$

This operator is elliptic if L is elliptic and is uniformly elliptic if L is uniformly elliptic. From this, an operator

$$\mathcal{P} := L + \sum_{i=1}^{n} b_i \frac{\partial}{\partial x_i} - \frac{\partial}{\partial t} \tag{1.39}$$

is parabolic if L is elliptic (and similarly for uniformly parabolic). Similar extensions exist for hyperbolic operators.

1.8.1. Maximum Principle

The maximum principle is one of the most fundamental and important concepts in the theory of partial differential equations. In general it states that, for a certain class of PDE's, the maximum (minimum) of the solution is obtained at the spatial or the temporal boundaries of the domain. This principle helps to prove uniqueness of solutions, as well as many other basic PDE properties. For example, if two curves are deforming according to a certain PDE and the first curve is strictly contained inside the second one, it is basically enough to show that the distance between the curves holds the conditions for the maximum principle to conclude that the curves will maintain their inclusion property. The same is true if we want to check if the number of self-intersections of a deforming curve increases or decreases with time or if we want to check if two images deforming under certain PDE keep the order between them or not (assuming that originally one image is smaller than the other at every pixel). We will see the importance of these properties later in the next chapter when introducing curvature based motions. In this section a number of basic maximum-principle-type results are given to make the reader familiar with the concept. A detailed study of this can be found in Ref. [321], an outstanding book on PDEs and highly recommended.

Let us illustrate the basic idea behind the maximum principle with a simple example. A continuous function $u(x) : [a, b] \to \mathbb{R}$ takes on its maximum at a point in the interval $[a, b]$. If u'' is continuous and the local maxima of u is achieved at a point $c \in [a, b]$, then from elementary calculus

$$u'(c) = 0, \quad u''(c) \leq 0. \tag{1.40}$$

Suppose that, in the open interval (a, b), u satisfies

$$u'' + g(x)u' > 0$$

for a bounded function $g(x)$. Then it is clear that relations (1.40) cannot be satisfied in (a, b), and the maximum can be attained at only the end points a or b. This is a case of maximum principle.

Let us now obtain a few results on the maximum principles for equations that we will use in the rest of this book. Once again, the description if far from being comprehensive, and it is just illustrative.

Let $u(x, y)$ be twice differentiable and consider the Laplacian Δ:

$$\Delta := \frac{\partial^2}{\partial x^2} + \frac{\partial^2}{\partial y^2}.$$

The Laplace equation is

$$\Delta u = 0$$

and functions u satisfying it in a domain D are called harmonic functions. If u has a maximum inside D, then its first derivatives vanish and the second derivatives are negative (or equal to zero). Therefore

$$\Delta u \leq 0$$

at a maximum. Therefore, if u satisfies

$$\Delta u > 0$$

at each point of the domain D, then u cannot attain its maximum inside D. The general result is the following theorem.

Theorem 1.9. *Let $\Delta u \geq 0$ in D. If u attains its maximum at any point of D, then u is constant in D.*

These results actually hold for general uniformly elliptic operators with bounded coefficients. Before proceeding with other results on the maximum principle, let us illustrate how this very simple result is already very helpful to prove uniqueness of solutions. Given two function $f(x, y)$ and $g(s)$, let us look for a function $u(x, y)$ that is twice differentiable in a bounded interval D, continuous in $D \cup \partial D$, and that satisfies

$$\Delta u = f(x, y) \quad \text{in } D,$$

with the boundary condition

$$u = g(s) \quad \text{on } \partial D.$$

Assume that u_1 and u_2 are solutions of the equation, with the same f and g. Defining $u := u_1 - u_2$, it is easy to show that

$$\Delta u = 0$$

inside D and 0 on ∂D. According to the maximum principle, u cannot have a maximum in the interior of D. Because u is continuous on $D \cup \partial D$ and $u = 0$ on ∂D, $u \leq 0$. Applying the same concept to $-u$, we have that $u \geq 0$. Hence $u = 0$ in D and $u_1 = u_2$ in D.

Parabolic, not elliptic, equations will be the main ones used throughout this book. Let us now illustrate a few examples of the maximum principle for parabolic equations (results for hyperbolic equations exist as well). Consider the function $u(x, t)$ satisfying

$$\frac{\partial^2 u}{\partial x^2} - \frac{\partial u}{\partial t} = f(x, t).$$

This is simply the heat equation, where f is the rate of heat removal. Suppose that u satisfies

$$L[u] := \frac{\partial^2 u}{\partial x^2} - \frac{\partial u}{\partial t} > 0$$

in a region E of the (x, t) plane. If u has a local maximum inside E, then $u_{xx} \leq 0$ and $u_t = 0$, thereby violating $L[u] > 0$. Let us formalize and generalize this now. Assume that $E := \{0 < x < a, 0 < t \leq \tau\}$ and let $S_1 := \{x = 0, 0 \leq t \leq \tau\}$, $S_2 := \{0 \leq x \leq a, t = 0\}$, and $S_3 := \{x = a, 0 \leq t \leq \tau\}$ be three of the four sides of E.

Theorem 1.10. *Suppose that $u(x, t)$ satisfies*

$$L[u] := \frac{\partial^2 u}{\partial x^2} - \frac{\partial u}{\partial t} \geq 0,$$

in E. Then the maximum of u on $E \cup \partial E$ must occur on one of the three sides S_1, S_2, S_3.

As in the elliptic case, these results can be generalized to uniformly parabolic operators with bounded coefficients, and also used to prove uniqueness of solutions. (The maximum principle results exist for nonlinear operators as well.) We should note that this result can also be made stronger. According to it, the maximum can also occur inside E, meaning that this is a weak maximum principle. The strong maximum principle shows that if the maximum also occurs inside E, then the solution is constant for a region.

1.8.2. Hyperbolic Conservation Laws and Weak Solutions

Consider a nonlinear PDE for $u(x, t) : \mathbb{R} \times [0, \tau) \rightarrow \mathbb{R}$ written in the form

$$u_t + [F(u)]_x = 0, \tag{1.41}$$

where F is a given function of u. Such differential equations are said to be in conservation form. A PDE that can be written in conservation form is a conservation law (see also [229]).

Consider, for example, the Burgers equation:

$$u_t + uu_x = 0.$$

If the solution u is smooth, this equation can be written in conservation form with $F(u) = \frac{1}{2}u^2$:

$$u_t + \left(\frac{1}{2}u^2\right)_x = 0.$$

A conservation law states that the time rate of change of the total amount of substance contained in some region is equal to the inward flux of that substance across the boundaries of that region. The total amount of the substance in a region $[a, b]$ is

$$\int_a^b u(x,t)\mathrm{d}x.$$

Let us now denote the flux by $F(u)$, having that the flux across the boundaries of $[a, b]$ is $-F(u)|_a^b$. Thus the conservation law may be written as

$$\frac{\mathrm{d}}{\mathrm{d}t}\int_a^b u(x,t)\mathrm{d}x = -F(u)|_a^b$$

or

$$\int_a^b u_t(x,t)\mathrm{d}x = -F(u)|_a^b.$$

Observing that

$$F(u)|_a^b = \int_a^b F_x(u)\mathrm{d}x,$$

we find that the integral form of the conservation law becomes

$$\int_a^b [u_t + F_x(u)]\mathrm{d}x = 0. \tag{1.42}$$

At every point where the derivatives exist, we obtain the differential form of conservation law (1.41).

If we carry out the differentiation in the conservation law, we obtain

$$u_t + a(u)u_x = 0, \tag{1.43}$$

where $a(u) := (\partial F/\partial u)$. If a is constant, the equation is quasi linear. Otherwise, it is nonlinear.

Consider the characteristic of the conservation law given by

$$\frac{\mathrm{d}x}{\mathrm{d}t} = a(u).$$

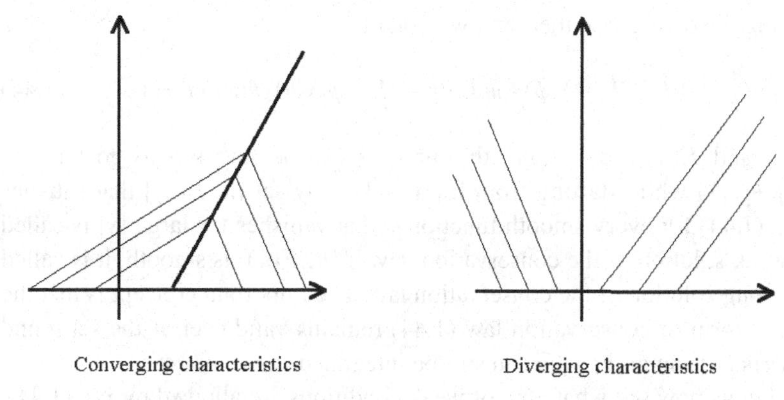

Fig. 1.2. Characteristic lines of a conservation law.

The left-hand side of Eq. (1.43) is the derivative of u in the characteristic direction. The function u is then constant along the characteristic, and so is $a(u)$. Therefore the characteristics of a conservation law are straight lines. Because the slopes of these straight lines depend on u, characteristics can intersect.

Let $u(x, 0) = f(x)$ be the initial condition for the conservation law. Let x_1 and x_2 be two points on the x axis at time $t = 0$; see Fig. 1.2. The slopes of these characteristics are $a_1 := a[u(x_1, 0)]$ and $a_2 := a[u(x_2, 0)]$. If these two characteristics intersect at a point (x, t), then the solution at this point must be equal to both $u(x_1, 0)$ and $u(x_2, 0)$, and unless we have a constant initial condition, the solution becomes multivalued at the point (x, t). This means that the differential form of the conservation law is not longer valid. We must then consider its integral form as given by Eq. (1.42).

Let us now define the operator $\vec{\nabla} := (\partial/\partial t, \partial/\partial x)$ and the vector $\vec{w} := [u, F(u)]$. The conservation law then becomes.

$$\vec{\nabla} \cdot \vec{w} = 0.$$

Let $\phi(x, t)$ be a smooth scalar function that vanishes for $|x|$ sufficiently large. Then

$$\phi \vec{\nabla} \cdot \vec{w} = 0,$$

and integrating over the $(-\infty, \infty)$ interval and for positive t's we have

$$\int_0^{\infty} \int_{-\infty}^{\infty} \phi \vec{\nabla} \cdot \vec{w} \mathrm{d}x \mathrm{d}t = 0.$$

Using the divergence theorem, we obtain

$$\int_0^\infty \int_{-\infty}^\infty (\vec{\nabla}\phi) \cdot \vec{w} \mathrm{d}x\mathrm{d}t + \int_{-\infty}^\infty \phi(x, 0) f(x)\mathrm{d}x = 0. \tag{1.44}$$

If u and $F(u)$ are smooth, this process can be reversed to go back to Eq. (1.41) while starting from Eq. (1.44). Any set $[u, F(u)]$ that satisfies Eq. (1.44) for every smooth function ϕ that vanishes for large $|x|$ is called a weak solution of the conservation law. If $[u, F(u)]$ is smooth, it is called a strong solution of the conservation law. The important concept is that the weak form of conservation law (1.44) remains valid even if the set u and $F(u)$ is not smooth, as $\vec{w}$ can still be integrated.

Let us now see what type of weak conditions are allowed by Eq. (1.44). For this, let Γ be a smooth curve across which u has a discontinuity (a jump); u has well-defined limits on both sides of Γ and is smooth away from Γ. Let P be any point on Γ and let D be a small disk centered at P. Assume that, inside D, the curve Γ is given by $x = \psi(t)$, and let D_1 and D_2 be the components of D that are determined by Γ. Let ϕ be a smooth test function. Then,

$$\iint_D (\vec{\nabla}\phi) \cdot \vec{w} \mathrm{d}x\mathrm{d}t = \iint_{D_1} (\vec{\nabla}\phi) \cdot \vec{w} \mathrm{d}x\mathrm{d}t + \iint_{D_2} (\vec{\nabla}\phi) \cdot \vec{w} \mathrm{d}x\mathrm{d}t = 0.$$

Let us select ϕ to be a smooth function that vanishes for ∂D, the boundary of D. Because u is smooth in each D_i, the divergence theorem gives

$$\iint_{D_i} (\vec{\nabla}\phi) \cdot \vec{w} \mathrm{d}x\mathrm{d}t = \iint_{D_i} \vec{\nabla} \cdot (\phi\vec{w})\mathrm{d}x\mathrm{d}t = \int_{\partial D_i} \phi(-u\mathrm{d}x + F\mathrm{d}t).$$

The lines' integrals are nonzero only along Γ. Thus, if $u_l := u[x(t) - 0, t]$ and $u_r := u[x(t) + 0, t]$, we have

$$\int_{\partial D_1} \phi(-u\mathrm{d}x + F\mathrm{d}t) = \int_{Q_1}^{Q_2} \phi[-u_l\mathrm{d}x + F(u_l)\mathrm{d}t],$$

$$\int_{\partial D_2} \phi(-u\mathrm{d}x + F\mathrm{d}t) = -\int_{Q_1}^{Q_2} \phi[-u_r\mathrm{d}x + F(u_r)\mathrm{d}t],$$

where Q_i are the points where Γ intersects ∂D. Therefore,

$$0 = \int_\Gamma \phi(-[u]\mathrm{d}x + [F(u)]\mathrm{d}t),$$

where $[u] := u_l - u_r$, the jump across Γ, and $[F(u)] := F(u_l) - F(u_r)$. Because ϕ is arbitrary, we conclude that

$$s[u] = [F(u)] \tag{1.45}$$

at each point of Γ, where $s := \mathrm{d}\phi/\mathrm{d}t$ is the speed of the discontinuity. We have then that the weak form of the conservation law implies Eq. (1.45), which ties together u on both sides of the discontinuity. Relation (1.45) is called the jump condition or the Rankine–Hugoniot condition.

Unfortunately, the jump condition is not enough to determine a unique solution to Eq. (1.44). To illustrate this, consider the stationary Burger's equation:

$$(u^2)_x = 0.$$

In the class of smooth solutions, $u \equiv$ constant, whereas in the class of weak solutions

$$u(x) = \begin{cases} c & 0 < x < p \\ -c & 0 < x < 2p, \end{cases}$$

$u(x + 2p) = u(x)$, $p > 0$ (a periodic Haar-type function). We need then to add conditions to obtain a physically relevant solution. A possible condition is what is called the entropy condition, which adds constraints on the discontinuity speed s. For every u between two states u_l and u_r, this conditions establishes that the relation

$$\frac{F(u) - F(u_l)}{u - u_l} \geq s \geq \frac{F(u) - F(u_r)}{u - r_r} \tag{1.46}$$

must hold. Weak solutions satisfying this entropy condition are unique for given initial data. A discontinuity is called a shock if the inequality holds. If equality holds, the discontinuity is called contact discontinuity.

By the definition of $a(u)$,

$$a(u_l) = \frac{\partial F}{\partial u}(u_l) = \lim_{u \to u_l} \frac{F(u) - F(u_l)}{u - u_l},$$

$$a(u_v) = \frac{\partial F}{\partial u}(u_r) = \lim_{u \to u_r} \frac{F(u) - F(u_r)}{u - u_r},$$

which means that the entropy condition can also be written as

$$a(u_l) \geq s \geq a(u_r).$$

Riemann Problems. A particular case of conservation laws is when the initial condition is a step function:

$$u(x, 0) = f(x) = \begin{cases} u_l, & x < 0 \\ v_r, & x > 0 \end{cases}.$$

This problem is called a Riemann problem. Solutions to a Riemann problem are self-similar:

$$u(x, t) = h(x/t),$$

where h is a piecewise continuous function of one variable. Solutions to the Riemann problem also hold the additivity property.

Solutions to the Riemann problem have the form of a wave propagating with speed s:

$$u(x, t) = f(x - st) = \begin{cases} u_l, & x/t < s \\ u_r, & x/t > s \end{cases}.$$

Suppose now that u_l and u_r are two states that violate the entropy condition:

$$a(u_l) < a(u_r).$$

Then we obtain a weak solution that violates the entropy condition. In this case, the solution is an expansion or rarefaction wave. The solution has the form

$$u(x, t) = w(x/t) = \begin{cases} u_l, & x/t \leq a(u_l) \\ h(x/t), & a(u_l) \leq x/t \leq a(u_r), \\ u_r, & x/t \geq a(u_r) \end{cases}$$

where h is a function such that $a[h(z)] = z$.

Viscosity Principle. The entropy condition receives a new intuitive meaning when we look at the viscosity principle as an alternative to determine the physically correct weak solution to the weak form of the conservation law. In this approach, we look for

$$u(x, t) := \lim{}_{\epsilon \to 0, \epsilon > 0} u_\epsilon(x, t),$$

where $u_\epsilon(x, t)$ satisfies

$$(u_\epsilon)_t + [F(u_\epsilon)]_x = \epsilon[\gamma(u_\epsilon)(u_\epsilon)_x]_x \tag{1.47}$$

for $\gamma(u_\epsilon) \geq 0$. This equation admits a unique solution, and the solution exists for all positive t. The solutions u_ϵ converge to a weak solution u of the conservation law. Moreover, this limit solution satisfies the entropy condition. The goal of the viscosity is then to smooth out the shock, in such a way that, when the viscosity vanishes, the correct weak solution is obtained.

The theory of conservation laws can be extended to systems of equations, but we will not be dealing with these in the book. We will see the use of conservation laws when numerical techniques are presented to solve curve and surface deformation problems in Chap. 2.

1.8.3. Viscosity Solutions

We have just seen how to deal with solutions of PDE that are not smooth. This is of course fundamental in image processing, as images have sharp edges (jumps) and curves have corners. But what happens if the PDE cannot be written in conservative form? How can we extend the notion of unique, physically meaningful, solutions for PDEs that are not conservation laws? We deal with this right now, briefly introducing the concept of viscosity solutions. A full description of the topic can be found in Ref. [105].

The basic idea in the theory of viscosity solutions is to study second-order partial differential equations of the form $F(x, u, \nabla u, \nabla^2 u) = 0$, where $u(x) : \mathbb{R}^n \to \mathbb{R}$ and ∇ and ∇^2 stand for the gradient and the matrix of second derivatives, respectively. The theory of viscosity solutions will provide existence and uniqueness theorems even when ∇u and $\nabla^2 u$ are not defined in the classical sense. That is, the function u does not have to be differentiable, discontinuities are allowed, and derivatives are considered in the weak sense. For the theory to work, the operator F must satisfy the monotonicity condition

$$F(x, r, p, X) \leq F(x, s, p, Y)$$

whenever $r \leq s$ and $Y \leq X$.

Later in this book will see specific examples of PDEs with relevance for image processing that hold this monotonicity condition. Let us now just give a few examples that are familiar from the general PDE literature. A simple example is the Laplace equation,

$$-\Delta u + c(x)u = f(x),$$

which holds the condition if $c > 0$. We can show this by noting that $F(x, r, p, X) = -\text{trace}(X) + c(x)r - f(x)$. This can be generalized to other linear elliptic equations.

Additional examples are first-order equations $F(x, u, \nabla u)$, which hold the property if $F(x, r, p)$ is nondecreasing in $r \in \mathbb{R}$. The equations of the type $F(x, u, \nabla u) = 0$ play an important role in the calculus of variations (see below), and they are called Hamilton–Jacobi equations (the operator is convex in (r, p)). Many image processing problems, such as shape from shading, have this form.

The last example is parabolic equations of the type

$$u_t + F(t, x, u, \nabla u, \nabla^2 u) = 0,$$

which will hold the monotonicity condition if the operator F holds it. (This example will cover a large portion of the PDEs introduced in this book.) For example, consider

$$u_t = \|\nabla u\| \operatorname{div}\left(\frac{\nabla u}{\|\nabla u\|}\right),$$

which, as we will later see, is one of the most important flows used in this book (div indicates divergence). Carrying out the differentiation, we obtain

$$u_t - \Delta u + \sum_{i,j=1}^{N} \frac{\partial^2 u}{\partial x_i \partial x_j} \frac{\partial u}{\partial x_i} \frac{\partial u}{\partial x_j} \|\nabla u\|^{-2} = 0.$$

Then,

$$F(x, p, X) = -\text{trace}\left[\left(I - \frac{p \times p}{\|p\|}\right) X\right],$$

which holds the required condition.

Let us now proceed with the basic definitions. We assume from now on that the monotonicity condition holds. Assume that u is twice differentiable and satisfies

$$F[x, u(x), \nabla u(x), \nabla^2 u(x)] \leq 0.$$

u is said to be a subsolution of $F = 0$ or a classical solution of $F \leq 0$. Suppose that there is a function ϕ that is also twice differentiable and that $\hat{x}$ is a local maximum of $u - \phi$. Then, $\nabla u(\hat{x}) = \nabla \phi(\hat{x})$ and $\nabla^2 u(\hat{x}) \leq \nabla^2 \phi(\hat{x})$. Therefore, from the monotonicity assumption,

$$F[\hat{x}, u(\hat{x}), \nabla \phi(\hat{x}), \nabla^2 \phi(\hat{x})] \leq F[\hat{x}, u(\hat{x}), \nabla u(\hat{x}), \nabla^2 u(\hat{x})] \leq 0.$$

We can then define that an arbitrary function u (not necessary derivable) is a subsolution of $F = 0$ if

$$F[\hat{x}, u(\hat{x}), \nabla \phi(\hat{x}), \nabla^2 \phi(\hat{x})] \leq 0$$

whenever ϕ is twice differentiable and $\hat{x}$ is a local maximum of $u - \phi$.

Having this, we can define that a viscosity subsolution of $F = 0$, or, equivalently, a viscosity solution of $F \leq 0$ is a function u that is upper semicontinuous and such that

$$F[x, u(x), p, X] \leq 0.$$

(See Ref. [105] for restrictions on the domain of x as well as on the pair (p, X), which are basically required for satisfying the Taylor expansion of u around $\hat{x}$.) Replacing $\leq$ with $\geq$ and upper semicontinuous by lower semicontinuous, we obtain the definition of viscosity supersolution. A function u is a viscosity solution if it is both a viscosity subsolution and a viscosity supersolution.

An intuitive way of looking at this is that we look to frame the (possibly) discontinuous solution u of $F = 0$ by twice-differentiable solutions from below and from above. We then consider the limit case in which these solutions from below and above coincide, giving us the viscosity solution.

The majority of the equations presented later in this book have a unique viscosity solutions. Then this framework allows us to freely discuss PDEs even when we know that discontinuities (e.g., edges) are present in the image and the image might not be differentiable.

1.9. Calculus of Variations and Gradient Descent Flows

A number of the PDEs used in image processing appear as means to minimize a given energy functional. In this section classical results on calculus of variations and gradient descent are described.

Given a one-dimensional function $u(x) : [0, 1] \to R$, the basic problem is to minimize a given energy **E**,

$$\mathbf{E}(u) := \int_0^1 F(u, u')\mathrm{d}x, \tag{1.48}$$

with given boundary conditions $u(0) = a$ and $u(1) = b$, and with $F : \mathbb{R}^2 \to \mathbb{R}$ given by the problem.

From classical calculus, we know that the extrema of a function $f(x) : \mathbb{R} \to \mathbb{R}$ are attained at the positions where $f' = 0$. Similarly, the extrema of the functional $\mathbf{E}(u)$ are obtained at points where $\mathbf{E}' = 0$, where $\mathbf{E}' = (\partial E/\delta u)$ is the first variation. We proceed now to find $\mathbf{E}'$.

The best u defeats any other candidate $u + v$, with $v(x) : [0, 1] \to \mathbb{R}$ and $v(0) = v(1) = 0$. If both v and v' are small, then

$$F(u + v, u' + v') = F(u, u') + v\frac{\partial F}{\partial u} + v'\frac{\partial F}{\partial u'}+, \ldots,$$

$$\mathbf{E}(u + v) = \mathbf{E}(u) + \int_0^1 \left(v\frac{\partial F}{\partial u} + v'\frac{\partial F}{\partial u'}\right)\mathrm{d}x+, \ldots,$$

yielding the necessary extrema (weak) condition:

$$\mathbf{E}'(u) = \int_0^1 \left(v\frac{\partial F}{\partial u} + v'\frac{\partial F}{\partial u'} \right) \mathrm{d}x = 0.$$

Integrating by parts, we obtain

$$\mathbf{E}'(u) = \int_0^1 \left[v\frac{\partial F}{\partial u} - v\frac{\mathrm{d}}{\mathrm{d}x}\left(\frac{\partial F}{\partial u'}\right) \right] \mathrm{d}x + v\,\frac{\partial F}{\partial u'}\bigg|_0^1 = 0.$$

Because the boundary terms vanish [$v(0) = v(1) = 0$] and the above relation is true for all v with vanishing boundary conditions, we obtain that

$$\frac{\partial F}{\partial u} - \frac{\mathrm{d}}{\mathrm{d}x}\left(\frac{\partial F}{\partial u'}\right) = 0$$

is the necessary (strong) condition for u to be an extremum of $\mathbf{E}(u)$. This is the Euler equation for a one-dimensional problem in the calculus of variations.

In a similar form we can obtain that, for an energy of the form

$$\mathbf{E}(u) := \int_0^1 F(u, u', u'')\mathrm{d}x, \tag{1.49}$$

the Euler equation is given by

$$\frac{\partial F}{\partial u} - \frac{\mathrm{d}}{\mathrm{d}x}\left(\frac{\partial F}{\partial u'}\right) + \frac{\mathrm{d}^2}{\mathrm{d}x^2}\left(\frac{\partial F}{\partial u''}\right) = 0. \tag{1.50}$$

The derivation for the 2D problem is completely analogous. Given a 2D function $u(x, y) : \Omega \in \mathbb{R}^2 \to \mathbb{R}$ and an energy $\mathbf{E}$, where

$$\mathbf{E} := \iint_\Omega F(u, \partial u/\partial x, \partial u/\partial y, \partial^2 u/\partial x^2, \partial^2 u/\partial y^2)\mathrm{d}x\mathrm{d}y, \tag{1.51}$$

the Euler equation is given by

$$\frac{\partial F}{\partial u} - \frac{\mathrm{d}}{\mathrm{d}x}\left(\frac{\partial F}{\partial u_x}\right) - \frac{\mathrm{d}}{\mathrm{d}y}\left(\frac{\partial F}{\partial u_y}\right) + \frac{\mathrm{d}^2}{\mathrm{d}x^2}\left(\frac{\partial F}{\partial u_{xx}}\right) + \frac{\mathrm{d}^2}{\mathrm{d}y^2}\left(\frac{\partial F}{\partial u_{yy}}\right) = 0. \tag{1.52}$$

Example. Let

$$F = \rho(\|\nabla u\|),$$

where $\rho(r) : \mathbb{R} \rightarrow \mathbb{R}$ is a given function and ∇u stands for the gradient of u. In detail, F is given by

$$F = \rho\left\{\left[\left(\frac{\partial u}{\partial x}\right)^2 + \left(\frac{\partial u}{\partial y}\right)^2\right]^{1/2}\right\}.$$

Therefore (we omit normalization constants in each step)

$$\begin{aligned}
\mathbf{E}' &= \frac{\mathrm{d}}{\mathrm{d}x}\left(\frac{\partial F}{\partial u_x}\right) + \frac{\mathrm{d}}{\mathrm{d}y}\left(\frac{\partial F}{\partial u_y}\right) \\
&= \frac{\mathrm{d}}{\mathrm{d}x}\left[\rho'(\|\nabla u\|)\frac{u_x}{\|\nabla u\|}\right] + \frac{\mathrm{d}}{\mathrm{d}y}\left[\rho'(\|\nabla u\|)\frac{u_y}{\|\nabla u\|}\right] \\
&= \operatorname{div}\left[\rho'(\|\nabla u\|)\frac{\nabla u}{\|\nabla u\|}\right].
\end{aligned}$$

In the particular case of $\rho(r) = r^2$, that is, a quadratic penalty term, we obtain $\rho' = 2r$, and the Euler equation is

$$\operatorname{div}(\nabla u) = 0$$

or

$$\Delta u = 0,$$

where Δu stands for the Laplacian of u.

Adding Constraints

In a number of cases, we are not allowed to freely search for the optimal u, but constraints are added. For example, we might be searching for a function u that minimizes the energy $\mathbf{E}(u)$ subject to a constraint on the average of u, that is, subject to

$$\frac{1}{|\Omega|}\int_{\Omega} u \mathrm{d}x = \alpha,$$

where α is a given constant. In this case, the constraint is added to the energy by use of Lagrange multiplier λ. The new energy becomes

$$\hat{\mathbf{E}} = \mathbf{E} + \lambda \cdot \text{constraint}.$$

For the example of given signal average, we have

$$\hat{\mathbf{E}} = \mathbf{E} + \lambda\left(\frac{1}{|\Omega|}\int_{\Omega} u \mathrm{d}x - \alpha\right),$$

and we can use for $\hat{\mathbf{E}}$ the same formulas we developed for $\mathbf{E}$. Note that the Lagrange multiplier λ is constant, as is α.

Gradient Descent Flows

One of the fundamental questions is how to solve the Euler equation, that is, how to solve the equation

$$\mathbf{E}'(u) = 0.$$

Only in a very limited number of simple cases does this problem gives an analytic and simple solution. In most image processing applications, directly solving this problem is not possible. One general technique of looking for a possible solution of $\mathbf{E}' = 0$ is to numerically solve (see Section 1.10) the PDE

$$\frac{\partial u}{\partial t} = \mathbf{E}'(u) \tag{1.53}$$

with given initial data u_0. The auxiliary variable t is denoted as the time-marching parameter.

When the steady estate of this equation is achieved,

$$\frac{\partial u}{\partial t} = 0,$$

we obtain

$$\mathbf{E}'(u) = 0,$$

and a solution to the Euler equation is obtained. Therefore, to solve the Euler equation, we start with some initial data u_0 and look for the steady-state solution of Eq. (1.53). This technique is denoted as the gradient descent. A number of questions need to be answered before this technique is used. For example, is the solution to the PDE unique and is the solution independent of the initial condition? In the case of nonconvex energies, of course, the solution will be greatly dependent on the initial data u_0.

Let us just conclude this section with the gradient descent flow for the example above, with $\rho(r) = r^2$. In this case, the Euler equation is $\Delta u = 0$, and then the corresponding gradient descent flow is

$$\frac{\partial u(x, y, t)}{\partial t} = \Delta u(x, y, t),$$

which is the heat flow introduced before (a parabolic equation).

1.10. Numerical Analysis

The goal of this section is to introduce the reader to the basic language of numerical analysis. Only generic concepts are introduced. In next chapter specific techniques are described that apply to some of the PDEs that are being used in image processing. We deal with only finite-difference methods, which are the most popular for the applications presented in this book. Finite elements have also been used in image processing and computer vision, and their description can be found, for example, in Ref. [187]. (The material in this section is inspired by and adapted from Ref. [240].)

Let us use a simple example to illustrate the ideas. Consider the function $u(x, t) : (-\infty, \infty) \times [0, \infty) \to \mathbb{R}$ satisfying

$$\frac{\partial u}{\partial t} + a\frac{\partial u}{\partial x} = 0, \tag{1.54}$$

with initial condition

$$u(x, 0) = u_0(x).$$

We discretize the (x, t) plane by choosing a mesh width $h = \Delta x$ and a time step $k = \Delta t$, and define the discrete mesh points (x_j, t_n) such that

$$x_j = jh, \quad j = , \ldots, -1, 0, 1, \ldots,$$
$$t_n = nk, \quad n = 0, 1, 2, \ldots.$$

It is also useful to define

$$x_{j+1/2} = x_j + h/2 = \left(j + \frac{1}{2}\right)h.$$

Throughout this section, as well as the rest of this book, we assume that both h and k are constant. Moreover, we also assume that these two parameters are related in some way, so that specifying one of them (usually k) is enough. The goal of numerical analysis is to compute approximations U_j^n to the solution $u(x_j, t_n)$ at the discrete grid points. The following notation will be used, throughout this section in particular and this book in general,

$$u_j^n := u(x_j, t_n),$$

to represent the pointwise values of the true solution, and

$$D_x^-(u) := u_j^n - u_{j-1}^n,$$
$$D_x^+(u) := u_{j+1}^n - u_j^n$$

to represent backward and forward derivatives respectively. For 2D problems, D_y^- and D_y^+ will be used as well.

For conservation laws it is often desirable to have U_j^n approximate the cell average, defined as

$$\bar{u}_j^n := \frac{1}{h}\int_{x_j-1/2}^{x_j+1/2} u(x, t_n)\mathrm{d}x, \tag{1.55}$$

rather than approximate the true value at a given mesh point.

The initial value U_j^0 of the discrete problem is obtained either by sampling or averaging of the initial value of the continuous problem $(u_0(x))$.

Because all the cells will have a constant value, this leads to the natural definition of a piecewise constant function $U_k(x, t)$:

$$U_k(x, t) := U_j^n,\ (x, t) \in [x_{j-1/2}, x_{j+1/2}) \times [t_n, t_{n+1}]. \tag{1.56}$$

Because the computation of Eq. (1.54) cannot be done in the full $(-\infty, \infty)$, we need to restrict the interval to some finite interval (b, c) and assume boundary conditions. Possible boundary conditions are periodic ones,

$$u(b, t) = u(c, t),$$

whereas other boundary conditions can be specified as well and incorporated into the numerical scheme.

The basic idea, now that we have both the initial data and boundary conditions, is to use a time-marching scheme, in which we compute U^{n+1} from U^n or, more generally, U^{n+1} from $U^n, U^{n-1}, \ldots, U^{n-r}$. The simplest way of solving this problem is to replace the derivatives in the PDE with appropriate finite-difference approximations. For example, replacing u_t with a forward-in-time approximation and u_x with a central-difference approximation, we obtain

$$\frac{U_j^{n+1} - U_j^n}{k} + a\frac{U_{j+1}^n - U_{j-1}^n}{2h} = 0.$$

Solving for U_j^{n+1}, we obtain

$$U_j^{n+1} = U_j^n - \frac{k}{2h}a\left(U_{j+1}^n - U_{j-1}^n\right). \tag{1.57}$$

This method is quite simple, defining an explicit system, but suffers from stability problems, requiring an impractically small time step, and even then it is not guaranteed that the result is correct. We should note that it is possible to define other explicit systems that do provide satisfactory results with reasonable time steps.

One important step forward is to use an implicit scheme, in which the central difference is also evaluated at $t = n + 1$:

$$\frac{U_j^{n+1} - U_j^n}{k} + a\frac{U_{j+1}^{n+1} - U_{j-1}^{n+1}}{2h} = 0.$$

Solving for U_j^{n+1}, we obtain

$$U_j^{n+1} = U_j^n - \frac{k}{2h}a\left(U_{j+1}^{n+1} - U_{j-1}^{n+1}\right). \tag{1.58}$$

Now, to solve for U^{n+1} from U^n, we consider these equations as a system of equations over all values of j. Having a bounded interval with N grid points, we obtain an $N \times N$ discrete system.

We can observe the grid points involved in the computation of U_j^{n+1}, which define the stencil of the method. The stencils for Eqs. (1.57) and (1.58) are a **T** and an inverted **T**, respectively. By using other stencils, we can obtain a wide variety of numerical methods for Eq. (1.54); see Ref. [240] for examples. In addition, methods that look at different past times, not just U^n but also $U^{n-1}, \ldots,$ can be devised, although they are not common for the kind of equations described in this book. Therefore we consider only methods that depend on U^n and write

$$U^{n+1} = \mathcal{H}_k(U^n),$$

meaning that the solution at time $n + 1$ depends on the solution at time n. Of course the value of U_j^{n+1} might depend on several components of the vector U^n, and this is expressed by

$$U_j^{n+1} = \mathcal{H}_k(U^n; j).$$

For the explicit method above we obtain

$$\mathcal{H}_k(U^n; j) = U_j^n - \frac{k}{2h}a\left(U_{j+1}^n - U_{j-1}^n\right).$$

The operator $\mathcal{H}_k$ can also be applied to functions. In particular, it can be applied to the piecewise constant function $U_k(x, t)$, yielding

$$U_k(x, t + k) = \mathcal{H}_k(U_k(x, t); x),$$

also a piecewise constant function. Finally, note that for linear discrete approximations, the operator $\mathcal{H}_k$ is just a matrix.

1.10.1. Convergence

The goal of numerical implementations is of course to approximate the solution u. We need then to analyze the error of the approximation. We can consider both the pointwise error

$$E_j^n := U_j^n - u_j^n$$

and the error relative to the cell average

$$E_j^n := U_j^n - \bar{u}_j^n.$$

We can actually just consider the error function

$$E_k(x,t) := U_k(x,t) - u(x,t).$$

We say that a method is convergent if for a given metric $\|\cdot\|$ we obtain

$$\|E_k(x,t)\| \to_{k\to 0} 0,$$

for given initial data u_0.

We then are left to define the metric to be used to measure convergence. One possibility is to consider the L_1 norm, which for a given function $v(x)$ is defined as

$$\|v(x)\|_1 = \int_{-\infty}^{\infty} |v(x)| \mathrm{d}x.$$

Another possibility is to use the L_∞ norm or the L_2 norm, given by

$$\|v(x)\|_\infty = \sup_x |v(x)|,$$

$$\|v(x)\|_2 = \left[\int_{-\infty}^{\infty} |v(x)|^2 \mathrm{d}x\right]^{1/2}.$$

Note that of course the L_∞ does not need to go to zero, and the method can still converge in the L_1 sense. The L_2 norm is frequently used for linear systems.

1.10.2. Local Truncation Error

The local truncation error $L_k(x,t)$ is a measure of how well the difference equation models the differential equation, locally. We define it by replacing the approximate solution U_j^n in the difference equation with the true solution $u(x_j, t_n)$. How well this true solution satisfies the difference equation gives an indication of how well the exact solution of the difference equation satisfies the differential equation.

Let us consider as an example the Lax–Friedrichs method, which is similar to the method developed in Eq. (1.57), except U_j^n is replaced with $\frac{1}{2}(U_{j-1}^n + U_j^n)$. This scheme is stable provided that k/h is small enough. Let us write the equation corresponding to this form:

$$\frac{1}{k}\left[U_j^{n+1} - \frac{1}{2}\left(U_{j-1}^n + U_{j+1}^n\right)\right] + \frac{1}{2h}a\left(U_{j+1}^n - U_{j-1}^n\right) = 0.$$

Let us proceed to perform the replacement:

$$\begin{aligned} L_k(x,t) :=& \frac{1}{k}\left\{u(x,t+k) - \frac{1}{2}[u(x-h,t) + u(x+h,t)]\right\} \\ &+ \frac{1}{2h}a[u(x+h,t) - u(x-h,t)]. \end{aligned}$$

We now assume smooth solutions and then perform a Taylor series expansion, obtaining ($u = u(x,t)$)

$$\begin{aligned} L_k(x,t) &= \frac{1}{k}\left[\left(u + ku_t + \frac{1}{2}k^2 u_{tt} + \cdots\right) - \left(u + \frac{1}{2}h^2 u_{xx} + \cdots\right)\right] \\ &\quad + \frac{1}{2h}a\left(2hu_x + \frac{1}{3}h^3 u_{xxx} + \cdots\right) \\ &= u_t + au_x + \frac{1}{2}\left(ku_{tt} - \frac{h^2}{k}u_{xx}\right) + O(h^2). \end{aligned}$$

Because we have assumed that u is the exact solution of the differential equation, $u_t + au_x = 0$. Using this and the fact that $u_{tt} = a^2 u_{xx}$, we obtain

$$L_k(x,t) = \frac{k}{2}\left(a^2 - \frac{h^2}{k^2}\right) u_{xx} + O(k^2) =_{k\to 0} O(k).$$

Recall that we have assumed that h/k is constant, so that h^2/k^2 is constant as the mesh is refined. With a bit of further analysis, it can actually be shown that

$$|L_k(x,t)| \leq Ck$$

for all $k < k_0$ and a constant C that depends on only the original data.

Generalizing this example, we have that the local truncation error is defined as

$$L_k(x,t) := \frac{1}{k}\{u(x,t+k) - \mathcal{H}_k(u(\cdot,t);x)\},$$

and the method is said to be consistent if

$$\|L_k(\cdot, t)\| \to_{k\to 0} 0.$$

The method is said to be of the order of p if for all sufficient smooth initial data with compact support there is some constant C such that

$$\|L_k(\cdot, t)\| \leq Ck^p$$

for all $k < k_0$ and $0 \leq t \leq T$.

1.10.3. Stability

Let us rewrite the definition of local truncation error in the form of

$$u(x, t + k) = \mathcal{H}_k(u(\cdot, t); x) + kL_k(x, t).$$

Because the numerical solution satisfies $U_k(x, t + k) = \mathcal{H}_k(U_k(\cdot, t); x)$, we can obtain a simple recursive relation for the error in the linear case:

$$E_k(x, t + k) = \mathcal{H}_k(E_k(\cdot, t); x) - kL_k(x, t).$$

The error at time $t + k$ is then composed of the local error $-kL_k$ introduced in this time step together with the cumulative error from previous steps. Applying this relation recursively, we obtain

$$E_k(\cdot, t_n) = \mathcal{H}_k^n E_k(\cdot, 0) - k \sum_{i=1}^{n} \mathcal{H}_k^{n-i} L_k(\cdot, t_{i-1}),$$

where $\mathcal{H}_k$ stands for the matrix representing the linear operator $\mathcal{H}_k(\cdot, t)$ and the superscripts represent powers of this matrix. We then say that the method is stable if for each time T there is a constant C and a value $k_0 > 0$ such that

$$\|\mathcal{H}_k^n\| \leq C$$

for all $nk \leq T$ and $k < k_0$. The method is of course stable if $\|\mathcal{H}_k\| \leq 1$.

We have then defined both the local in time consistency criteria and the global stability criteria. We have also seen before the definition of convergence (the error goes to zero). The Lax equivalent theorem connects these concepts and says that for a consistent, linear method, stability is necessary and sufficient for convergence.

1.10.4. The CFL Condition

Courant, Friedrichs, and Lewy (CFL) wrote one of the first papers on finite-difference methods for PDEs [103]. They used finite-difference methods for proving the existence of solutions of certain PDEs. The idea is to define a sequence of approximate solutions by means of finite-difference equations, prove that they converge as the grid is refined, and then show that the limit function must satisfy the PDE. While proving convergence of this sequence, they realized that a necessary stability condition for any numerical method is that the domain of dependence of the finite-difference method should include the domain of dependence of the corresponding PDE as $k, h \to 0$. This condition is known as the CFL condition.

The domain of dependence $D(x, t)$ of a PDE is the set of values x of $u_0(x)$ on which the solution $u(x, t)$ depends. For conservation laws, for example, the solution $u(x, t)$ might depend on u_0 at only a single point, as we have seen, and the domain of dependence is just this point. The numerical domain of dependence $D_k(x, t)$ is similarly defined. It is the set of points x for which the initial data $u_0(x)$ could possibly affect the numerical solution at (x, t).

We will use an example to illustrate the CFL condition. Consider the numerical implementation of Eq. (1.57) for Eq. (1.54). The value of $U_k(x_j, t_n)$ depends on the values of U_k at time t_{n-1} at the points x_{j+q}, $q = -1, 0, 1$. These values in turn depend on U_k at time t_{n-2} at the points x_{j+q}, $q = -2, -1, 0, 1, 2$. We see that the solution at t_n depends on the initial data at $x_{j+q}, q = -n, \ldots, n$, and so the numerical domain of dependence satisfies

$$D_k(x_j, t_n) \subset \{x : |x - x_j| \leq nh\},$$

or, in terms of a fixed point $(\bar{x}, \bar{t})$,

$$D_k(\bar{x}, \bar{t}) \subset \{x : |x - \bar{x}| \leq (\bar{t}/k)h\}.$$

If the domain is refined while $k/h = r$ is kept fixed, in the limit the domain of dependence fills out the entire set,

$$D_0(\bar{x}, \bar{t}) = \{x : |x - \bar{x}| \leq \bar{t}/r\}.$$

The CFL condition requires that

$$D(\bar{x}, \bar{t}) \subset D_0(\bar{x}, \bar{t}).$$

Because for linear systems of the form of Eq. (1.54), the domain of dependence is given by the points $\bar{x} - \lambda\bar{t}$ for all λ eigenvalues of the corresponding system matrix (or a for a scalar equation), the CFL condition gives

$$|(\bar{x} - \lambda\bar{t}) - \bar{x}| \leq \bar{t}/r$$

or

$$\left|\frac{\lambda k}{h}\right| \leq 1,$$

again for all the possible eigenvalues of the system.

The CFL condition is necessary, but not sufficient. The fact that it is necessary is quite clear. If the domain of the equation is not included in the numerical domain, changing the initial condition outside the numerical domain while inside the continuous domain will change the true solution but not the numerical one. Therefore the numerical solution will not converge to the true one for all possible smooth initial data. For certain numerical schemes, such as the Lax–Friedrichs scheme discussed in Subsection 1.10.2, it is possible to show that the CFL condition is actually sufficient for stability, but this is not true for all possible stencils.

1.10.5. Upwind Methods

Consider the one-sided method for solving Eq. (1.54):

$$U_j^{n+1} = U_j^n - \frac{ak}{h}\left(U_j^n - U_{j-1}^n\right).$$

If

$$0 \leq \frac{ak}{h} \leq 1,$$

then it can be shown that the method is stable. If $a > 0$, this method is called a first-order upwind method, because the one-sided stencil points in the upwind or upstream direction, the direction from which characteristic information propagates. An analogous upwind method exists for $a < 0$. Upwind methods will be useful for a number of PDEs presented later in this book (see next chapter), specially for equations that have discontinuous solutions.

1.10.6. Comments

Some of the basic concepts and language of numerical analysis have been introduced. Unfortunately (or fortunately, if you do research in numerical analysis), the picture does not end here. It is very easy to see that the following Riemannian problem, already studied when hyperbolic conservation laws were introduced,

$$u_t + au_x = 0,$$

with initial condition

$$u_0(x) = \begin{cases} 1, & x < 0 \\ 0, & x > 0 \end{cases},$$

cannot be approximated with any of the simple stencils introduced above. This is because the solution is discontinuous, and simple finite-difference approximations across the discontinuity will blow as $h \to 0$. Discontinuities not just appear in images (edges!), but one of the key advantages of the use of PDEs is that those discontinuities can be preserved. In addition to this, the concepts above were described for only linear systems, whereas all the important PDEs introduced later in this book are nonlinear. This introduces additional difficulties into the numerical implementation. Numerical techniques that work for nonlinear systems and are devised to preserve discontinuities are described later in Chapter 2. Further information can be found, for example, in Ref. [240].

Exercises

1. Compute the Euclidean arc length ds for an ellipse.
2. Compute the formula for the curvature of a curve $\mathcal{C}(p)$, where p is not the Euclidean arc length. Compute this formula for the case $p = x$ (the curve is expressed as a function).
3. Compute the curvature of a circle, an ellipse, a parabola, and a hyperbola.
4. Show that if $\vec{\mathcal{T}} = (\cos\theta, \sin\theta)$, then $(\mathrm{d}\vec{\mathcal{T}}/\mathrm{d}\theta) = \vec{\mathcal{N}}$ and $(\mathrm{d}\vec{\mathcal{N}}/\mathrm{d}\theta) = -\vec{\mathcal{T}}$.
5. Prove the equation for the curvature of a level curve.
6. Compute the affine curvature for a curve $\mathcal{C}(p)$, where p is not the affine arc length. Compute this also in the case $p = x$ (the curve is expressed as a function).
7. Compute the affine curvature of a conic (ellipse, hyperbola, and parabola) and verify that they are constant.
8. Compute the affine curvature of a curve given in implicit representation.
9. Derive the Cartan matrix for the Euclidean differential geometry of space curves.

10. Compute the basic geometric characteristics of the Euclidean and the affine evolutes (tangent, normal, arc length, curvature).

11. Compute the Euclidean arc length, curvature, and torsion of the 3D curve given by $(a \cos p, a \sin p, bp)$, where a and b are constants and p parameterizes the curve.

12. Prove that surfaces of revolution are regular surfaces.

13. Compute the first fundamental form coefficients for a helix and a sphere.

14. Show that a necessary and sufficient condition for a connected regular curve $\mathcal{C}(p)$ on $\mathcal{S}$ to be a line of curvature (all its tangents are principal directions) is given by $\vec{\mathcal{N}}'(p) \parallel \mathcal{C}'(p)$ (the variations of the normals to the surface restricted to the curve are parallel to the tangents to the curve at every point).

15. Exemplify all the concepts described in this chapter for conservation laws for Bruger's equation with a step initial condition.

16. Compute the Euler equations and solve them for the following functionals:

a. Shortest path between two points $(0, a)$ and $(1, b)$, given by the energy $\mathbf{E}(u) = \int_0^1 \sqrt{1 + (u')^2}$.

b. Same data as in the previous exercise, but the path should enclose a given area.

c. $E(u) = \iint [1 + (u_x)^2 + (u_y)^2]^{1/2} \mathrm{d}x\mathrm{d}y$ (this is the formula for surface area).

17. Numerically implement the scalar PDE $u_x + u_t = 0$ with smooth initial conditions. Use the different stencils described in this chapter.

18. Repeat the exercise for a step function as initial condition. Show that the methods are not stable.

CHAPTER TWO

Geometric Curve and Surface Evolution

In this chapter the theory of geometric curve and surface evolution is introduced. First present the basic concepts are presented, and then a number of examples of intrinsic evolutions are given. Chapter 3 will deal with evolutions that depend not only on the intrinsic properties of the curve or surface, but on external forces as well. We should note that the use of curve and surface evolution flows goes beyond image processing and computer vision; see, for example, Ref. [361] for a collection of applications in other fields, as well as the reports in Ref. [290].

2.1. Basic Concepts

The basic concepts of the theory of curve and surface evolution are now presented. We start with curve evolution; the basic concepts described now are easily extended to surfaces.

We consider curves to be deforming in time. Let $\mathcal{C}(\tilde{p}, t) : S^1 \times [0, T) \to \mathbb{R}^2$ denote a family of closed (embedded) curves, where t parameterizes the family and $\tilde{p}$ parameterizes the curve. Assume that this family of curves obeys the following PDE:

$$\frac{\partial \mathcal{C}(\tilde{p}, t)}{\partial t} = \alpha(\tilde{p}, t)\vec{\mathcal{T}}(\tilde{p}, t) + \beta(\tilde{p}, t)\vec{\mathcal{N}}(\tilde{p}, t), \tag{2.1}$$

with $\mathcal{C}_0(p)$ as the initial condition. Here $\vec{\mathcal{N}}$ stands for the inward unit normal. Throughout this book we use $\vec{\mathcal{N}}$ both for the inner and the outer unit normals, and the exact direction will be clear from the context. This equation has the most general form and means that the curve is deforming with α velocity in the tangential direction and β velocity in the normal direction. Note that for a general velocity $\vec{\mathcal{V}}$, $\alpha = \langle \vec{\mathcal{V}} \cdot \vec{\mathcal{T}} \rangle$ and $\beta = \langle \vec{\mathcal{V}} \cdot \vec{\mathcal{N}} \rangle$. If we are just interested in the geometry of the deformation, but not in its parameterization, this flow can be further simplified following the result of Epstein and Gage [123]:

Lemma 2.1. *If β does not depend on the parameterization, meaning that is a geometric intrinsic characteristic of the curve, then the image of $\mathcal{C}(\tilde{p}, t)$ that satisfies Eq. (2.1) is identical to the image of the family of curves $\mathcal{C}(p, t)$ that satisfies*

$$\frac{\partial \mathcal{C}(p, t)}{\partial t} = \beta(p, t) \vec{\mathcal{N}}(p, t). \tag{2.2}$$

In other words, the tangential velocity does not influence the geometry of the deformation, just its parameterization.

Proof: Let $\tilde{p} = \tilde{p}(p, \tau)$, with $t = \tau$ and $\partial \tilde{p} / \partial p > 0$. Applying the chain rule, we have

$$\begin{aligned} \mathcal{C}_\tau &= \mathcal{C}_p p_\tau + \mathcal{C}_t t_\tau = \mathcal{C}_p p_\tau + \mathcal{C}_t \\ &= \mathcal{C}_p p_\tau + \alpha \vec{\mathcal{T}} + \beta \vec{\mathcal{N}} = (s_p p_\tau + \alpha) \vec{\mathcal{T}} + \beta \vec{\mathcal{N}} \\ &= \tilde{\alpha} \vec{\mathcal{T}} + \beta \vec{\mathcal{N}}, \end{aligned}$$

where we have used the fact that, for the Euclidean arc length s, $\mathcal{C}_p = s_p \vec{\mathcal{T}}$. By changing the parameterization we have then modified the tangential velocity without altering the normal one. To show that we can make $\tilde{\alpha} = 0$, it remains to show that the equation $s_p p_\tau + \alpha = 0$ has a solution. See Ref. [123] for this. □

This result can also be proved locally as follows: Let $\gamma(x, t)$ locally represent the curve, where x and y are Cartesian coordinates. Then

$$\vec{\mathcal{T}} = \frac{1}{\sqrt{1 + \gamma_x^2}} (1, \gamma_x),$$

$$\vec{\mathcal{N}} = \frac{1}{\sqrt{1 + \gamma_x^2}} (-\gamma_x, 1),$$

$$\kappa = \frac{\gamma_{xx}}{\left(\sqrt{1 + \gamma_x^2}\right)^3}.$$

Now, clearly

$$\begin{aligned} y_t &= \gamma_x x_t + \gamma_t, \\ \gamma_t &= y_t - \gamma_x x_t \\ &= \frac{1}{\sqrt{1+\gamma_x^2}}(\alpha\gamma_x + \beta) - \gamma_x \frac{1}{\sqrt{1+\gamma_x^2}}(\alpha - \beta\gamma_x) \\ &= \frac{1}{\sqrt{1+\gamma_x^2}}\beta(1+\gamma_x^2) = \beta\sqrt{1+\gamma_x^2}, \end{aligned}$$

and we see again that the tangential component α does not affect the deformation of the curve. The level-set approach described in Section 2.2 will provide yet an additional proof.

Basically, with Eq. (2.2) we do not know exactly where each point $\mathcal{C}(p)$ is moving; this depends on the parameterization. We just know how the whole curve, as a geometric object, deforms. (We will see in the rest of this chapter how it is not always convenient, from the analysis point of view, to remove the tangential component of the velocity.) This is illustrated in Fig. 2.1. The art in curve evolution then becomes the search for the function β that solves a given problem. Finally, note that a flow of the form

$$\frac{\partial \mathcal{C}}{\partial t} = \alpha\vec{\mathcal{T}}$$

will leave the curve unchanged; only the parameterization is changing.

This section concludes with a comment on surface deformations. Following similar arguments as for the planar curve case, the most general

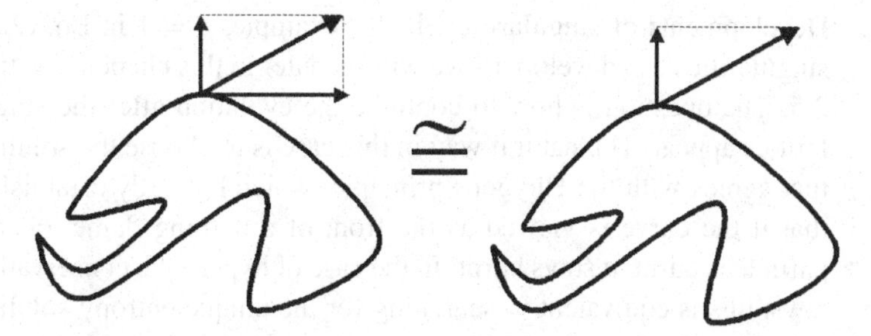

Fig. 2.1. Most general form of a curve evolution flow. The velocity is decomposed into its tangential and normal components, the former not affecting the geometry of the flow, just its parameterization.

geometric deformation for a family of surfaces $\mathcal{S} : \mathbb{R}^2 \to \mathbb{R}^3$ is given by

$$\frac{\partial \mathcal{S}}{\partial t} = \beta \vec{\mathcal{N}}, \tag{2.3}$$

where the geometric velocity is now in the direction of the 3D normal $\vec{\mathcal{N}}$.

2.2. Level Sets and Implicit Representations

As we saw in Chap. 1, curves and surfaces can be represented in an implicit form, as level sets of a higher-dimensional surface. In this section we discuss the evolution of curves and surfaces given by implicit representations or level sets. Only the basic concepts are presented here; details can be found, for example, in Refs. [262, 290, and 361].

The pioneering and fundamental work on the deformation of level sets was introduced in Ref. [294] (see also Ref. [358], and see Ref. [274] for an early introduction of level-set flows) to solve basic problems with the numerical implementation of flows such as Eqs. (2.2) and (2.3). A number of problems need to be solved when we are implementing these type of flows:

1. Accuracy and stability. The numerical algorithm must approximate the continuous flow, and it must be robust. A simple Lagrangian approximation, based on moving particles along the curve or surface, requires an impractically small time step to achieve stability. The basic problem is that the marker particles on the evolving shape can come very close together or very far away during the deformation. These can be solved by a frequent redistribution of the markers, altering the motion of the curve in a nonobvious way.
2. Developments of singularities. If, for example, $\beta = 1$ in Eq. (2.2), singularities can develop, as we will see later in this chapter, Section 2.5. The question is how to continue the evolution after the singularities appear. The natural way in this case is to choose the solution that agrees with the Hüygens principle, which basically establishes that if the curve is viewed as the front of a burning flame, once a particle is burnt, it stays burnt. In the case of hyperbolic conservation laws, this is equivalent to searching for the unique entropy solution (see Chap. 1). In general, we need a scheme that finds the right weak solution when singularities are present in the flow.
3. Topological changes. When a curve or surface is deforming, its topology can change (split or merge). Tracking the possible topological

changes with marker particles is an almost impossible task, or at least incredibly hard to implement and time consuming to run.

All these problems lead to the development of level-set techniques, which are presented now. The technique for 2D deforming curves is described; the extension to higher-dimensional surfaces is straightforward.

Let, as above, $\mathcal{C}(p, t) : S^1 \times [0, T) \to \mathbb{R}^2$ denote a family of closed (embedded) curves, where t parameterizes the family and p parameterizes the curve. Assume that this family of curves obeys the following PDE:

$$\frac{\partial \mathcal{C}(p, t)}{\partial t} = \vec{\mathcal{V}} = \beta(p, t)\vec{\mathcal{N}}(p, t), \tag{2.4}$$

with $\mathcal{C}_0(p)$ as the initial condition.

Let us represent this curve as the level set of an embedding function $u(x, y, t) : \mathbb{R}^2 \times [0, T) \to \mathbb{R}$:

$$\mathcal{L}_c(x, y, t) := \{(x, y, t) \in \mathbb{R}^2 : u(x, y, t) = c\}, \tag{2.5}$$

where $c \in \mathbb{R}$ is a given constant. The initial curve $\mathcal{C}_0$ is represented by an initial function u_0. For example, we can consider the signed distance function $d(x, y)$ from a point on the plane to the curve $\mathcal{C}_0$ (negative in the interior and positive in the exterior of $\mathcal{C}_0$), and $u_0 = d(x, y) + c$. We now have to find the evolution of u such that

$$\mathcal{C}(x, y, t) \equiv \mathcal{L}_c(x, y, t),$$

that is, the evolution of $\mathcal{C}$ coincides with the evolution of the level sets of u (we will see that all the level sets obey the same curve evolution flow, independently of c). By differentiating Eq. (2.5) with respect to t, we have

$$\nabla u(\mathcal{L}, t) \cdot \frac{\partial \mathcal{L}}{\partial t} + \frac{\partial u}{\partial t}(\mathcal{L}, t) = 0.$$

(Note that we can omit the specification of the level-set value c, as it has been automatically eliminated from the equation above after the derivative was taken.)

As seen in Chap. 1,

$$\frac{\nabla u}{\|\nabla u\|} = -\vec{\mathcal{N}},$$

where $\vec{\mathcal{N}}$ is the normal to the level set $\mathcal{L}$ (the sign depends on the assumed convention for the direction of the normal). This equation combines information from the function u (left-hand side) with information from the planar curve (right-hand side).

To have $\mathcal{C} \equiv \mathcal{L}$, we must have

$$\frac{\partial \mathcal{L}}{\partial t} = \vec{\mathcal{V}} = \beta \vec{\mathcal{N}}.$$

Combining the last three equations, we have

$$\begin{aligned} 0 &= \nabla u \cdot \vec{\mathcal{V}} + u_t = \nabla u \cdot \beta \vec{\mathcal{N}} + u_t \\ &= \nabla u \cdot \beta \left(-\frac{\nabla u}{\|\nabla u\|} \right) + u_t = -\beta \|\nabla u\| + u_t, \end{aligned}$$

obtaining

$$\frac{\partial u}{\partial t} = \beta \|\nabla u\|. \tag{2.6}$$

Recapping, when a function is moving according to Eq. (2.6), its level sets, all of them, are moving according to Eq. (2.4). A number of comments are worth mentioning regarding this formulation:

1. We see once again that this time, because of the dot product in the derivation above, only the normal component of the velocity affects the flow.
2. Related to the point above, the level-set formulation is a parameterization free formulation, it is written in a fixed (x, y) coordinate system. For this reason, this is called an Eulerian formulation.
3. A number of questions must be asked when the above derivation above is introduced: (1) Is the deformation independent of the initial embedding u_0? (2) What happens when the curve is not smooth, and classical derivatives cannot be computed? Fortunately, these questions have been answered in the literature, e.g., in Refs. [90, 126 , 127, and 372]. For a large class of initial embeddings, the evolution is independent of them. The classical solution of Eq. (2.6), if it exists, coincides with the classical solution of Eq. (2.4). In addition, and of special significance when singularities develop, when the theory of viscosity solutions is used, the existence and uniqueness of the solutions to the level-set flow can be proved for velocities β as for those used in this book. This automatically gives a generalization of the curve flows of Eq. (2.4) when the curve becomes singular and notions such as normal are not well defined. This coincides with the unique entropy condition in the case of hyperbolic conservation laws. This will be specified for the different curve flows introduced throughout this book.

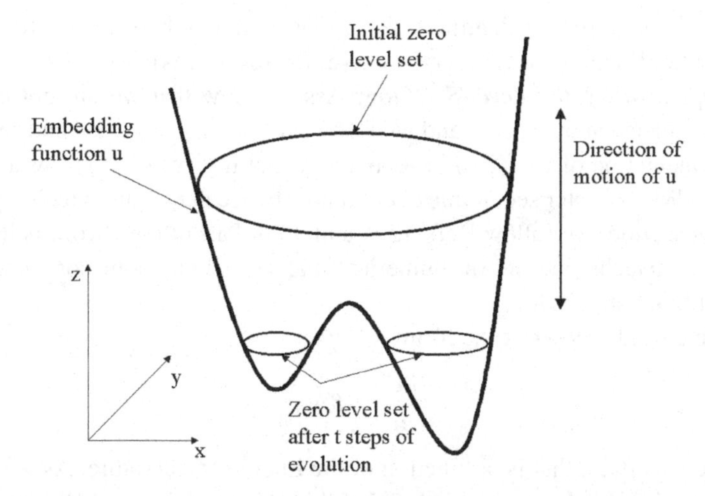

Fig. 2.2. Embedding the curve in a higher-dimensional function automatically solves topological problems. Although the curve can be changing topology, the surface moves up and down on a fixed coordinate system without altering its topology. The topological change is just discovered when the corresponding level set of the function is computed.

4. Although the curve can change its topology, the topology of u is fixed; see Fig. 2.2. This means that there is no need to track topological changes, the evolution of u is implemented, and the changes in topology are discovered when the corresponding level set is computed.

2.2.1. Numerical Implementation of Level-Set Flows

To complete the picture, we need to show how to numerically implement Eq. (2.6). This is not a straightforward case, as, for example, we have to make sure that in the case of singularities the correct weak solution is picked by the numerical implementation. A brief description of the basic numerical techniques here is given here; for details see Refs. [290, 294, and 361].

Before the numerical implementation of the level-set flow is presented, note that we also need to have an algorithm to compute (zero) level sets of discrete functions. Many ideas are available for doing this. In the 2D case, for example, a zero level set is considered as passing through every cell that has values of different signs in its four corners. Once those cells are identified, zero-order or more accurate interpolations can be used to find the exact place where the zero level set crosses the cell. The same technique is applied for 3D cases by means of the popular marching-cubes technique.

Most of the popular scientific software have already built in functions to compute 2D and 3D level sets (isocurves or isosurfaces).

Implementing the Level-Set Flow. Assume now that we are not especially worried about speed, and we just want to find an accurate and robust implementation of Eq. (2.6) in such a way that it gives the right weak solution. We will later see in this Section how to accelerate these techniques. In this section, we follow Refs. [294 and 358]. Part of their work is based on general techniques for the numerical analysis of hyperbolic conservation laws [228, 240, 371].

We consider flows of the form

$$\frac{\partial \mathcal{C}}{\partial t} = \beta \vec{\mathcal{N}},$$

where β is $\beta(\kappa)$, that is, a function of the Euclidean curvature. As we will see later in this chapter, Section 2.5 and in this book in general, these are the kinds of velocities that are important from the point of view of image processing and computer vision. Moreover, we restrict ourself to velocities of the form

$$\beta(\kappa) = 1 + \epsilon\kappa,$$

where ϵ is some positive real number. This velocity not only is fundamental for image analysis, as we will later see, but it is also sufficient to show the basic numerical problems and techniques used in curve flows.

If $\mathcal{C}(p, t) = [x(p, t), y(p, t)]$, the above flow can be rewritten as

$$x_t = \beta(\kappa)\left[\frac{y_p}{\left(x_p^2 + y_p^2\right)^{1/2}}\right],$$

$$y_t = \beta(\kappa)\left[\frac{-x_p}{\left(x_p^2 + y_p^2\right)^{1/2}}\right],$$

and the curvature can also be expressed as

$$\kappa = \frac{y_{pp}x_p - x_{pp}y_p}{\left(x_p^2 + y_p^2\right)^{3/2}}.$$

(Recall that we are considering only closed curves.)

As explained above, we embed the curve as the level set of the function u deforming according to

$$u_t = (1 + \epsilon\kappa)\|\nabla u\|$$

and implement this higher-dimensional flow. This equation looks very much like a Hamilton–Jacobi equation with viscosity, $u_t = \|\nabla u\|$ being the Hamilton–Jacobi and $\epsilon\kappa$ the viscosity; see Chap. 1. We will use this connection to devise an appropriate numerical scheme.

Assume now that the moving front is a curve in the plane that remains a graph for at least some short period of time. Let $f(x)$ be the initial graph and both f and f' periodic in $[0, 1]$. Let γ be the height of the propagating function, $\gamma(x, 0) = f(x)$. The normal at (x, γ) is $(-\gamma_x, 1)$, and the equation of motion becomes (computations were explicitly done before, in Section 2.1 when showing the effect of the tangential velocity)

$$\gamma_t = \beta(\kappa)\left(1 + \gamma_x^2\right)^{1/2}.$$

For the specific β that we are considering, and with $\kappa = \{(\gamma_{xx})/[(1 + \gamma_x^2)^{3/2}]\}$, we obtain

$$\gamma_t = \left(1 + \gamma_x^2\right)^{1/2} + \epsilon\frac{\gamma_{xx}}{1 + \gamma_x^2}.$$

Let us now compute the evolution of the slope $\phi = \gamma_x$. Differentiating both sides of the equation with respect to x, we obtain

$$\phi_t = \left[\left(1 + \gamma_x^2\right)^{1/2}\right]_x + \epsilon\left[\frac{\gamma_{xx}}{1 + \gamma_x^2}\right]_x.$$

Therefore the evolution of the slope ϕ of γ is a hyperbolic conservation law with viscosity (see Chap. 1), and the machinery of these kinds of flows can then be applied. As expressed in Chap. 1, if $\epsilon > 0$, the parabolic part of the flow diffuses steep gradients and enforces smooth solutions. However, for $\epsilon = 0$, discontinuous solutions can arise from smooth initial data. Of all the possible solutions of this case, we are of course interested in the one that is the limit of the smooth solution when $\epsilon \to 0$, or in other words, the entropy solution. We must then construct a numerical scheme that picks up this entropy solution from all possible weak solutions.

Let us see then how to construct the numerical scheme for a general one-dimensional (1D) conservation law of the form

$$\phi_t + [H(\phi)]_x = 0.$$

Let ϕ_i^n indicate the value of ϕ at the point (pixel) i at the time n. An algorithm to approximate the solution of this equation is said to be in conservation form if it can be written in the form

$$\phi_i^{n+1} = \phi_i^n - \Delta t/\Delta x\left(g_{i+1/2}^n - g_{i-1/2}^n\right),$$

where the numerical flux function $g_{i+1/2} = g(\phi_{i-p+1}, \ldots, \phi_{i+q+1})$ must be Lipschitz and satisfy the consistency requirement $g(\phi, \ldots, \phi) = H(\phi)$. A scheme is called monotone if the right-hand side of the equation above is a nondecreasing function of all its arguments. A conservative monotone scheme obeys the entropy condition [371], and this is then the key concept of the numerical implementation. All that is left to do is to select the function g. Many selections are possible, and in the particular case in which $H(\phi) = h(\phi^2)$, which is of particular interest in our equations, one such possible selections is

$$g\left(\phi_i^n, \phi_{i+1}^n\right) = h\left[(\min\{\phi_i, 0\})^2 + \left(\max\left\{\phi_{i+1}^n, 0\right\}\right)^2\right].$$

Note that in the case of

$$u_t = \|\nabla u\|,$$

for 1D signals we have

$$u_t = \sqrt{u_x^2},$$

and, considering $\phi := u_x$ and $H(\cdot) = -\sqrt{(\cdot)^2}$, we obtain an equation of exactly this form.

This extends to two dimensions, and if $H(\cdot, \cdot) = h(\cdot^2, \cdot^2)$ then

$$\begin{aligned} g = h\Big[&\left(\min\left\{D_x^- u_{ij}^n, 0\right\}\right)^2 + \left(\max\left\{D_x^+ u_{ij}^n, 0\right\}\right)^2 \\ &+ \left(\min\left\{D_y^- u_{ij}^n, 0\right\}\right)^2 + \left(\max\left\{D_y^+ u_{ij}^n, 0\right\}\right)^2\Big], \end{aligned}$$

and the 2D flow

$$u_t = \|\nabla u\|$$

is approximated as

$$u_{ij}^{n+1} = u_{ij}^n + \Delta t g.$$

The functions of the min and the max are interchanged according to the sign of the velocity to guarantee an upwind scheme. A general upwind method for an equation of the form $u_t = \beta\|\nabla u\|$ for $\beta\|\nabla u\|$ convex is given by (see Refs. [294 and 361] for details and extensions to higher-order schemes and nonconvex Hamiltonians)

$$u_{ij}^{n+1} = u_{ij}^n - \Delta t[\max(-\beta_{ij}, 0)\Delta^+ + \min(-\beta_{ij}, 0)\Delta^-],$$

where

$$\Delta^+ := [\max(D_x^-, 0)^2 + \min(D_x^+, 0)^2 + \max(D_y^-, 0)^2 + \min(D_y^+, 0)^2]^{1/2},$$
$$\Delta^- := [\max(D_x^+, 0)^2 + \min(D_x^-, 0)^2 + \max(D_y^+, 0)^2 + \min(D_y^-, 0)^2]^{1/2}.$$

We have then approximated the first part of the level-set flow with velocity $\beta = 1 + \epsilon\kappa$ (more sophisticated conservative and monotone schemes are developed in the literature, but this is one of the most popular ones for the equations throughout this book). Because, as we will see later in this chapter, Section 2.5, the curvature motion does not produce singularities (it is like the viscosity part in the hyperbolic conservation law), all curvature-dependent components are implemented just by central derivatives,

$$\frac{\partial u}{\partial x} \equiv \frac{u_{i+1,j}^n - u_{i-1,j}^n}{2\Delta x},$$
$$\frac{\partial^2 u}{\partial x^2} \equiv \frac{u_{i+1,j}^n - 2u_{i,j}^n + u_{i-1,j}^n}{(\Delta x)^2},$$
$$\frac{\partial^2 u}{\partial x \partial y} \equiv \frac{u_{i+1,j+1}^n + u_{i-1,j-1}^n - u_{i+1,j-1}^n - u_{i-1,j+1}^n}{4(\Delta x)^2},$$

and the same for derivatives with respect to y. (Recall the expression for the curvature of the level sets from Chap. 1). This gives then the basic implementation of the level-set flow.

Local Schemes and Extension Velocities. The direct implementation of Eq. (2.6) presented above is not particularly fast. In particular, because of the embedding, 1 order of magnitude is added to the computational complexity of the curve evolution flow. That is, for 2D curve evolution, we move from $O(N)$, a linear complexity, to $O(N^2)$ (N stands for the size of the grid). In three dimensions, we have to move a full volume, thereby arriving at a complexity of $O(N^3)$. This problem can be solved by the narrow band, introduced by Chopp [93] and later used and extended by Malladi et al. [250,251] and by Adalsteinsson and Sethian [2, 361], and the local methods developed by Osher and colleagues [262, 308]. The basic idea is to operate on only a surrounding band around the region of interest, that is, around the level set being tracked (e.g., the zero level set). This reduces the complexity to $O(kN)$ in the 2D case and $O(kN^2)$ in the 3D case, where k stands for width of the narrow band and N is, as before, the size of the grid in each direction. With this technique, Eq. (2.6) is solved only inside the narrow band, and this band is updated each time the curve (surface) approaches the border of the band.

In the narrow-band method, the technique requires a bit more of programming than the direct implementation because of the band-rebuilding process, but speeds up the computations considerably. Note that if, for example, the embedding function is a distance function (see below), a narrow band of width ϵ around the zero level set, which is computed on the (x, y) (or (x, y, z) in the 3D) plane, is just given by those points on the embedding function with an absolute value less than or equal to ϵ [262, 308]. That is, to perform a local computation in the level-set framework, all we need is to maintain the embedding function as a signed distance function and update only those points (pixels) where the embedding function is less than a certain ϵ. This tremendously simplifies the computation of the narrow band. Therefore, if the narrow-band method is used, there is not an imperative need to maintain the embedding function as a distance function, although the band is calculated from the location of points in the domain [2]. On the other hand, in local methods, while the embedding function is maintained as a distance function, the local band of computations is directly computed from the function itself [308]. This second approach is simpler, if we can in a reliable and fast way keep the distance function.

To complete the picture, we must indicate how to keep the embedding function as a signed distance function. This embedding function can be, for example, computed as the steady state of (e.g., Refs. [308 and 378])

$$\frac{\partial u}{\partial t} = \text{sign}(u)(1 - \|\nabla u\|), \tag{2.7}$$

which gives the signed distance function at the steady-state solution [378]. Then, every few steps of the level-set flow, the embedding function u is rebuilt with this equation.

In addition, the level-set technique can be modified to maintain the embedding function as a signed distance function without the need to interleave (Eq. 2.7) every time. This was developed in Refs. [159 and 426]. The basic idea is (following Ref. [159]) to search for a new deformation

$$\frac{\partial u}{\partial t} = \tilde{\beta}$$

such that we add to the requirement that the velocity of the zero level set of u must be $\beta\vec{\mathcal{N}}$, the requirement $\|\nabla u\| = 1$. This additional constraint forces the relation $\nabla u \cdot \nabla\tilde{\beta} = 0$, meaning that the new velocity does not vary along the characteristics of u. Recalling that the characteristics of a distance function are straight lines, we find that the new velocity is given by

$$\tilde{\beta}(\mathcal{X}) = \beta(\mathcal{X} - u\nabla u),$$

where $\mathcal{X}$ is the Cartesian coordinate of the point (pixel) in two or three dimensions. Note that this equation means that the new velocity $\tilde{\beta}$ for a point $\mathcal{X}$ is computed from the old velocity β at the point $\mathcal{X} - u\nabla u$. Therefore the equation is not necessarily, even for simple velocities β, a PDE.

Other reinitialization techniques, as well as other techniques to rebuild the narrow band, can be found in Ref. [361].

We conclude the discussion on the narrow band and local methods by noting that because the CFL condition needs to be satisfied only inside the narrow band, the algorithm might enjoy further computational speedups.

The last remark we must make is on extending β beyond the level set of interest (normally the zero level set). In general, for a given problem, β is given for only the deforming curve or surface and, in order to implement the level-set flow $u_t = \beta\|\nabla u\|$, we must define the values of β for all the space, or at least for the narrow band around the level set of interest. Many times there is a natural way to extend β. For example, in cases such as curvature motion the velocity is naturally extended to be the curvature also at the other level sets. In other cases there is no natural extension, and decisions have to be made. A number of techniques have been proposed to do this, e.g., see Refs. [3, 89, 250, 308, 361, and 427]. One possibility is to extend β so that it is constant normal to each level curve of u. In other words, $\nabla\beta \cdot \nabla u = 0$ (note the similarity of this to the condition stated above on the velocity $\tilde{\beta}$ that guarantees the preservation of u as a distance function; redistancing is equivalent to extending the velocity field perpendicular to the level sets; see Ref. [426]). We can achieve this once again by searching for the steady-state solution of the flow

$$\frac{\partial\beta}{\partial t} = \operatorname{sign}(u_0)(\nabla\beta \cdot \nabla u_0).$$

Of course, all these techniques (local schemes, preservation of the distance function, extension of velocity fields) can and should be combined [308].

Fast Techniques for Positive Velocities. It is well known in graph theory that the optimal path (trajectory or minimal weighted distance) between two nodes in a graph with positive edges can be found by the Dijkstra algorithm. In the case of a positive velocity β, the curve evolution flow

$$\mathcal{C} = \beta\vec{\mathcal{N}}$$

can also be implemented with the same complexity $[O(N \log N)]$, with an extremely small constant, where N are the number of grid points. This very important concept was developed by Tsitsiklis [393] whith motivations from optimal control, and independently by Helmsen et al. [180] and Sethian [359] following upwind schemes formulations, and extended and applied to numerous fields by Sethian in Refs. [360 and 361]. Additional developments (and connections between the above mentioned approaches) can be found in the work of Osher and Helmsen [292], and studies on efficient implementations appear in Ref. [317]. Extensions to triangulated surfaces are reported in Refs. [207 and 215]. To present this technique, we follow the elegant development in Ref. [361]. (Once again, we deal with only evolving curves, and the technique is extended in a straightforward fashion to surfaces.)

Suppose we graph the evolving curve $\mathcal{C}$ above the (x, y) plane, that is, let $T(x, y)$ be a function representing the time at which the curve crosses the point (x, y). Because $\beta > 0$, $T(x, y)$ is indeed a function, and the curve crosses each planar point only once (at most). $T(x, y)$ is called the time-of-arrival function. Figure 2.3 illustrates this idea.

The time-of-arrival function satisfies

$$\|\nabla T(x, y)\|\beta = 1, \tag{2.8}$$

and the solution of this equation also solves the curve evolution flow. This is a Hamilton–Jacobi equation, which becomes the well-known Eikonal equation for velocities that depend on only the position (x, y). We have then transformed the curve evolution problem into a stationary one; see also Refs. [50,128 ,129, and 289]. Before showing how to solve this equation, let us summarize the results so far, comparing the time-of-arrival formulation

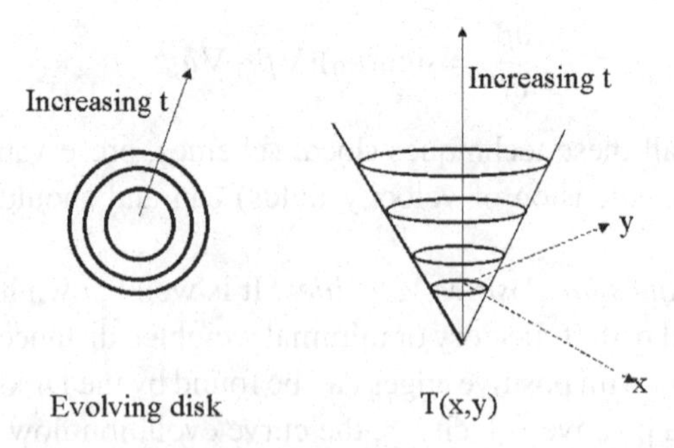

Fig. 2.3. A disk evolving with unit speed, together with the corresponding time-of-arrival function $T(x, y)$.

with the level-set formulation presented above:

Level-Set Formulation	Time-of-Arrival Formulation
$u_t = \beta \|\nabla u\|$	$\|\nabla T(x, y)\| \beta = 1$
$\mathcal{C}(t) = \{(x, y) : u(x, y, t) = 0\}$	$\mathcal{C}(t) = \{(x, y) : T(x, y) = t\}$
β arbitrary	$\beta > 0$

The static Hamilton-Jacobi equation (2.8) is implemented with an upwind, viscosity solution approximation of the gradient, leading to

$$\left[\max\left(D_{ij}^{-x}T, 0\right)^2 + \min\left(D_{ij}^{+x}T, 0\right)^2 + \max\left(D_{ij}^{-y}T, 0\right)^2 + \min\left(D_{ij}^{+y}T, 0\right)^2\right]^{1/2} = 1/\beta_{ij}^2.$$

There are several techniques to solve this discrete equation [129, 327], one that achieves the optimal computational complexity [360, 361, 393]. The key idea is to observe that, in the above equation, information flows in only one direction, from small values of T to large values (for those familiar with Dijkstra, this is exactly the same case as in this famous graphics algorithm). The algorithm then reads as follows [361]:

1. Initialize
   a. Let *alive* be the set of all grid points at which the value of T is zero.
   b. Let *narrowband* be the set of all grid points in the narrow band (around the front). Set these points to $\sqrt{dx^2 + dy^2}/\beta$.
   c. Let *faraway* be the set of all the rest of grid points. Set $T_{i,j} = \infty$ for these points.
2. Marching forward
   a. Let $(i_{\min}, j_{\min})$ be the point in *narrowband* with the smallest T.
   b. Add the point $(i_{\min}, j_{\min})$ to *alive* and remove it from *narrowband*.
   c. Tag as neighbors any points $(i_{\min} - 1, j_{\min})$, $(i_{\min} + 1, j_{\min})$ $(i_{\min}, j_{\min} - 1)$ $(i_{\min}, j_{\min} + 1)$ that are not *alive*. If the neighbor is in *faraway*, remove it from that set and add it to the *narrowband* set.
   d. Recompute the values of T at all the neighbors according to the upwind equation above.
   e. Return to 2a.

Details on the validity of the algorithm are given in the cited references. To conclude, we must note that it is fundamental to find the smallest value of T in an efficient way, and this is done with the well-known min-heap algorithm, thereby producing a total computational cost of $O(N \log N)$, where N is the number of grid points.

2.2.2. Region Tracking on Level-Set Techniques

Let us now consider again the level-set flow for an m-dimensional closed surface $\mathcal{S}$, $m \geq 3$. Let's represent the initial surface $\mathcal{S}(0)$ as the zero level set of $u(\mathcal{X}, t) : \mathbb{R}^m \times [0, \tau) \to \mathbb{R}$, i.e., $\mathcal{S}(0) \equiv \{\mathcal{X} \in \mathbb{R}^m : u(\mathcal{X}, 0) = 0\}$. From the level-set approach, if the surface is deforming according to

$$\frac{\partial \mathcal{S}(t)}{\partial t} = \beta \vec{\mathcal{N}}_{\mathcal{S}}, \tag{2.9}$$

where $\vec{\mathcal{N}}_{\mathcal{S}}$ is the unit normal to the surface, then this deformation is represented as the zero level set of $u(\mathcal{X}, t) : \mathbb{R}^m \times [0, \tau) \to \mathbb{R}$ deforming according to

$$\frac{\partial u(\mathcal{X}, t)}{\partial t} = \beta(\mathcal{X}, t) \| \nabla u(\mathcal{X}, t) \|, \tag{2.10}$$

where $\beta(\mathcal{X}, t)$ is computed on the level sets of $u(\mathcal{X}, t)$.

In a number of applications, it is important not just to know how the whole surface deforms, but also how some of its regions do. Because the parameterization (tangential velocity) is missing, this is not possible in a straightforward level-set approach. This problem is related to the aperture problem in optical flow computation, and it is also the reason why the level-sets approach can deal with only parameterization-independent flows that do not contain tangential velocities. Although, as we have seen, tangential velocities do not affect the geometry of the deforming shape, they do affect the point correspondence in the deformation. For example, with a straight level-set approach, it is not possible to determine where a given point $\mathcal{X}_0 \in \mathcal{S}(0)$ is at certain time t. One way to solve this problem is to track isolated points with a set of ordinary differential equations (ODEs), and this was done, for example, in grid generation; see Ref. [361]. This is a possible solution if we are just interested in tracking a number of isolated points. If we want to track regions, for example, then using particles brings us back to a Lagrangian formulation and some of the problems that actually motivated the level-set approach. For example, what happens if the region splits during the deformation? What happens if the region of interest is represented by particles that start to come too close together in some parts of the region and travel too far from each other in others?

An alternative solution is now proposed to the problem of region tracking on surface deformations implemented by means of level sets [27]. The basic idea is to represent the boundary of the region of interest $\mathcal{R} \in \mathcal{S}$ as the intersection of the given surface $\mathcal{S}$ and an auxiliary surface $\hat{\mathcal{S}}$, both of them given as zero level sets of $m+1$ dimensional functions Φ and $\hat{\Phi}$, respectively. The tracking of the region $\mathcal{R}$ is given by tracking the intersection of these two surfaces, that is, by the intersection of the level sets of Φ and $\hat{\Phi}$. See [208, 213] for tracking curves on functions.

Note that we could also track, in theory, the boundary of $\mathcal{R}$ with the Ambrosio–Soner approach for level sets in the high codimension described below. This approach is difficult to implement numerically and in some cases it is not fully appropriate for numerical 3D curve evolution. Then there is a need for techniques such as the one described now. Although we use the technique described below to track regions of interest on deforming surfaces, with the region deformation dictated by the surface deformation, the same general approach here presented of simultaneously deforming n hypersurfaces ($n \geq 2$) and looking at the intersection of their level sets can be used for the numerical implementation of generic geometric deformations of curves and surfaces of high codimension [54].

Note also that the use of multiple level-set functions was used in the past for problems such as the motion of junctions [260, 261]. Both the problem and its solution are different from the ones we describe now.

The Algorithm. Assume that the deformation of the surface $\mathcal{S}$, given by Eq. (2.9), is implemented with the level-set algorithm, i.e., Eq. (2.10). Let $\mathcal{R} \in \mathcal{S}$ be a region we want to track during this deformation and $\partial\mathcal{R}$ be its boundary. Define a new function $\hat{u}(\mathcal{X}, 0) : \mathbb{R}^m \to \mathbb{R}$ such that the intersection of its zero level set $\hat{\mathcal{S}}$ with $\mathcal{S}$ defines $\partial\mathcal{R}$ and then $\mathcal{R}$. In other words,

$$\partial\mathcal{R}(0) := \mathcal{S}(0) \cap \hat{\mathcal{S}}(0) = \{\mathcal{X} \in \mathbb{R}^n : \mathbf{u}(\mathcal{X}, 0) = \hat{u}(\mathcal{X}, 0) = 0\}.$$

The tracking of $\mathcal{R}$ is done by the simultaneous deformation of u and $\hat{u}$. The auxiliary function $\hat{u}$ deforms according to

$$\frac{\partial \hat{u}(\mathcal{X}, t)}{\partial t} = \hat{\beta}(\mathcal{X}, t)\|\nabla \hat{u}(\mathcal{X}, t)\|, \tag{2.11}$$

and then $\hat{\mathcal{S}}$ deforms according to

$$\frac{\partial \hat{\mathcal{S}}}{\partial t} = \hat{\beta}\vec{\mathcal{N}}_{\hat{\mathcal{S}}}. \tag{2.12}$$

We then have to find the velocity $\hat{\beta}$ as a function of β. To track the region of interest, $\partial\mathcal{R}$ must have exactly the same geometric velocity in both Eqs. (2.10) and (2.11). The velocity in Eq. (2.10) (or Eq. (2.9)) is given by the problem in hand and is $\beta\vec{\mathcal{N}}_{\mathcal{S}}$. Therefore the velocity in Eq. (2.12) will be the projection of this velocity into the normal direction $\vec{\mathcal{N}}_{\hat{\mathcal{S}}}$ (recall that tangential components of the velocity do not affect the geometry of the flow), that is, for (at least) $\partial\mathcal{R}$,

$$\hat{\beta} = \beta\vec{\mathcal{N}}_{\mathcal{S}} \cdot \vec{\mathcal{N}}_{\hat{\mathcal{S}}}.$$

Outside of the region corresponding to $\mathcal{R}$, the velocity $\hat{\beta}$ can be any function that connects smoothly with the values in $\partial\mathcal{R}$. (To avoid the creation of spurious intersections during the deformation of Φ and $\hat{\Phi}$, these functions can be reinitialized every few steps, as is frequently done in the level-set approach.)

This technique, for the moment, requires finding the intersection of the zero level sets of u and $\hat{u}$ at every time step in order to compute $\hat{\beta}$. To avoid this, we choose a particular extension of $\hat{\beta}$ outside of $\partial\mathcal{R}$, and simply define $\hat{\beta}$ as the projection of $\beta\vec{\mathcal{N}}_{\mathcal{S}}$ for all the values of $\mathcal{X}$ in the domain of Φ and $\hat{\Phi}$. Note that although $\mathcal{S}$ and $\hat{\mathcal{S}}$ do not occupy the same regions in the m-dimensional space, their corresponding embedding functions u and $\hat{u}$ do have the same domain, making this velocity extension straightforward. Therefore the auxiliary level-set flow is given by

$$\frac{\partial \hat{u}}{\partial t}(\mathcal{X}, t) = \left[\beta(\mathcal{X}, t)\frac{\nabla u(\mathcal{X}, t)}{\|\nabla u(\mathcal{X}, t)\|} \cdot \frac{\nabla \hat{u}(\mathcal{X}, t)}{\|\nabla \hat{u}(\mathcal{X}, t)\|}\right]\|\nabla \hat{u}(\mathcal{X}, t)\|, \quad (2.13)$$

and the region of interest $\mathcal{R}(t)$ is given by the portion of the zero level set that belongs to $u(\mathbf{X}, t) \cap \hat{u}(\mathbf{X}, t)$:

$$\partial\mathcal{R}(t) = \{\mathcal{X} \in \mathbb{R}^n : u(\mathcal{X}, t) = \hat{u}(\mathcal{X}, t) = 0\}. \quad (2.14)$$

Examples and Comments. Examples of the proposed technique are now presented. We should note the following: (1) The fast techniques described above, such as narrow-band, fast-marching, and local methods, can also be used with the technique proposed here to evolve each one of the surfaces; (2) in the examples below, we compute a zero-order type of intersection between the implicit surfaces, meaning that we consider part of the intersection as the full vortex where both surfaces go through (giving a jagged boundary). More accurate intersections can be easily computed with subdivisions, as in marching cubes.

Four examples are given in Fig. 2.4, one per row. In each example, the first figure shows the original surface with the marked regions to be tracked

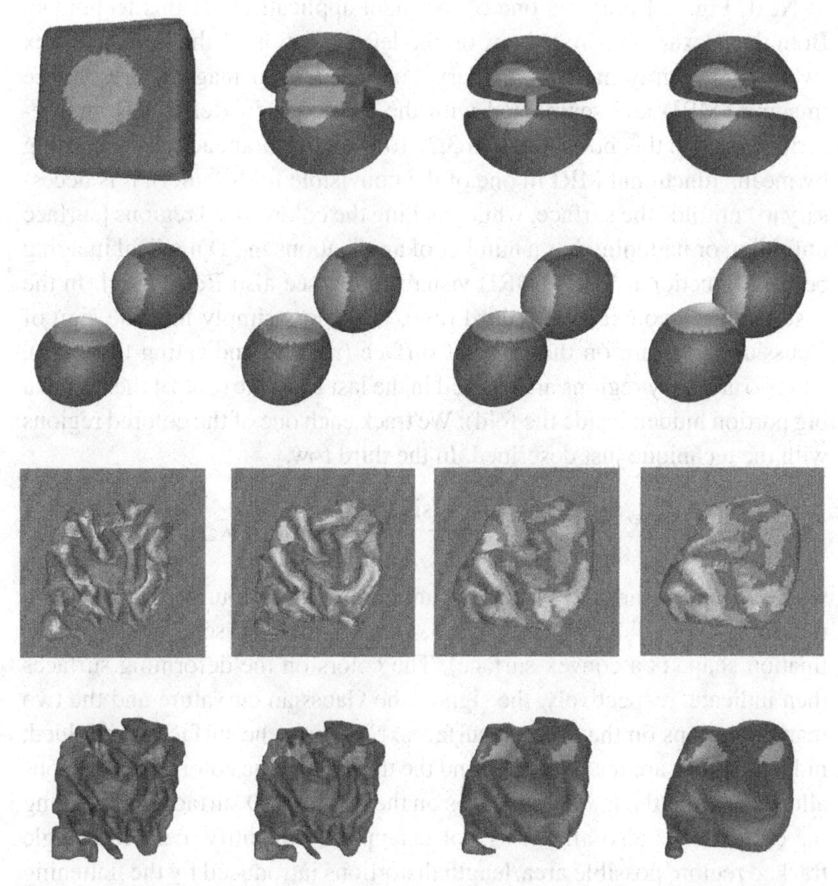

Fig. 2.4. Examples of the region-tracking algorithm (brighter regions are the ones being tracked). The two top rows show toy examples, demonstrating possible topological changes on the tracked region. The next rows show unfolding the cortex and tracking the marked regions with a curvature-based flow and a 3D morphing one, respectively.

(brighter regions), followed by different time steps of the geometric deformation and region tracking.

Figure 2.4 first shows two toy examples in the first two rows. We track the painted regions on the surfaces while they are deforming with a morphing-type velocity [23, and 26]. ($\beta(\mathbf{X}, t)$ is simply the difference between the current surface $\Phi(\mathbf{X}, t)$ and a desired goal surface $\Phi(\mathbf{X}, \infty)$, two separate surfaces and two merged balls, respectively, thereby morphing the initial surface toward the desired one [26].) Note how the region of interest changes topology (splits on the left example and merges on the next one).

Next, Fig. 2.4 presents one of the main applications of this technique. Both these examples first show, on the left, a portion of the human cortex (white-matter/gray-matter boundary), obtained from magnetic resonance imaging (MRI) and segmented with the technique in Ref. [384] and described later in this book, Section 3.2. To visualize brain activity recording by means functional MRI in one of the nonvisible folds (sulci), it is necessary to "unfold" the surface, while tracking the color-coded regions [surface unfolding or flattening has a number of applications in 3D medical imaging beyond Functional MRI (fMRI) visualization; see also Ref. [182]]. In the first of these two examples (third row), the colors simply indicate sign of Gaussian curvature on the original surface (roughly indicating the sulci), and two arbitrary regions are marked in the last example (one of them with a big portion hidden inside the fold). We track each one of the colored regions with the technique just described. In the third row,

$$\beta(\mathbf{X}, t) = \frac{\text{sign}(\kappa_1) + \text{sign}(\kappa_2)}{2}\min(|\kappa_1|, |\kappa_2|),$$

where κ_1 and κ_2 are the principal curvatures. In the fourth row, we use a morphing-type velocity as before [23, and 26] (in this case, the desired destination shape is a convex surface). The colors on the deforming surfaces then indicate, respectively, the sign of the Gaussian curvature and the two marked regions on the original surfaces. Note how the surface is unfolded, hidden regions are made visible, and the tracking of the color-coded regions allow us to find the matching places on the original 3D surface representing the cortex. This also allows us, for example, to quantify, per each single tracked region, possible area/length distortions introduced by the flattening process. To track all the marked regions simultaneously in these two examples, we select the zero level set of $\hat{\Phi}$ to intersect the zero level set of Φ at all these regions. If we have regions with more than two color codes to track, as will frequently happen in fMRI, we just use one auxiliary function $\hat{\Phi}$ per color (region).

The same technique can be applied to visualize lesions that occur on the hidden parts of the cortex. After unfolding, the regions become visible, and the region tracking allows us to find their position on the original surface. When level-set techniques are used to deform two given shapes, one toward the other (a 3D cortex to a canonical cortex, for example), this technique can be used to find the region-to-region correspondence. This technique then solves one of the basic shortcomings of the very useful level-set approach. Other approaches for 3D MRI visualization can be found in Ref. [182].

Level Sets in High Codimension. In the subsection above, we basically dealt with the tracking of a curve in three dimensions. This curve is on the surface $\mathcal{S}$ and is deforming with a velocity β intrinsic to the surface, but not necessarily intrinsic to the curve. We can now raise the question of how to extend the level-set technique to curves in three dimensions and, in general, to high codimension surfaces that move with intrinsic velocities (or, in some cases, with the combination of intrinsic and external velocities). This problem was addressed in Ref. [9]. The basic idea is to embed the high codimension surface into a squared distance function and then evolve this function with normal velocity equal to the $m - k$ smallest principal curvatures, where k is the codimension of the surface in $\mathbb{R}^m$. The curve is then the zero level set of a positive-everywhere function. Ambrosio and Soner [9] showed the existence of a unique weak solution for the function and showed that that the level-set solution coincides with the classical solution for the deforming hypersurface, when this exists. This approach is difficult to implement numerically and in some cases is not fully appropriate for numerical 3D curve evolution (Osher, personal communication; Faugeras, personal communication). A modification of this technique was used by Lorigo et al. [234] to extend the segmentation flows presented in Chap. 3 to detect curves and tubular sections in three dimensions.

2.3. Variational Level Sets

We have seen that the basic idea behind the level-set method is to represent a curve or surface in implicit form, as the (zero) level sets or isophote of a higher-dimensional function, and then track the deformation of the curve or surface by means of the deformation of this embedding function. In a number of curve and surface evolution applications, the curve/surface is deforming in order to minimize a given energy, that is, the shape motion is a gradient descent flow such as those described in Chap. 1. A natural question then is whether we can represent this variational/energy formulation in an implicit from and then use level-set techniques to numerically implement the corresponding gradient descent flow. The key idea is to define a variational formulation on an $n + 1$ dimensional function, while we are actually interested in only minimizing the value of the energy at the n-dimensional zero level set. This technique was developed in Refs. [426 and 427], and it is briefly described here.

Let us assume that we are given a variational formulation for a curve or surface deformation problem:

$$\mathbf{E}_\Gamma = \int_\Gamma L(\Gamma)\mathrm{d}s,$$

where Γ stands for either the curve $\mathcal{C}$ or the surface $\mathcal{S}$, and ds stands for the corresponding element of length or area, that is, we formulate a given problem as the minimization of $\mathbf{E}_\Gamma$. As we have seen before, in Section 1.9, this can be done by means of gradient descent, obtaining a Lagrangian flow for Γ. We then embed Γ in a higher-dimensional function u as before and replace $\mathbf{E}_\Gamma$ with

$$\mathbf{E}_u = \int_\Omega L(\mathcal{X})\delta(u)\|\nabla u\|\,\mathrm{d}\mathcal{X},$$

where Ω is the (fixed) 2D or 3D Euclidean space, $\mathcal{X}$ stands for a coordinate vector in this space, and $\delta(\mathcal{X})$ is the Dirac delta function (note that $\delta(\mathcal{X})\|\nabla u\|$ is the length/area element of the zero level set of u). Then, instead of computing the gradient descent of $\mathbf{E}_\Gamma$, we compute the gradient descent of $\mathbf{E}_u$ and use all the level-set machinery, obtaining an Eulerian flow. The gradient descent flow is obtained with standard calculus of variations. Figure 2.5, courtesy of the authors of Ref. [427], gives an example of the use of this technique to interpolate unorganized 3D data. In this case, L depends on the distance from the given unorganized data to the surface. The data can be points, curves, or surfaces, or a combination of them.

2.4. Continuous Mathematical Morphology

So far general techniques have been described to implement and study curve and surface evolution flows. As stated above, once these techniques are available, the art of curve and surface evolution is on the design of the velocity $\beta\vec{\mathcal{N}}$ to accomplish certain tasks. The first specific example, related to the design of $\beta\vec{\mathcal{N}}$ for basic morphological operations, is now presented. We now show that, when we are considering convex structuring elements, the continuous mathematical morphology on Euclidean space is nothing else but a curve evolution flow with a velocity $\beta\vec{\mathcal{N}}$ that depends on the structuring element. This leads to a new definition of discrete mathematical morphology based on the numerical implementation of the continuous flow.

Traditionally, mathematical morphology is defined in a set-theoretical setting [173, 255, 355], and morphological operators are defined over sets

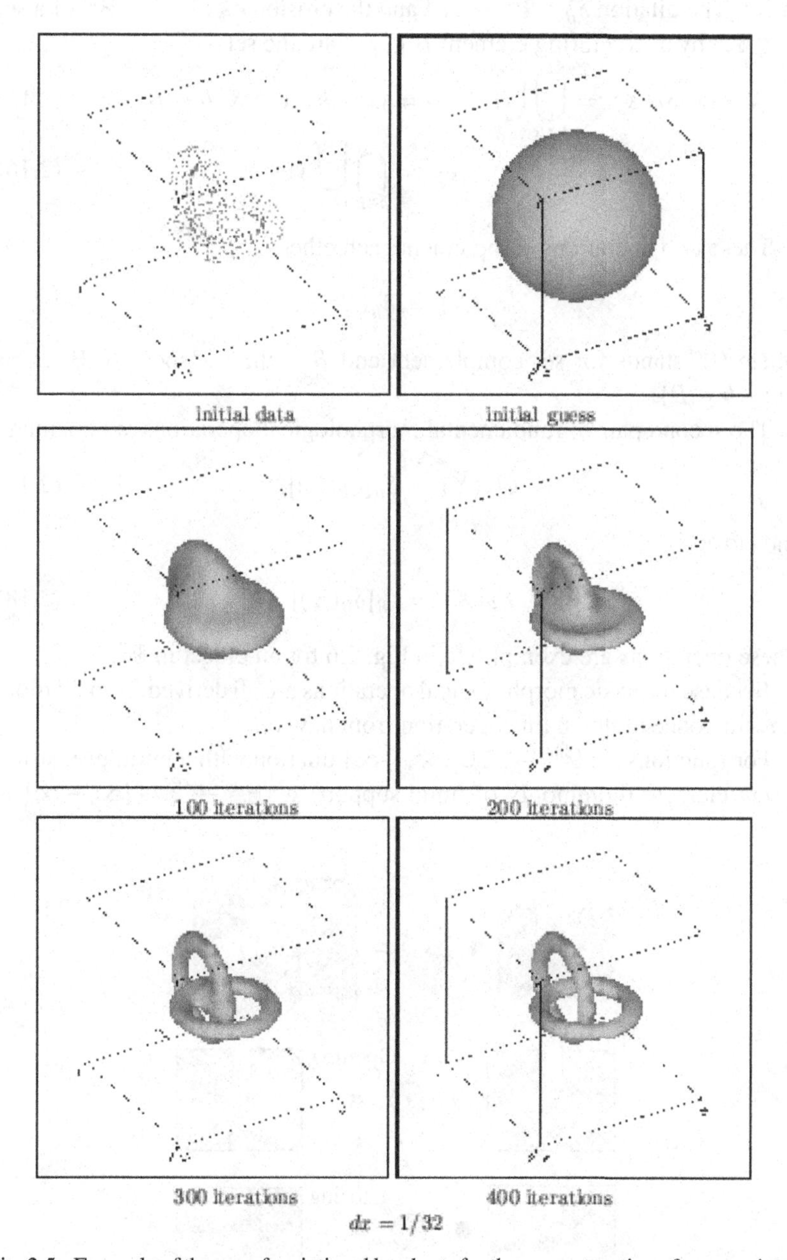

Fig. 2.5. Example of the use of variational level sets for the reconstruction of unorganized points. Observe how the two tori are reconstructed, without any prior knowledge of the connection between the points or the topology of the object. (Figure courtesy of Zhao et al.)

in $\mathbb{R}^m$. The dilation $\delta_B : \mathbb{R}^m \to \mathbb{R}^m$ and the erosion $\epsilon_B : \mathbb{R}^m \to \mathbb{R}^m$ of a set $X \subset \mathbb{R}^m$ by a structuring element $B \subset \mathbb{R}^m$ are the sets

$$\delta_B(X) := \bigcup_{b \in B} \bigcup_{x \in X} x + b = \{x + b : x \in X, b \in B\}, \tag{2.15}$$

$$\epsilon_B := \bigcap_{b \in B} \bigcup_{x \in X} x - b. \tag{2.16}$$

These two operations complement each other:

$$\epsilon_B(X) = [\delta_{\hat{B}}(X^c)]^c,$$

where $(\cdot)^c$ stands for set complement and $\hat{B}$ is the transpose of B, $\hat{B} := \{b : -b \in B\}$.

The second pair of fundamental morphological operations are opening

$$\mathcal{O}_B(X) := \delta_B[\epsilon_B(X)], \tag{2.17}$$

and closing

$$\mathcal{E}_B(X) := \epsilon_B[\delta_B(X)]. \tag{2.18}$$

These operations are exemplified in Fig. 2.6 for an object in $\mathbb{R}^2$.

Because the basic morphological operations are all derived from dilation, we will concentrate on this operation from now on.

For functions $u : \mathbb{R}^m \to \mathbb{R} \cup \{\infty, -\infty\}$ dilation with a multilevel structuring element (commonly of finite support) $g : \mathbb{R}^m \to \mathbb{R} \cup \{\infty, -\infty\}$ is

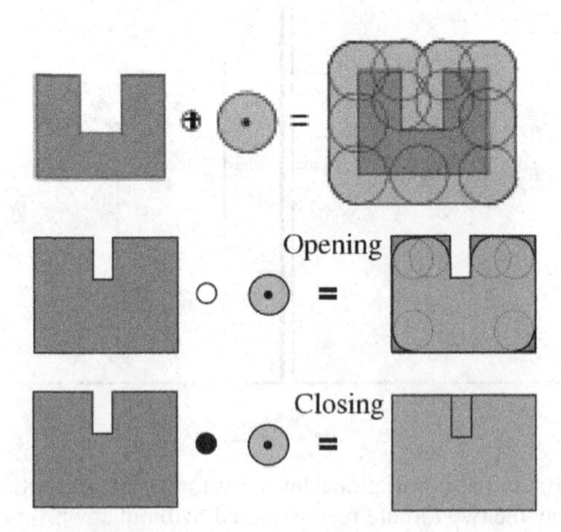

Fig. 2.6. Examples of dilation (first row), opening, and closing. (Figures courtesy of Ronny Kimmel.)

given by

$$\delta_g(u) := \sup_{y \in \mathbb{R}^m} \{u(x-y) + g(y)\}. \tag{2.19}$$

This is also a set operation, considering the concept of umbra.

A particular and useful case is when $g = -\infty$ outside of a finite set G and zero inside the set (flat structuring element). Dilation then becomes

$$\delta_G(u) := \sup_{y \in G}\{u(x-y)\}. \tag{2.20}$$

.

One of the most interesting properties of dilation is that $\delta_G(u)$ in Eq. (2.20) can be obtained as a collection of binary (set) morphological operators. More precisely, the dilation of f with a flat structuring element g is the union of the dilation of each one of the level sets of u with a structuring element given by G, the support of g. In other words, dilation and thresholding (level-sets decomposition) commute. In general, operations that commute with level-set decompositions are denoted as morphological operations, and they will have a very significant role in this book, and will be important in this section as well.

A second very important property of dilation is that it holds an associative property for convex elements. If B is convex, then

$$\delta_{(t_1+t_2)B}(X) = \delta_{t_1 B}[\delta_{t_2 B}(x)],$$
$$\delta_{(t_1+t_2)g}(f) = \delta_{t_1 g}[\delta_{t_2 g}(x)],$$
$$\delta_{(t_1+t_2)G}(X) = \delta_{t_1 G}[\delta_{t_2 G}(x)],$$

where $t \in \mathbb{R}^+$ and $tB := \{tx : x \in B\}$, $tg(x) = tg(x/t)$. For example, performing the dilations with a disk of radius r is equivalent to performing r/r_0 consecutive dilations of radius r_0.

Although we can give all the above definitions for discrete Euclidean spaces just by replacing $\mathbb{R}$ with $\mathbb{Z}$ (see Fig. 2.7), many of the nice geometric properties of continuous morphology are missing in its discrete version. Part of the problem is due to the difficulty in defining digital disks off all possible radii. One possible way to address this problem is to use accurate distance transforms [42, 108]. Another direction, which actually accurately implements distance transforms, is to address continuous mathematical morphology from the point of view of curve evolution and discrete mathematical morphology as the numerical implementation of the continuous flow. We proceed to develop this approach now, following Ref. [343] (see also Ref. [49]).

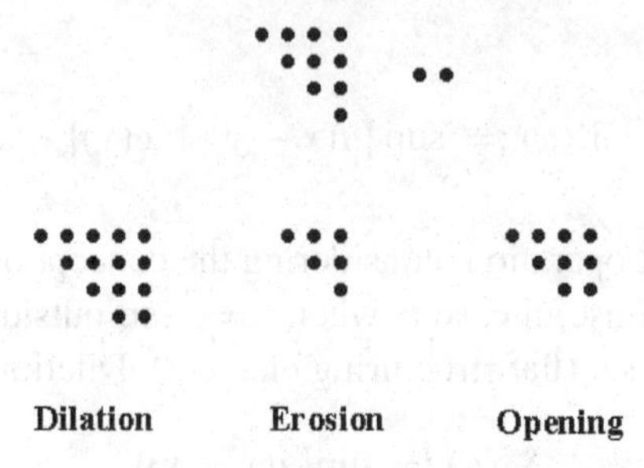

Fig. 2.7. Examples of basic discrete morphological operations.

2.4.1. The Continuous Morphology Velocity

According to the results at the beginning of this chapter, all possible geometric curve evolution flows are given by the equation

$$\frac{\partial \mathcal{C}}{\partial t} = \beta \vec{\mathcal{N}},$$

where $\mathcal{C}$ is the deforming curve. Assuming that $\mathcal{C}$ is the boundary of a planar set X, we need to find the velocity β such that the solution to the above flow gives $\delta_B(X)$.

Let ∂B be the boundary of the convex structuring element B. Because B is convex, ∂B can be represented by a function $\vec{r}(\theta) : [0, 2\pi] \to \mathbb{R}^2$. Dilation can be interpreted as a generalization of the Hüygens principle: Each point on the boundary $\mathcal{C}$ of X is the source of a new local wave, whose shape is given by the structuring element. The new wave front is obtained by the envelopes of the local waves; see Fig. 2.8. The velocity of each point is then given by the maximal projection of the structuring element on the

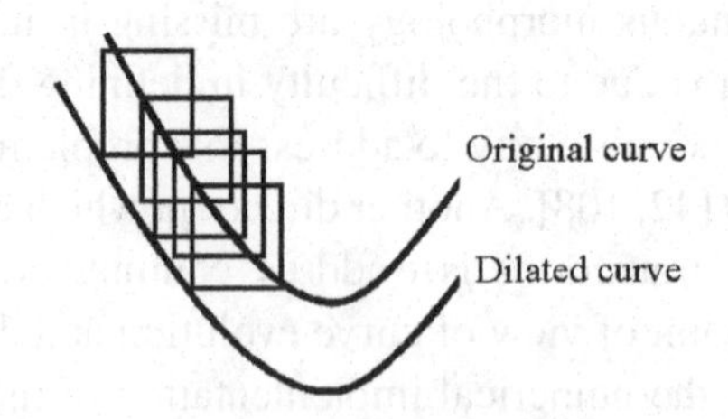

Fig. 2.8. Generalized Hüygens principle.

normal direction:

$$\beta = \sup_{\theta}\{\vec{r}(\theta) \cdot \vec{\mathcal{N}}\}. \tag{2.21}$$

This velocity can be explained as follows: Each point p in the curve $\mathcal{C}$, boundary of X, moves, according to the Hüygens principle, toward ∂B (centering at p) (recall that, because of the convexity of B, the dilation can be performed at infinitesimal steps). This motion generates a set of infinite possible velocities vectors, one for each point on ∂B, that is, $\vec{r}(\theta)$. Because the tangential component of those vectors does not affect the geometry of the flow, we can consider only their normal components, $\vec{r}(\theta) \cdot \vec{\mathcal{N}}$. The supremum of these projections is the effective velocity.

We then have that, for a convex structuring element B represented by $\vec{r}(\theta)$,

$$\frac{\partial \mathcal{C}}{\partial t} = \sup_{\theta}\{\vec{r}(\theta) \cdot \vec{\mathcal{N}}\}\vec{\mathcal{N}},$$

and, considering a level-set formulation, with $u(x, y) : \mathbb{R}^2 \to \mathbb{R}$, we have

$$\frac{\partial u}{\partial t} = \sup_{\theta}\{\vec{r}(\theta) \cdot \nabla u\}.$$

Note that because all the level sets of u are moving with the velocity corresponding to the binary dilation with B, the function u itself is dilating (flat dilation with a function g having B as its support). This is due to the commutativity between dilation and level-set decomposition.

It is easy to show that the following velocities are obtained for popular structuring elements:

$$\sup_{\theta}\{\vec{r} \cdot \nabla u\} = \begin{cases} \|\nabla u\|, & B = \text{disk} \\ \max\{|u_x|, |u_y|\}, & B = \text{diamond}. \\ |u_x| + |u_y|, & B = \text{square} \end{cases}$$

The equation corresponding to a disk means that the curve is moving with constant Euclidean velocity. This flow can also be used to find the skeleton of a shape [41]; see Section 2.5. The other equations, of course, mean constant velocity with other metrics (here we see again the necessity of convexity in order to define a metric). Examples are given in Fig. 2.9.

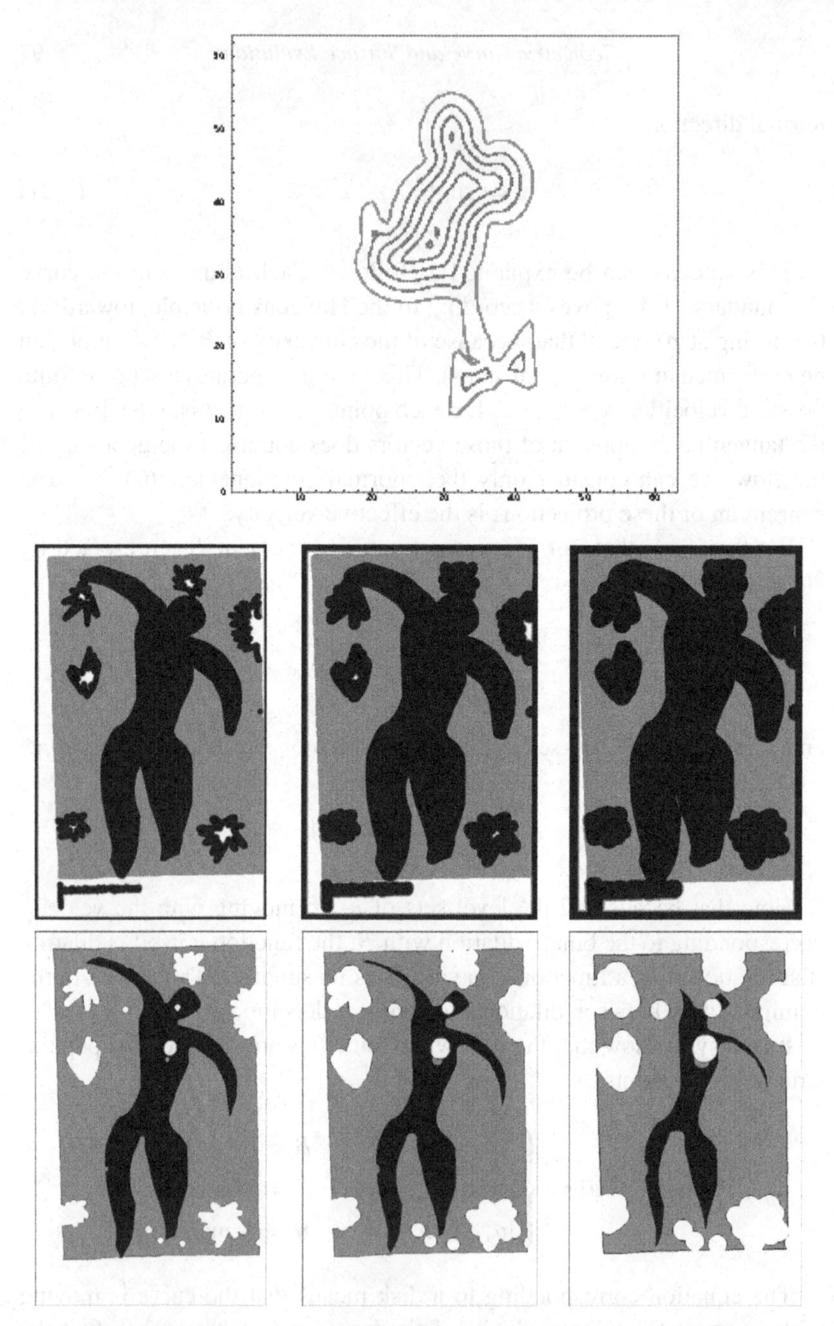

Fig. 2.9. Examples of continuous morphology by means of curve evolution. The erosion of a curve is shown in the first row. Steps of image erosion (with a disk) are shown in the second row and steps of image dilation in the third. Note of course that when the bright objects are eroded, the dark objects (background) are dilated, and vice versa. Also note that several objects are eroded/dilated at the same time, as the process is performed on the gray-valued image, which is equivalent to the level-set process (binary erosion/dilation).

2.5. Euclidean and Affine Curve Evolution and Shape Analysis

2.5.1. Constant-Velocity Flow

We have already seen the basic Euclidean invariant flow,

$$\mathcal{C}_t = \vec{\mathcal{N}}, \tag{2.22}$$

or, in level-set form,

$$u_t = \|\nabla u\|, \tag{2.23}$$

which moves the curve with constant inward velocity and is equivalent to the erosion of the enclosed shape with a disk. Inverting the direction of the flow, we obtain a morphological dilation. As we have seen, this flow can create singularities and change the topology of the evolving curve; see Fig. 2.9 (recall that erosion with a disk is equivalent to constant Euclidean motion). Moreover, the singularities created by this flow define the skeleton of the shape [41, 204, 216, 239, 364, 365, 422]. The skeleton, introduced by Blum [41], is the center of bitangent disks inside the shape. In other words, it is the center of disks that have a common point and tangent with at least two points of the shape, with the condition that the disk be completely inside the shape. Figure 2.10 is an example. Considering the outward flow, we can obtain the bitangent disks that are completely outside the shape. The constant-velocity flow is then implementing Blum's prerifire transform: If fire is turned on the boundary of the shape and this fire is allowed to travel inward with constant velocity, the skeleton is given by the places where the fire fronts collide.

It is easy to show that, e.g. [205], as expected, the normal and tangent to the evolving curve does not change direction with this flow, that is,

$$\frac{\partial \vec{\mathcal{T}}}{\partial t} = \frac{\partial \vec{\mathcal{N}}}{\partial t} = 0.$$

The curvature changes according to (examples of how to perform these computations are given later when presenting the affine heat flow)

$$\frac{\partial \kappa}{\partial t} = \kappa^2,$$

whose solution is

$$\kappa(p, t) = \frac{\kappa(p, 0)}{1 - \kappa(p, 0)},$$

which is singular already at $t = 1/\kappa(p, 0)$. Note that, as is well known

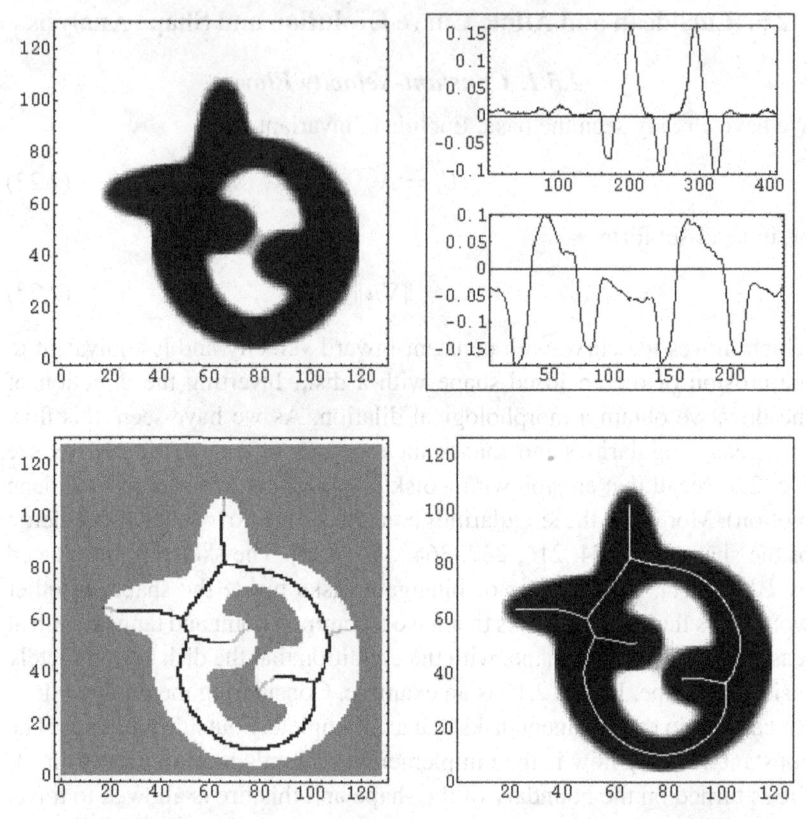

Fig. 2.10. Examples of the skeleton constructed with the constant-velocity curve flow. The first row shows the original binary shape and its curvature (for both inner and outer contours). The second row shows a discrete computation (left) and the curve evolution one (right). (Figure courtesy of Ronny Kimmel.)

from the skeleton theory, the first singularity corresponds to the maxima curvature, and it happens exactly at the corresponding center of curvature.

We now have a flow and, as we will later see in Section 2.8, the simplest one, which produces shocks. The next step is to find the simplest flow that smooths any simple planar curve without producing singularities (we will later also discuss the constant-affine-velocity flows). This is presented now.

2.5.2. Geometric Heat Flows

We consider, again planar simple curves $\mathcal{C}(p, t) = [x(p, t), y(p, t)]$ that are deforming in time, where p now stands for an arbitrary parameterization and t parameterizes the family. (For work on polygons see [52].) The simplest way to smooth the curve is to apply to each one of its coordinates the linear heat flow

$$\frac{\partial \mathcal{C}}{\partial t} = \frac{\partial^2 \mathcal{C}}{\partial p^2} \tag{2.24}$$

with a given initial curve $\mathcal{C}(p, 0)$. This flow, which is equivalent to smoothing each one of the coordinate curves with a Gaussian filter

$$\mathcal{G}(p, t) := \sqrt{\frac{1}{4\pi t}} \exp\left\{\frac{-p^2}{4t}\right\}$$

or

$$\begin{bmatrix} x_t \\ y_t \end{bmatrix} = \begin{bmatrix} x_{pp} \\ y_{pp} \end{bmatrix},$$

was the first equation introduced to define a scale space for planar curves. The basic idea behind the concept of scale space is to have multiple representations of the shape, starting from the original shape $\mathcal{C}(p, 0)$, going toward coarser and coarser representations $\mathcal{C}(p, t)$ when t, the scale parameter, increases [218, 413]. It is a crucial requirement in the scale-space theory that the information in $\mathcal{C}(p, t_1)$ is more than the information in $\mathcal{C}(p, t_2)$ whenever $t_1 < t_2$. Another important requirement is that $\mathcal{C}(p, t_2)$ can be obtained from $\mathcal{C}(p, t_1)$ without requiring or even knowing $\mathcal{C}(p, 0)$. This is the semigroup or causality property.

The Gaussian filtering or, equivalently, the linear heat flow is the only linear filter that defines a scale space [18, 423] (information is considered to be given by the number of zero crossings of the signal or, equivalently, the number of local extrema). It can also be derived from basic principles from physics [142, 218–221, 324]. This leads to the acceptance of this filter and its wide use for shape analysis; see, e.g., Refs. [263 ,264, and 324].

In spite of these very attractive properties, the linear heat flow (or Gaussian filtering) has a number of fundamental problems. First, we note that if the curve is parameterized by a different parameterization $w = w(p, t)$, $(\partial w/\partial p) > 0$, then, in spite of the fact that geometrically $\mathcal{C}(p, 0)$ and $\mathcal{C}(w, 0)$ are identical, the solution to the linear heat flow gives different results, that is, $\mathcal{C}(p, t) \neq \mathcal{C}(w, t)$ for $t > 0$, and two different implementations will get two different scale-space representations (the flow is not intrinsic to the curve, but depends on its parameterization).

A possible attempt to solve this is to parameterize the original curve with the Euclidean arc-length parameterization v such that $\|(\partial \mathcal{C}/\partial v)\| = 1$ or the affine arc length, if an affine invariant flow is required. (In this chapter we use v for the Euclidean arc length and s for the affine one.) This will make the flow intrinsic, but will violate the semigroup property, as the arc length is a time-dependent parameterization, and then a reparameterization will be required. In addition, order is not preserved, and if two initial curves are one inside the other, this is not necessarily the case after this arc-length parameterization technique is used. This is especially problematic for shapes with holes, for example. Problems with the numerical implementation of the heat flow were found as well [242], mainly that the corresponding discrete flow violates the basic causality principle. We need therefore to look for a different way to obtain an intrinsic geometric scale space.

In spite of its problems, Gaussian filtering leaves very important messages. Not only does it introduce the concept of scale space, but it also hints at the possible solution to all its problems. As pointed out by Hummel [191], other PDE formulations, not just the linear heat flow equivalent to Gaussian filtering, can lead to scale spaces. All that we need to request is that the PDE holds the maximum principle discussed in Chap. 1. This will guarantee that information is decreasing when t increases. We are then in the search for a PDE that holds the maximum principle and does not suffer from the problems above, and we do this from the theory of curve evolution, that is, we search for the appropriate velocity $\beta\vec{\mathcal{N}}$ for the curve deformation flow

$$\mathcal{C}_t = \beta\vec{\mathcal{N}}.$$

We start with a brief presentation of the Euclidean case, moving then to a more detailed description of the affine case. Other groups, such as the projective one, are discussed in Refs. [131, 132, 134, 135, 284, and 286] and results are very briefly described at the end of this chapter.

A natural choice to obtain an intrinsic Euclidean invariant scale space is to consider $\beta = \kappa$ (we later actually show that this is the simplest possible selection), leading to the Euclidean geometric heat flow:

$$\frac{\partial \mathcal{C}}{\partial t} = \kappa\vec{\mathcal{N}}, \tag{2.25}$$

with the original closed curve as the initial condition.

Because $\mathcal{C}_{vv} = \kappa\vec{\mathcal{N}}$, the Euclidean heat flow can be written as

$$\frac{\partial \mathcal{C}}{\partial t} = \frac{\partial^2 \mathcal{C}}{\partial v^2}, \tag{2.26}$$

which looks very much like the linear heat flow, with the important difference that v is a time-dependent parameterization, making the equation nonlinear (and interesting!).

This flow is very well studied in the literature, also for nonsimple curves and for spatial curves, and has a number of amazing properties. Any non-convex curve becomes convex in finite time, enclosing a finite area [162], and then it converges into a round point [150], that is, a point that, when dilated, is a disk (a convex curve remains convex under this flow). Moreover, the total curvature, given by

$$\oint |\kappa| \mathrm{d}v,$$

decreases. The number of vertices and the number of local maxima of κ and y (when we are considering a local Cartesian system) decrease, and the local minima increase. The number of inflection points also decreases. Last, the order of curves enclosing is preserved. All these properties clearly show that this is a perfect candidate for a geometric, Euclidean invariant scale space. Figure 2.11 shows examples.

The evolution of the main geometric characteristics of the curve when it evolves according to Eq. (2.25) is given by

$$\begin{aligned}
\vec{\mathcal{T}}_t &= -\kappa_s \vec{\mathcal{N}}, \\
\vec{\mathcal{N}}_t &= \kappa_s \vec{\mathcal{T}}, \\
\text{area}_t &= -2\pi, \\
\text{length}_t &= -\oint \kappa^2 \mathrm{d}s,
\end{aligned}$$

Fig. 2.11. Steps of the Euclidean geometric heat flow. Note how the objects are becoming smoother.

and

$$\kappa_t = \kappa_{ss} - \kappa^3.$$

The Euclidean heat flow can of course be combined with the constant velocity flow to yield

$$\mathcal{C}_t = (1 + \epsilon\kappa)\vec{\mathcal{N}}.$$

As we have seen before, this is the flow introduced by Osher and Sethian to study the numerical analysis of curve evolution theory. It was introduced to the computer vision community in Refs. [204 and 206], in which its importance for shape analysis was studied. Note that in the level-set formulation, this flow is

$$u_t = (1 + \epsilon\kappa)\|\nabla u\|,$$

where, as we have seen before, the curvature of the level set is given by

$$\kappa = \frac{u_{xx}u_y^2 - 2u_x u_y u_{xx} + u_{yy}u_x^2}{(\|\nabla u\|)^3}.$$

This level-set flow is well defined and well posed also for nonsmooth curves, based on the theory of viscosity solutions [5]. This also means that the geometric heat flow is well defined also for nonsmooth curves. We do this by embedding the curve in the level-set framework and considering the unique viscosity solution. Recall that, for a large class of embedding functions, the evolution is independent of the exact form of the embedding. Therefore the theory of viscosity solutions not only formalizes the level-set framework but also allows for extensions of flows such as the heat flow to nonsmooth curves. This will be true for all the curve and surface evolution flows studied in this book.

Let us conclude the discussion of the Euclidean geometric heat flow with a property that not only makes it very important in physics but that makes the flow a star in shape segmentation as well. The flow is the gradient descent flow of the Euclidean length

$$L := \oint \mathrm{d}v,$$

and for this reason it is also called the Euclidean shortening flow.

Affine Geometric Heat Flow

It is time now to extend the Euclidean geometric heat flow to the affine group. In other words, we want a flow that has the same smoothing characteristics of the Euclidean geometric heat flow, but that is invariant to affine transformations as well. We restrict ourself to the special affine group and follow the developments in Refs. [15 and 346–348] (a completely different approach for affine invariant scale spaces is given in Ref. [338, 341] whereas the basic affine flow and its uniqueness, from the point of view of the viscosity theory of PDEs, was also derived in the seminal paper of Ref. [5]). See also [11].

Following Eq. (2.26), we can obtain an affine invariant flow by means of

$$\frac{\partial \mathcal{C}}{\partial t} = \frac{\partial^2 \mathcal{C}}{\partial s^2}, \tag{2.27}$$

derivatives being taken now with respect to the affine arc length as introduced in Chap. 1. Because the tangential component of the velocity does not affect the flow, and

$$\vec{\mathbf{N}} = \mathcal{C}_{ss} = \kappa^{1/3}\vec{\mathcal{N}} - \frac{\kappa_v}{3\kappa^{5/3}}\vec{\mathcal{T}},$$

the affine invariant flow of Eq. (2.27) is geometrically equivalent to

$$\frac{\partial \mathcal{C}}{\partial t} = \kappa^{1/3}\vec{\mathcal{N}}. \tag{2.28}$$

Note that because affine differential geometry is not defined for nonconvex curves the flow of Eq. (2.27), the affine geometric heat flow, is not defined for nonconvex curves, whereas Eq. (2.28), which is geometrically affine invariant, is defined for all planar curves. Therefore, by projecting the affine velocity into the Euclidean normal, we have made the flow well defined. Moreover, in spite of the fact that the flow is now also affine invariant, it has the same complexity (same number of derivatives) as the Euclidean heat flow. If we want to have a flow that is not just geometrically invariant, but is invariant with respect to the parameterization as well, we need to keep the tangential component and consider

$$\frac{\partial \mathcal{C}}{\partial t} = \begin{cases} \frac{\partial^2 \mathcal{C}}{\partial s^2}, & \text{noninflection points} \\ 0, & \text{inflection points} \end{cases}. \tag{2.29}$$

A number of properties of the affine heat flow are presented below, showing that it indeed defines a scale space. In Ref. [5] it is proved that this is the unique affine flow that holds all the key properties of a scale space for planar curves. This is also derived from the results at the end of this chapter, which also show that this is the simplest possible affine invariant curve flow. Note that this contrasts with the Euclidean case, in which the simplest flow, constant velocity, did not define a scale space and created singularities.

As in the Euclidean case, the affine geometric heat flow deforms any simple nonconvex curve into a convex one and from there into an elliptic point. All the deformation is smooth, without creating singularities or self-intersections. The same basic properties, such as reducing the total curvature and reducing the number of inflections, hold for this flow as well. All this means that this flow defines a geometric affine invariant scale space. In the rest of this section details are given on the flow for convex curves, that is, on Eq. (2.27). The nonconvex case requires deeper knowledge of PDEs, not provided in this book. Details can be found in Ref. [15]. Full details on the convex case can be found in Refs. [346–348].

Basic Evolution of Geometric Characteristics. Evolution equations are now presented for both affine and Euclidean invariants when the curve is deforming according to the affine heat flow. With

$$g := [\mathcal{C}_p, \mathcal{C}_{pp}]^{1/3}$$

as the affine metric, the following change in the order of derivation will be useful later:

$$\begin{aligned}\frac{\partial}{\partial t}\frac{\partial}{\partial s} &= \frac{\partial}{\partial t}\left(\frac{1}{g}\frac{\partial}{\partial p}\right)\\ &= -\frac{g_t}{g^2}\frac{\partial}{\partial p} + \frac{1}{g}\frac{\partial}{\partial t}\frac{\partial}{\partial p} = -\frac{g_t}{g}\frac{\partial}{\partial s} + \frac{\partial}{\partial s}\frac{\partial}{\partial t}.\end{aligned}$$

We now compute the affine metric evolution. We first note that

$$\begin{aligned}\frac{\partial g^3}{\partial t} &= \left[\frac{\partial \mathcal{C}_p}{\partial t}, \mathcal{C}_{pp}\right] + \left[\mathcal{C}_p, \frac{\partial \mathcal{C}_{pp}}{\partial t}\right]\\ &= \left[\frac{\partial \mathcal{C}_t}{\partial p}, \mathcal{C}_{pp}\right] + \left[\mathcal{C}_p, \frac{\partial^2 \mathcal{C}_t}{\partial p^2}\right].\end{aligned}$$

However,

$$\begin{aligned}\left[\frac{\partial \mathcal{C}_t}{\partial p}, \mathcal{C}_{pp}\right] &= \left[\frac{\partial \mathcal{C}_{ss}}{\partial p}, \mathcal{C}_{pp}\right]\\ &= [g\mathcal{C}_{sss}, \mathcal{C}_{pp}] = g[\mathcal{C}_{sss}, g^2\mathcal{C}_{ss} - g^2\mathcal{C}_p p_{ss}]\\ &= -g^3[\mathcal{C}_{ss}, \mathcal{C}_{sss}] + g[\mathcal{C}_{sss}, -g^2\mathcal{C}_p p_{ss}]\\ &= -g^3\mu + g[\mathcal{C}_{sss}, -g^3\mathcal{C}_s p_{ss}] = -g^3\mu,\end{aligned}$$

$$\frac{\partial^2 \mathcal{C}_t}{\partial p^2} = \frac{\partial^2 \mathcal{C}_{ss}}{\partial p^2} = \frac{\partial}{\partial p}\left(\frac{\partial \mathcal{C}_{ss}}{\partial p}\right)$$
$$= \frac{\partial}{\partial p}(g\mathcal{C}_{sss}) = g\frac{\partial}{\partial s}(g\mathcal{C}_{sss})$$
$$= g^2\mathcal{C}_{ssss} + gg_s\mathcal{C}_{sss}.$$

Further,

$$\left[\mathcal{C}_p, \frac{\partial^2 \mathcal{C}_t}{\partial p^2}\right] = [g\mathcal{C}_s, g^2\mathcal{C}_{ssss} + gg_s\mathcal{C}_{sss}].$$
$$= -g^3\mu.$$

Therefore

$$\frac{\partial g^3}{\partial t} = -2g^3\mu$$
$$\frac{\partial g}{\partial t} = -\frac{2}{3}g\mu, \tag{2.30}$$

which gives the required evolution of the affine metric.

We can compute in a similar form the affine tangent evolution and affine normal evolution:

$$\frac{\partial \vec{\mathbf{T}}}{\partial t} = -\frac{1}{3}\mu\vec{\mathbf{T}}, \tag{2.31}$$
$$\frac{\partial \vec{\mathbf{N}}}{\partial t} = \frac{1}{3}\mu\vec{\mathbf{N}} - \frac{1}{3}\mu_s\vec{\mathbf{T}}. \tag{2.32}$$

Similarly

$$\frac{\partial \vec{\mathbf{N}}_s}{\partial t} = \mu\vec{\mathbf{N}}_s - \frac{1}{3}\mu_{ss}\vec{\mathbf{T}}. \tag{2.33}$$

We now come to a crucial computation, namely the affine curvature evolution. First note that

$$\frac{\partial \mu}{\partial t} = \frac{\partial}{\partial t}[\mathcal{C}_{ss}, \mathcal{C}_{sss}] = \left[\frac{\partial \mathcal{C}_{ss}}{\partial t}, \mathcal{C}_{sss}\right] + \left[\mathcal{C}_{ss}, \frac{\partial \mathcal{C}_{sss}}{\partial t}\right].$$

Consequently, by using the formulas above, we see that the affine curvature evolves according to the following reaction-diffusion equation:

$$\frac{\partial \mu}{\partial t} = \frac{4}{3}\mu^2 + \frac{1}{3}\mu_{ss}. \tag{2.34}$$

Proposition 2.1. *If $\mu(p, 0) > 0$, then $\mu(p, t) > 0$.*

Proof: The proof of this proposition is a straightforward application of the maximum principle described in Chap. 1. Let $\mu_{\min}(t) := \inf\{\mu(p,t)\}$. Suppose that at some $t > 0$, $\mu_{\min}(t) = \beta, 0 < \beta < \mu_{\min}(0)$. Let $t_0 := \inf\{t : \mu_{\min}(t) = \beta\}$. Then, β is achieved for the first time at $\mu(p_0, t_0)$. We have that $\mu_t \leq 0$, $\mu_{pp} \geq 0$, $\mu_p = 0$, and $\mu = \beta > 0$. Further,

$$\mu_{ss} = \mu_{pp}(p_s)^2 + \mu_p p_{ss} \geq 0.$$

Then we have that the left-hand side of Eq. (2.34) is negative, whereas the right-hand side is positive, which is not possible. This proves that $\mu_{\min}(t)$ is a nondecreasing function. Therefore $\mu(p,t) \geq \mu_{\min}(t) \geq \mu_{\min}(0) > 0$. □

Remark: Using the strong maximum principle [150, 321, 328] we can prove that segments of the initial curve $\mathcal{C}(p, 0)$ such that $\mu(p, 0) = 0$ disappear immediately as the curve evolves, and the succeeding curves are strictly elliptically curved ($\mu > 0$).

Because $\mathcal{C}(t)$ is a convex curve (see Theorem 2.1), we can express the affine curvature evolution equation (2.34) by using the parameter θ, where θ is the angle between the affine tangent vector $\vec{\mathbf{T}}$ and the x axis (i.e., $\vec{\mathbf{T}} = (\cos\theta, \sin\theta)\|\vec{\mathbf{T}}\|$). If we change the time parameter from t to τ, we obtain

$$\frac{\partial \mu}{\partial t} = \frac{\partial \mu}{\partial \tau} + \frac{\partial \mu}{\partial \theta}\frac{\partial \theta}{\partial t}.$$

Because

$$\frac{\partial \vec{\mathbf{T}}}{\partial t} = -\frac{1}{3}\mu \vec{\mathbf{T}},$$

θ does not change with time, i.e.,

$$\frac{\partial \theta}{\partial t} = 0.$$

Therefore

$$\frac{\partial \mu}{\partial t} = \frac{\partial \mu}{\partial \tau},$$

and we can replace τ with t. Also,

$$\mu_{ss} = \frac{\mu_{\theta\theta}}{g^2} - \mu_\theta \frac{g_\theta}{g^3}$$

where g is the affine metric for the parameter θ, i.e., $g = [\mathcal{C}_\theta, \mathcal{C}_{\theta\theta}]^{1/3} = \kappa^{-2/3}$

(κ is the Euclidean curvature). Then we obtain

$$\frac{\partial \mu}{\partial t} = \frac{1}{3}\kappa^{4/3}\mu_{\theta\theta} + \frac{2}{9}\kappa^{1/3}\kappa_\theta\mu_\theta + \frac{4}{3}\mu^2. \tag{2.35}$$

We next compute the evolution of the affine perimeter $L := \oint \mathrm{d}s = \oint g\mathrm{d}p$. Indeed, noting that

$$\frac{\partial L}{\partial t} = \frac{\partial}{\partial t}\oint g\mathrm{d}p = \oint \frac{\partial g}{\partial t}\mathrm{d}p$$

we see from Eq. (2.30) that

$$\frac{\partial L}{\partial t} = -\frac{2}{3}\oint g\mu\mathrm{d}p = -\frac{2}{3}\oint \mu\mathrm{d}s, \tag{2.36}$$

and in Ref. [348] it was proved that this expression is negative, meaning that the affine geometric heat flow is an affine length-shortening process.

We now write an expression for the Euclidean curvature evolution. Let κ be the Euclidean curvature and $\rho := 1/\kappa$ be the Euclidean radius of curvature. Therefore

$$\rho = \frac{\langle \mathcal{C}_p, \mathcal{C}_p\rangle^{3/2}}{[\mathcal{C}_p, \mathcal{C}_{pp}]} = \frac{\langle g\mathcal{C}_s, g\mathcal{C}_s\rangle^{3/2}}{g^3}.$$

Then

$$\rho = \langle \mathcal{C}_s, \mathcal{C}_s\rangle^{3/2} = \|\vec{\mathbf{T}}\|^3. \tag{2.37}$$

Let $\mathcal{C} = (x, y)^t$. Then

$$\begin{aligned}
\rho &= \left(x_s^2 + y_s^2\right)^{3/2}, \\
\frac{\partial \rho}{\partial t} &= \frac{3}{2}\left(x_s^2 + y_s^2\right)^{1/2}\left(2x_s\frac{\partial x_s}{\partial t} + 2y_s\frac{\partial y_s}{\partial t}\right) \\
&= 3\|\vec{\mathbf{T}}\|\langle \vec{\mathbf{T}}, \vec{\mathbf{T}}_t\rangle = 3\|\vec{\mathbf{T}}\|\langle \vec{\mathbf{T}}, -\frac{1}{3}\mu\vec{\mathbf{T}}\rangle \\
&= -\mu\|\vec{\mathbf{T}}\|\langle \vec{\mathbf{T}}, \vec{\mathbf{T}}\rangle = -\mu\|\vec{\mathbf{T}}\|^3.
\end{aligned}$$

Hence

$$\frac{\partial \rho}{\partial t} = -\mu\rho, \tag{2.38}$$

$$\frac{\partial \kappa}{\partial t} = -\frac{1}{\rho^2}\frac{\partial \rho}{\partial t},$$

$$\frac{\partial \kappa}{\partial t} = \frac{\mu}{\rho} = \mu\kappa. \tag{2.39}$$

This equation was also used to numerically compute the affine curvature μ [136].

Theorem 2.1. *The curve $\mathcal{C}(t)$ remains convex as it evolves according to the affine heat flow.*

Proof: In the case in which $\mu(0) \geq 0$, $\mu(t) > 0$ $(t > 0)$ immediately. Then from the evolution equation of the Euclidean curvature (2.39), we see that $\kappa(t)$ is a nondecreasing function, and because $\kappa(0) > 0$ ($\mathcal{C}(0)$ is convex), the curve remains convex.

If $\mu(0)$ is negative on some segments, by using the maximum principle, we can easily show that $\mu_{\min}(t)$ is a nondecreasing function ($\mu_{\min}(t)$ is defined as in Proposition 2.1). More precisely, define $\hat{\mu}(t) := \mu(t) - \mu_{\min}(0)$. Then $\hat{\mu}(0) \geq 0$, and, with the same argument as that in the proof of Proposition 2.1, $\hat{\mu}_{\min}(t)$ is a nondecreasing function, and thus so is $\mu_{\min}(t)$. Therefore, in the worst case (see Eq. (2.39)), $\kappa(t)$ decreases exponentially as $\kappa(0)\exp[\mu_{\min}(0)\, t]$, never becoming zero ($\kappa(0) > 0$). □

We now examine the Euclidean perimeter evolution. (See also Refs. [12, 13, 150, 162, and 204].) First, the Euclidean perimeter is defined as

$$\iota := \oint \left(x_p^2 + y_p^2\right)^{1/2} \mathrm{d}p. \tag{2.40}$$

Let

$$h = \left(x_p^2 + y_p^2\right)^{1/2} = g\left(x_s^2 + y_s^2\right)^{1/2}.$$

Then

$$\begin{aligned}\frac{\partial h}{\partial t} &= g_t \|\vec{\mathbf{T}}\| + g\|\vec{\mathbf{T}}\|^{-1}\langle \vec{\mathbf{T}}, \vec{\mathbf{T}}_t\rangle \\ &= -\frac{2}{3} g\mu\|\vec{\mathbf{T}}\| - \frac{1}{3} g\|\vec{\mathbf{T}}\|^{-1}\mu\|\vec{\mathbf{T}}\|^2 \\ &= -g\mu\|\vec{\mathbf{T}}\| = -\mu h,\end{aligned}$$

$$\frac{\partial \iota}{\partial t} = \frac{\partial}{\partial t}\oint h \mathrm{d}p = \oint \frac{\partial h}{\partial t}\mathrm{d}p.$$

Therefore

$$\frac{\partial \iota}{\partial t} = -\oint g\mu\|\vec{\mathbf{T}}\|\mathrm{d}p. \tag{2.41}$$

Remark: If μ is nonnegative, we see immediately from Eq. (2.41) that the Euclidean curve perimeter decreases as a function of time. If μ is somewhere negative, with the curve parameter p taken as the Euclidean arc length, the affine curvature can be expressed as a function of the Euclidean curvature:

$$\mu = \kappa^{4/3} - \frac{5}{9}\kappa^{-8/3}(\kappa_p)^2 + \frac{1}{3}\kappa^{-5/3}\kappa_{pp}.$$

Using this formula in the integral in Eq. (2.41) and integrating by parts, we find

$$\oint g\mu\|\vec{\mathbf{T}}\|\mathrm{d}p$$

is positive. Therefore the Euclidean perimeter decreases with time, even if μ is somewhere negative.

Define the Euclidean curvature metric Υ as $\Upsilon := \kappa h$ [203]. Then we have the following elementary result:

Lemma 2.2. *The quantity Υ is conserved in the affine curve evolution.*

Proof: Left as an exercise. □

We now discuss the evolution of the area $A^{(r)}$. First,

$$A^{(r)} = \frac{1}{2}\oint [C, C_p]\mathrm{d}p, \tag{2.42}$$

$$\frac{\partial A^{(r)}}{\partial t} = \frac{1}{2}\oint \frac{\partial}{\partial t}[C, C_p]\mathrm{d}p = \frac{1}{2}\oint [C_t, C_p]\mathrm{d}p + \frac{1}{2}\oint [C, (C_p)_t]\mathrm{d}p,$$

$$\oint [C_t, C_p]\mathrm{d}p = \oint [C_{ss}, gC_s]\mathrm{d}p = -\oint g\mathrm{d}p = -L.$$

Then

$$\begin{aligned}
\oint [C, (C_p)_t]\mathrm{d}p &= \oint [C, (C_t)_p]\mathrm{d}p = \oint [C, (C_{ss})_p]\mathrm{d}p \\
&= \oint [C, gC_{sss}]\mathrm{d}p = \oint [C, C_{sss}]\mathrm{d}s = \oint (xy_{sss} - yx_{sss})\mathrm{d}s \\
&= (xy_{ss} - yx_{ss})|_0^L - \oint (x_s y_{ss} - y_s x_{ss})\mathrm{d}s \\
&= -\oint [C_s, C_{ss}]\mathrm{d}s = -\oint \mathrm{d}s = -L.
\end{aligned}$$

Consequently,

$$\frac{\partial A^{(r)}}{\partial t} = -L. \tag{2.43}$$

Remark: If we take p (the curve parameter) as equal to v (the Euclidean arc length), we obtain $g = \kappa^{1/3}$. Then,

$$L = \oint \kappa^{1/3} \mathrm{d}v.$$

Because the curve is convex (and remains convex in the affine curve evolution process), L is positive and the area decreases as a function of time.

We now turn to the affine isoperimetric ratio evolution. More precisely, from the affine isoperimetric inequality introduced in Chap. 1, we have the following result about the affine isoperimetric ratio:

Theorem 2.2. *For ovals, the affine isoperimetric ratio*

$$\frac{L^3(t)}{A^{(r)}(t)}$$

is a nondecreasing function of time.

Proof: First note that

$$\frac{\partial}{\partial t}\left[\frac{L^3}{A^{(r)}}\right] = \frac{3L^2 L_t A^{(r)} - L^3 A_t^{(r)}}{[A^{(r)}]^2}$$

$$= \frac{L^2}{A^{(r)}}\left[3L_t - \frac{LA_t^{(r)}}{A^{(r)}}\right].$$

Then we have to prove that

$$3L_t - \frac{LA_t^{(r)}}{A^{(r)}} \geq 0$$

or

$$-3L_t \leq -\frac{LA_t^{(r)}}{A^{(r)}}.$$

From Eqs. (2.36) and (2.43) above, we obtain

$$-3L_t = 2\oint \mu \mathrm{d}s,$$

$$-\frac{LA_t^{(r)}}{A^{(r)}} = \frac{L^2}{A^{(r)}}.$$

Hence we must show that

$$2\oint \mu ds \le \frac{L^2}{A^{(r)}},$$

which is precisely the affine isoperimetric inequality presented in Chap. 1.

□

Remark: The above inequality is strictly negative if the curve is not an ellipse [348]. Therefore the isoperimetric ratio increases when $\mathcal{C}(t)$ is not an ellipse.

Remark: If $\mu(p, t_0) =$ constant for some $t_0 \ge 0$, i.e., $\mathcal{C}(p, t_0)$ is an ellipse, then

$$2\oint \mu ds = \frac{L^2}{A^{(r)}},$$

$$\frac{\partial}{\partial t}\left[\frac{L^3}{A^{(r)}}\right] = 0.$$

This means that for all $t \ge t_0$, $\mathcal{C}(p, t)$ remains an ellipse:

$$\frac{L^3(t_0)}{A^{(r)}(t_0)} = \frac{L^3(t)}{A^{(r)}(t)} = 8\pi^2.$$

Some of the above ideas are illustrated by the following example.

Example. An explicit example of affine curve evolution. Let $\mathcal{C}(p, 0) = \mathcal{C}_0(p)$ (the initial curve) be an ellipse with affine curvature $\mu(p, 0) = \mu_0$ (constant) and Euclidean curvature $\kappa(p, 0)$. In our discussion below we will sometimes set $\mu(t) = \mu(\cdot, t)$, when the curve parameter is understood. $\mathcal{C}_0$ is given by (see Ref. [53]):

$$\mathcal{C}_0 = \begin{bmatrix} a\ \cos(\mu^{1/2} s) \\ b\ \sin(\mu^{1/2} s) \end{bmatrix}, \quad ab\mu^{3/2} = 1, \ \ a, b > 0.$$

From the last Remark, we know that the curve remains an ellipse as it evolves. Then $\mu_s(t) = \mu_{ss}(t) = 0$, and the evolution for the affine curvature is given by Eq. (2.34):

$$\mu_t(t) = \frac{4}{3}\mu^2(t).$$

In this case, $\mu(t)$ can be explicitly calculated to be

$$\mu(t) = \frac{\mu_0}{1 - \frac{4}{3}\mu_0 t}.$$

We see from this equation that the curve exists for all $t \in [0, \frac{3}{4\mu_0})$.

Next the Euclidean curvature evolution is given by Eq. (2.39):

$$\kappa_t = \mu\kappa = \frac{\mu_0}{1 - \frac{4}{3}\mu_0 t}\kappa.$$

Then

$$\begin{aligned}
\frac{\kappa_t}{\kappa} &= \frac{\mu_0}{1 - \frac{4}{3}\mu_0 t}, \\
[\ln(\kappa)]_t &= \frac{\mu_0}{1 - \frac{4}{3}\mu_0 t}, \\
[\ln(\kappa)]_t &= -\frac{3}{4}\left[\ln\left(1 - \frac{4}{3}\mu_0 t\right)\right]_t, \\
\ln[\kappa(p,t)] - \ln[\kappa(p,0)] &= -\frac{3}{4}\left[\ln\left(1 - \frac{4}{3}\mu_0 t\right) - \ln(1)\right],
\end{aligned}$$

and so

$$\kappa(p,t) = \frac{\kappa(p,0)}{\left(1 - \frac{4}{3}\mu_0 t\right)^{3/4}}. \tag{2.44}$$

Similarly, we may compute that

$$\begin{aligned}
g(p,t) &= g(p,0)\left(1 - \frac{4}{3}\mu_0 t\right)^{1/2}, \\
\vec{\mathbf{T}}(p,t) &= T(p,0)\left(1 - \frac{4}{3}\mu_0 t\right)^{1/4}, \\
\vec{\mathbf{N}}(p,t) &= \frac{N(p,0)}{\left(1 - \frac{4}{3}\mu_0 t\right)^{1/4}}, \\
L(t) &= L(0)\left(1 - \frac{4}{3}\mu_0 t\right)^{1/2}, \\
A^{(r)}(t) &= \frac{\pi}{\mu^{3/2}(t)} = A(0)\left(1 - \frac{4}{3}\mu_0 t\right)^{3/2},
\end{aligned}$$

where all the initial values can be expressed as functions of μ_0 and $\mathcal{C}_0$.

From the above (see Eq. (2.44)), we see that $\mathcal{C}_0$ and $\mathcal{C}(t)$ are related by a similarity motion [51], i.e., a homothetic expansion with ratio $\alpha(t) =$

$(1 - \frac{4}{3}\mu_0 t)^{3/4}$, and perhaps a rotation. Symmetry considerations show that rotation is not possible. Therefore the ellipse evolves in a homothetic manner. The radii $r_1(t)$ and $r_2(t)$ of the ellipse $\mathcal{C}(t)$ are given by

$$r_1(t) = \alpha(t)a,$$
$$r_2(t) = \alpha(t)b.$$

If, instead of looking at the curves $\mathcal{C}(t)$, we look at their normalized versions $c(t)$ [such that $c(t)$ encloses area π], which are defined as

$$c(t) := \sqrt{\frac{\pi}{A^{(r)}(t)}}\mathcal{C}(t),$$

we see (as expected) that $c(t) \equiv c(0)$ for all $t \in [0, \frac{3}{4\mu_0})$.

Curve Convergence. In this section, we show that the family of convex curves $\mathcal{C}(t)$, evolving according to the affine heat flow, converges to an elliptical point in the Hausdorff metric. (This means that a certain associated family of dilated curves converges to an ellipse. See below for the precise details.) The technique we use is similar to that of Gage [148], who shows that Euclidean curve shortening makes convex curves circular. Instead of dealing with the convergence of the family of curves, we deal with the convergence of the laminae enclosed by them. The term ellipse is used for both the curve and its corresponding lamina. We assume throughout this section that our curves are regarded as mappings $S^1 \to \mathbb{R}^2$ are C^3 and are closed (i.e., the mappings are periodic). Once again, when the context is clear, we set $\mathcal{C}(t) = \mathcal{C}(\cdot, t)$.

First, the definition of the Hausdorff metric is reviewed and the Blaschke selection theorem is stated for completeness. See Ref. [227] for a complete discussion.

Definition 2.1. *Let S be a nonempty convex subset of $\mathbb{R}^2$. Then for given $\delta > 0$, the parallel body S_δ is defined to be*

$$S_\delta := \bigcup_{s \in S} K(s, \delta)$$

where $K(s, \delta) = \{x : d(x, s) \le \delta\}$, and $d(\cdot, \cdot)$ is the ordinary Euclidean plane distance.

Definition 2.2. *Let S and R be nonempty compact convex subsets of $\mathbb{R}^2$. Then the Hausdorff distance between S and R is defined as*

$$D(S, R) := \inf\{\delta : S \subset R_\delta \text{ and } R \subset S_\delta\}$$

We see from the definition that if $S \subset R$, then $D(S, R) = \inf\{\delta : R \subset S_\delta\}$, i.e., is the infimum of the values of δ such that the boundary ∂R is included in the annulus $(S_\delta \backslash S) \cup \partial S$ ($\backslash$ denotes complement).

We now list the following standard definitions.

Definition 2.3.

1. *A sequence $\{S_i\}$ of compact convex subsets of $\mathbb{R}^2$ is said to converge (in the Hausdorff metric) to the set S if*
$$\lim_{i\to\infty} D(S_i, S) = 0.$$
2. *Let $\mathcal{C}$ be the collection of nonempty compact subsets of $\mathbb{R}^2$. A subcollection $\mathcal{M}$ of $\mathcal{C}$ is uniformly bounded if there exists a circle in $\mathbb{R}^2$ that contains all the members of $\mathcal{M}$.*
3. *A collection of subsets $\{S_t\}$ is decreasing if $S_t \subseteq S_\tau$ for all $t \geq \tau$. In the case of curve evolution, the collection of laminae $H(t)$ associated with the curves $\mathcal{C}(t)$ is called decreasing if $H(t) \subseteq H(\tau)$ for all $t \geq \tau$.*

We state the following well-known theorem (see Ref. [227] for a proof).

Theorem 2.3 (Blaschke Selection Theorem). *Let $\mathcal{M}$ be a uniformly bounded infinite subcollection of $\mathcal{C}$. Then $\mathcal{M}$ contains a sequence that converges to a member of $\mathcal{C}$.*

Next we list several elementary results for future use.

Lemma 2.3. *A decreasing infinite collection of compact subsets in $\mathbb{R}^2$ converges in the Hausdorff metric to a set in $\mathbb{R}^2$.*

Proof: Because a decreasing infinite collection $S_t \subset \mathbb{R}^2$ is uniformly bounded, we see from the Blaschke selection theorem that the collection contains a sequence S_i that converges to $S_\infty \subset \mathbb{R}^2$ in the Hausdorff metric ($S_\infty \subseteq S_i$ for all i, and $S_\infty = \bigcap_{i=0}^{\infty} S_i$).

We have to prove now that all the collection converges to S_∞, i.e., that for all $\delta > 0$ there exists $t_0(\delta)$ such that $D(S_t, S_\infty) < \delta$ for all $t \geq t_0(\delta)$. Because the sequence $\{S_i\}$ converges, for all $\delta > 0$ there exists $i_0(\delta)$ such that $D(S_i, S_\infty) > \delta$ for all $i \geq i_0(\delta)$. From the decreasing property of the collection, we have $S_{i+1} \subseteq S_t \subseteq S_i$ for all $t \in [i, i+1]$. Then, if we choose $t_0(\delta) = i_0(\delta)$, we obtain $D(S_t, S_\infty) < \delta$ for all $t \geq t_0(\delta)$. $\square$

Corollary 2.1. *If a decreasing infinite collection of compact subsets in $\mathbb{R}^2$ has a sequence that converges to a set S, then all the collection converges to S.*

Using these results, we proceed to prove the convergence of the family of convex plane curves $\mathcal{C}(t)$ that satisfies the affine heat flow. We prove both the convergence of the affine isoperimetric ratio to the value of an ellipse $(8\pi^2)$ and the convergence (in the Hausdorff metric) of the laminae enclosed by the $\mathcal{C}(t)$ to an ellipse. We assume that the family $\mathcal{C}(t)$ of closed C^3 curves satisfying the affine heat flow is defined for the time interval $[0, T)$.

Lemma 2.4. *For convex laminae and their associated convex boundary curves, as above, L (affine perimeter), $A^{(r)}$ (area), μ (affine curvature), r_0 (inradius), and R_0 (circumradius), are continuous functionals with respect to the topology induced by the Hausdorff distance.*

Proof: The result is a direct consequence of properties of the Hausdorff metric together with the bounded variation of smooth convex curves. □

Lemma 2.5. *If* $\lim_{t\to T} A^{(r)}(t) = 0$ [*where* $A^{(r)}(t)$ *denotes the area enclosed by* $\mathcal{C}(t)$], *then*

$$\limsup_{t\to T}\left\{L(t)\left[2\oint \mu ds - \frac{L^2}{A^{(r)}}(t)\right]\right\} \geq 0. \tag{2.45}$$

Proof: From the evolution equations developed above [see Eqs. (2.36), and (2.43)], we have

$$\left[\frac{L^3}{A^{(r)}}\right]_t = -\frac{L^2}{A^{(r)}}(t)\left[2\oint \mu ds - \frac{L^2}{A^{(r)}}(t)\right] \geq 0.$$

Suppose that in a neighborhood of T

$$L(t)\left[2\oint \mu ds - \frac{L^2}{A^{(r)}}(t)\right] < -\epsilon,$$

for some $\epsilon > 0$. Then

$$\left[\frac{L^3}{A^{(r)}}\right]_t \geq \frac{L(t)\epsilon}{A^{(r)}(t)} = -\epsilon(\ln A)_t.$$

By integrating the latter inequality we obtain

$$\frac{L^3}{A^{(r)}}(t) \geq \frac{L^3}{A^{(r)}}(t_1) + \epsilon \ln\left[A^{(r)}(t_1)\right] - \epsilon \ln\left[A^{(r)}(t)\right]. \tag{2.46}$$

The left-hand side of inequality (2.46) $\leq 8\pi^2$ by the affine isoperimetric inequality, but the right-hand side goes to infinity (positive) as $A^{(r)}(t)$ goes to zero. This contradiction proves the lemma. □

For the next result, we need to define a certain functional on the space of closed convex C^3 curves. We consider the closed convex sets defined by $\mathcal{C}$ and $-\mathcal{C}_{ss}$ and use an extension of the isoperimetric inequality presented in Chap. 1 [348], which follows from the Minkowsky inequality for areas:

$$L^2 \geq 2A^{(r)} \oint \mu \mathrm{d}s + \frac{1}{4}\left(\oint \mu \mathrm{d}s\right)^2 (R_0 - r_0)^2. \tag{2.47}$$

Here r_0 and R_0 are the inradius and the circumradius, respectively. (The inradius of a set S_0 with respect to a second set S_1 is the largest real number r_0 such that a translate of $r_0 S_1$ is inside S_0, and this holds similarly for the circumradius.) Equality holds if and only if $\mathcal{C}$ is an ellipse. Equivalently,

$$L^2 \geq 2A^{(r)} \oint \mu \mathrm{d}s \left[1 + \frac{1}{8A^{(r)}}\left(\oint \mu \mathrm{d}s\right)(R_0 - r_0)^2\right]. \tag{2.48}$$

Because μ, $A^{(r)}$, r_0, and R_0 are all functions of $\mathcal{C}$, we can define the non-negative functional

$$F(\mathcal{C}) := \frac{1}{8A^{(r)}}\left(\oint \mu \mathrm{d}s\right)(R_0 - r_0)^2. \tag{2.49}$$

Note that we can express relation (2.48) as

$$2[1 + F(\mathcal{C})] \oint \mu \mathrm{d}s \leq \frac{L^2}{A^{(r)}}. \tag{2.50}$$

Finally, if $F(\mathcal{C}) = 0$, then $r_0 = R_0$, and so $\mathcal{C}$ is an ellipse [348] ($\mathcal{C}$ and $\mathcal{C}_{ss}$ are homothetic). Conversely, if $\mathcal{C}$ is an ellipse, then certainly $F(\mathcal{C}) = 0$.

We now have the following lemma.

Lemma 2.6. *The notation is the same as that used above. Let $\mathcal{C}_i$ ($i \geq 0$) be a sequence of closed convex curves such that*

$$\lim_{i\to\infty} F(\mathcal{C}_i) = 0.$$

Let H_i be the lamina enclosed by $\mathcal{C}_i$ for each $i \geq 0$. If the H_i form a decreasing sequence, then the sequence $\{H_i\}$ converges to an ellipse in the Hausdorff metric.

Proof: Because the H_i form a decreasing sequence, the entire sequence $\{H_i\}$ lies in a bounded region of the plane. Then the Blaschke selection theorem ensures that a subsequence $\{H_{i_k}\}$ converges to a limit H_∞. But from Lemma 2.4, $A^{(r)}$, μ, r_0, and R_0 are all continuous functionals of convex sets. Hence the functional F is also continuous and $F(H_\infty) = \lim(H_{i_k}) = 0$. Therefore H_∞ is an ellipse. Because the sequence H_i is decreasing, the entire sequence $\{H_i\}$ converges to the ellipse in the Hausdorff metric. □

Theorem 2.4. *The family of convex closed curves $\mathcal{C}(t)$ that satisfies the affine heat flow for $t \in [0, T)$ and with $\lim_{t \to T} A^{(r)}(t) = 0$ satisfies*

$$\lim_{t \to T} \frac{L^3}{A^{(r)}}(t) = 8\pi^2.$$

Moreover, $\mathcal{C}(t)$ converges in Hausdorff metric to an elliptical point as $A^{(r)}(t)$ goes to zero ($t \to T$).

Proof: From relation (2.50) we have

$$2 \oint \mu \mathrm{d}s - \frac{L^2}{A^{(r)}} \leq -2F \oint \mu \mathrm{d}s.$$

Then

$$L \left(2 \oint \mu \mathrm{d}s - \frac{L^2}{A} \right) \leq -2F \oint \mathrm{d}s \oint \mu \mathrm{d}s \leq 0. \tag{2.51}$$

From Lemma 2.5, there exists a subsequence $\mathcal{C}(t_i)$ of curves for which the left-hand side of inequality (2.51) is greater than or equal to zero when $A^{(r)}(t)$ goes to zero. Therefore $F[\mathcal{C}(t_i)]$ tends to zero.

We have seen that the affine curve evolution process shrinks the curves as time progresses. Therefore, if $H(t)$ is the lamina enclosed by $\mathcal{C}(t)$, $H(t) \subseteq H(\tau)$ if $t > \tau$. Thus the collection $\{H(t)\}$ is decreasing and in particular the sequence $\{H(t_i)\}$. From Lemma 2.6, it follows that the sequence of laminae $H(t_i)$ enclosed by the curves $\mathcal{C}(t_i)$ converges to an ellipse H_∞ in the Hausdorff metric. From the continuity of L and $A^{(r)}$, the isoperimetric ratio $L^3/A^{(r)}$ converges to $8\pi^2$ for this sequence. Because $L^3/A^{(r)}$ is a nondecreasing function of time in the affine evolution process [and is increasing for nonelliptical $\mathcal{C}(t)$], the ratio $L^3/A^{(r)}$ converges to $8\pi^2$ (the value for an ellipse) for the whole collection $\{\mathcal{C}(t)\}$.

To complete the proof, we have to show that the family $H(t)$ [the laminae enclosed by the $\mathcal{C}(t)$] converges to H_∞ as $t \to T$. This follows from the fact that the collection $\{H(t)\}$ is decreasing. □

Instead of dealing with the convergence of the family of curves $\{\mathcal{C}(t)\}$ and their associated laminae $\{H(t)\}$, we can consider the family of normalized curves

$$c(t) := \sqrt{\frac{\pi}{A^{(r)}}} \mathcal{C}(t)$$

and their associated laminae $h(t)$ (with area π) (this was done by Gage in Ref. [148] for the Euclidean case). This will allow us to make precise the notion of an elliptical point.

Definition 2.4. *Two curves a and b in the plane (or their associated laminae) are equivalent under the group of motions $\mathcal{G}$ (denoted by $a \equiv_{\mathcal{G}} b$) if there exists $G \in \mathcal{G}$ such that $a = Gb$.*

We assume that all the curves are centered around the origin of the Cartesian axes, so we do not need to add a translation vector to the motion.

Examples.

1. $\mathcal{G}$ is the group of rotations. A given disk is equivalent only to itself. Then the equivalence class of the disk contains only one member. Let $\mathcal{D}$ denote the set of disks with area π (only one member, namely the unit disk).
2. $\mathcal{G}$ is the group of similarities [51]. All disks are in the same equivalence class.
3. $\mathcal{G}$ is the group $SL_2(\mathbb{R})$. All ellipses with the same area are in the same equivalence class. Let $\mathcal{E}$ denote the set of ellipses with area π. [All elements of $\mathcal{E}$ are equivalent under $SL_2(\mathbb{R})$.]

Definition 2.5. *A collection $\{S_t\}$ of compact convex sets in $\mathbb{R}^2$ converges in the Hausdorff metric to a collection of sets $\mathcal{S}$, if $\{S_t\}$ is composed of subsequences $\{S_{t_{i_k}}\}$, each one converging in the Hausdorff metric to an element of $\mathcal{S}$.*

Because the function $F(\cdot)$ defined in Eq. (2.49) is an affine invariant, if we have that $\lim_{i\to\infty} F(\mathcal{C}_i) = 0$, then $\lim_{i\to\infty} F(c_i) = 0$. Using the same arguments as those given in Theorem 2.4 and assuming that $\{h_i\}$ is bounded (instead of assuming that $\{H_i\}$ is decreasing), we can easily prove that the sequence $\{h_i\}$ is composed of subsequences $\{h_{i_{k_j}}\}$, each one converging to an ellipse of area π ($F(c_\infty) = 0$ if and only if c_∞ is an ellipse). Therefore $\{h_i\}$ converges to elements of $\mathcal{E}$.

Note that, in the Euclidean case, Gage [148] proves that the sequence of normalized laminae converges to the unit disk, i.e., to the unique element of $\mathcal{D}$ (see Lemma 2 in Ref. [148]). In the affine case, all the ellipses in the set $\mathcal{E}$ are equivalent, and so convergence to $\mathcal{E}$ in the affine evolution is analogous to convergence to $\mathcal{D}$ in the Euclidean evolution.

More precisely, we check now the convergence of the normalized curves $\{c(t)\}$ when $\{\mathcal{C}(t)\}$ satisfies the affine heat flow. Because $H(t) \subseteq H(\tau)$ if $t > \tau$, we can choose one of the points in the set

$$\bigcap_{0\le t<T} H(t)$$

as the origin of the homothetic expansion [148] and obtain from $\{\mathcal{C}(t)\}$ a family $\{c(t)\}$ of bounded and normalized curves.

Using similar arguments as those given in the proof of Theorem 2.4 and the above discussion, we find that the affine isoperimetric ratio $L^3/A^{(r)}$ converges to $8\pi^2$, i.e., the value for an ellipse. We see therefore that $\{h(t)\}$ converges to the set $\mathcal{E}$, i.e., it is composed of subsequences, each one converging to an element of $\mathcal{E}$. Because all the members of the set $\mathcal{E}$ are equivalent under $SL_2(\mathbb{R})$ (i.e., from the point of view of affine geometry), in this sense $\{h(t)\}$ converges to an ellipse, and the family of curves $\mathcal{C}(t)$ converges to an elliptical point.

Affine Evolution Existence. This section deals with the existence and the smoothness of solutions of the affine heat flow. The methodology is similar to that used for proving the existence of solutions of the Euclidean curve-shortening process (see Refs. [149, 150, 162 , 163, and 204]). Once more, when the context is clear, we will set $\mathcal{C}(t) = \mathcal{C}(\cdot, t)$ and $\mu(t) = \mu(\cdot, t)$.

Lemma 2.7. *If* μ *is bounded in the interval* $[0, T)$, *then*

$$\frac{\partial^n \mu}{\partial s^n}, \quad n \geq 0,$$

is also bounded in the interval.

Proof: From the affine curvature evolution equation (2.34),

$$\frac{\partial \mu}{\partial t} = \frac{4}{3}\mu^2 + \frac{1}{3}\mu_{ss}$$

and the relation between time and affine arc-length derivatives,

$$\frac{\partial}{\partial t}\frac{\partial}{\partial s} = \frac{2}{3}\mu\frac{\partial}{\partial s} + \frac{\partial}{\partial s}\frac{\partial}{\partial t},$$

we obtain

$$\begin{aligned}\frac{\partial}{\partial t}\left(\frac{\partial \mu}{\partial s}\right) &= \frac{2}{3}\mu\mu_s + \frac{8}{3}\mu\mu_s + \frac{1}{3}\mu_{sss}\\ &= \frac{10}{3}\mu\left(\frac{\partial \mu}{\partial s}\right) + \frac{1}{3}\frac{\partial^2}{\partial s^2}\left(\frac{\partial \mu}{\partial s}\right).\end{aligned}$$

This equation bounds the rate of growth of μ_s to exponential. Therefore μ_s is bounded for finite time.

In general,

$$\frac{\partial}{\partial t}\left(\frac{\partial^n \mu}{\partial s^n}\right) = \frac{8+2n}{3}\mu\left(\frac{\partial^n \mu}{\partial s^n}\right) + \frac{1}{3}\frac{\partial^2}{\partial s^2}\left(\frac{\partial^n \mu}{\partial s^n}\right) + \text{bounded terms}.$$

Thus $\partial^n \mu/\partial s^n$ does not grow faster than exponentially, and so it is bounded for finite time. □

Lemma 2.8. *If μ is bounded for $t \in [0, T)$, $\partial^n \kappa / \partial s^n$ $(n \geq 0)$ is also bounded in the interval.*

Proof: The proof is as in Lemma 2.7 by use of the Euclidean curvature evolution equation (2.39). □

The evolution of $\mu(t)$ (Eq. (2.34)) is governed by a reaction-diffusion equation [369]. If we take the initial condition

$$\mu(0) = \mu_0,$$

where μ_0 is bounded and smooth, then standard results in the theory of reaction-diffusion equations guarantee short-term existence, boundedness, uniqueness, and smoothness of the solutions of Eq. (2.34) [369]. Short-term existence of the solution of Eq. (2.34) may also be obtained from standard results on parabolic equations [181, 321]. We will use these results about reaction-diffusion equations in what follows; the interested reader is referred to any of the aforementioned references for the details.

We now state and prove the following:

Theorem 2.5 (Short-Term Existence). *Let $\mathcal{C}(\cdot, 0)$ be a smooth, convex, embedded closed curve in the plane $\mathbb{R}^2$. Then there exists an $\epsilon > 0$, and a (classical) solution $\mathcal{C} : S^1 \times [0, \epsilon) \rightarrow \mathbb{R}^2$ for the affine heat flow.*

Proof: The result follows directly from the short-term existence results presented in Refs. [12, 94, 149, 150, 162 , 163, and 204]. (In some cases, e.g., in Refs. [94, 150, and 162], long-term existence is also proved, but short-term existence is enough for our purposes.)

More precisely, if v denotes the Euclidean arc length, then from Eq. (1.10) we obtain

$$\mathcal{C}_{ss} = \frac{\mathcal{C}_{vv}}{g^2(v)} - \mathcal{C}_v \frac{g_v(v)}{g^3(v)}.$$

Note that if $\mathcal{T}$ and $\mathcal{N}$ denote the Euclidean unit tangent and normal, respectively, then $\mathcal{C}_v = \mathcal{T}$, $\mathcal{C}_{vv} = \kappa\mathcal{N}$, and $g = \kappa^{1/3}$. Therefore the affine curve evolution equation can be expressed as follows:

$$\mathcal{C}_t = \kappa^{1/3}\mathcal{N} - \frac{\kappa_v}{3\kappa^{5/3}}\mathcal{T}. \tag{2.52}$$

As we have already seen before, in Lemma 2.1 the tangential component of the velocity vector affects only the parameterization of the family of curves in the evolution, not their shape [1, 149]. So the existence of the family of curves is determined by the normal component of the velocity ($\kappa^{1/3}\mathcal{N}$).

For equations of the form of Eq. (2.52) short-term existence may be derived immediately from the results of Refs. [12, 94, and 204]. For example, Chow [94] proves, by using the same technique as that in Ref. [150], short-term existence, uniqueness, and smoothness of the solution of a system defined by convex hypersurfaces deforming by means of an exponential function of their Gaussian curvature (see Theorem 2.1 in Ref. [94]). Theorem 2.5 may be obtained as a special case of this result for 1D curves and an exponent equal to $\frac{1}{3}$. □

Theorem 2.6. *There exists a time T such that for $t \in [0, T)$, the affine curve evolution process,*

$$\begin{cases} \frac{\partial \mathcal{C}}{\partial t} = \mathcal{C}_{ss} \\ \mathcal{C}(0) = \mathcal{C}_0\,,\ \mathcal{C}_0 \text{ smooth and closed} \end{cases}, \tag{2.53}$$

is equivalent to looking for a solution of the following PDE problem:

$$\begin{cases} \mu \in C^{3+\alpha,1+\alpha}(S^1 \times [0, T)) \\ \frac{\partial \mu}{\partial t} = \frac{4}{3}\mu^2 + \frac{1}{3}\mu_{ss} \\ \mu(0) = \mu(\mathcal{C}_0) = \mu_0 \text{ (bounded, smooth, and periodic)}. \end{cases} \tag{2.54}$$

Proof: Let ϵ_1 be the time given in Theorem 2.5 and ϵ_2 be the one given by the short-term solution of PDE problem (2.54). Define $T := \min\{\epsilon_1, \epsilon_2\}$, and consider the time interval $[0, T)$. From the affine curvature evolution equation (2.34), if a curve $\mathcal{C}(t)$ satisfies the affine curve evolution (2.53), the corresponding affine curvature $\mu(t)$ satisfies PDE problem (2.54).

Let $\mu^*(t)$ be a solution of PDE problem (2.54). From the preceding results on reaction-diffusion equations, this solution exists and is unique in $[0, T)$. The affine curvature $\mu^*(t)$ defines, up to an affine transformation, unique curve $X^*(t)$ [53]. Therefore the family of solutions $\mu^*(t), t \in [0, T)$ determines a unique family of curves $X^*(t), t \in [0, T)$. Because $\mu^*(t)$ is also the affine curvature of $\mathcal{C}(t)$, which satisfies the affine curve evolution process given by Eqs. (2.53), $\mathcal{C}(t)$ and $X^*(t)$ are equal (up to an affine motion) for all $t \in [0, T)$. □

We proceed now to prove that $\mu(t)$, which satisfies PDE problem (2.54) in the interval $[0, T)$, is bounded as long as the area $A^{(r)}(t)$ enclosed by the correspondent curve $\mathcal{C}(t)$, which satisfies the affine evolution process, is bounded away from zero. A result similar to the following lemma was proved by Grayson in Ref. [162] for the Euclidean curve evolution process. We prove it now for the affine evolution process.

Lemma 2.9. *If $\mu(t)$ satisfies affine curvature evolution equation (2.34) for $[0, T)$, and*

$$\oint \mu(t)\mathrm{d}s$$

is bounded, then

$$\oint \left(\frac{\partial \mu}{\partial s}\right)^2 \mathrm{d}s$$

is also bounded in the interval.

Proof: Using affine curvature evolution equation (2.34) and the relation between the derivatives by time and affine arc length, we obtain

$$\begin{aligned}
\frac{\partial}{\partial t}\oint \left(\frac{\partial \mu}{\partial s}\right)^2 \mathrm{d}s &= \oint 2\left(\frac{\partial \mu}{\partial s}\right)\frac{\partial}{\partial t}\left(\frac{\partial \mu}{\partial s}\right) + \frac{g_t}{g}\left(\frac{\partial \mu}{\partial s}\right)^2 \mathrm{d}s \\
&= \oint 2\left(\frac{\partial \mu}{\partial s}\right)\frac{\partial}{\partial s}\left(\frac{\partial \mu}{\partial t}\right) - \frac{g_t}{g}\left(\frac{\partial \mu}{\partial s}\right)^2 \mathrm{d}s \\
&= \oint 2\left(\frac{\partial \mu}{\partial s}\right)\frac{\partial}{\partial s}\left(\frac{\partial \mu}{\partial t}\right) + \frac{2}{3}\mu\left(\frac{\partial \mu}{\partial s}\right)^2 \mathrm{d}s.
\end{aligned}$$

Integrating by parts and using Eq. (2.34), we obtain

$$\begin{aligned}
\frac{\partial}{\partial t}\oint \left(\frac{\partial \mu}{\partial s}\right)^2 \mathrm{d}s &= \oint -2\left(\frac{\partial \mu}{\partial t}\right)\left(\frac{\partial^2 \mu}{\partial s^2}\right) + \frac{2}{3}\mu\left(\frac{\partial \mu}{\partial s}\right)^2 \mathrm{d}s \\
&= \oint -2\left(\frac{4}{3}\mu^2 + \frac{1}{3}\frac{\partial^2 \mu}{\partial s^2}\right)\left(\frac{\partial^2 \mu}{\partial s^2}\right) + \frac{2}{3}\mu\left(\frac{\partial \mu}{\partial s}\right)^2 \mathrm{d}s \\
&= \oint -\frac{2}{3}(\mu_{ss})^2 - \frac{8}{3}\mu^2\mu_{ss} + \frac{2}{3}\mu(\mu_s)^2 \mathrm{d}s \\
&= \oint -\frac{2}{3}(\mu_{ss})^2 + 6\mu(\mu_s)^2 \mathrm{d}s.
\end{aligned}$$

Because $\oint \mu ds$ is bounded by hypothesis, we can find U such that

$$\int_{\mu>U} \mu \mathrm{d}s < \frac{1}{9L},$$

where L is the initial affine perimeter. The part of the curve with $\mu < U$ is bounded as

$$\int_{\mu<U} 6\mu(\mu_s)^2 \mathrm{d}s < 6U \oint (\mu_s)^2 \mathrm{d}s.$$

Where $\mu > U$, we have

$$\int_{\mu>U} 6\mu(\mu_s)^2 \mathrm{d}s < 6 \sup(\mu_s)^2 \int_{\mu>U} \mu \mathrm{d}s < \frac{2}{3L} \sup(\mu_s)^2$$

The supremum of a differentiable function that is somewhere zero is less than or equal to the integral of the absolute value of its derivative. It is well known (see, e.g., Ref. [53]) that an oval has at least six sextactic points, i.e., points where $\mu_s = 0$. Hence,

$$\sup(\mu_s)^2 \leq \left[\oint |\mu_{ss}| \mathrm{d}s\right]^2 \leq L(t) \oint (\mu_{ss})^2 \mathrm{d}s \leq L(0) \oint (\mu_{ss})^2 \mathrm{d}s.$$

Then

$$\oint 6\mu(\mu_s)^2 < 6U \oint (\mu_s)^2 \mathrm{d}s + \frac{2}{3} \oint (\mu_{ss})^2 \mathrm{d}s$$

and

$$\frac{\partial}{\partial t} \oint \left(\frac{\partial \mu}{\partial s}\right)^2 \mathrm{d}s < 6U \oint \left(\frac{\partial \mu}{\partial s}\right)^2 \mathrm{d}s.$$

Therefore the growth of $\oint \left(\frac{\partial \mu}{\partial s}\right)^2 \mathrm{d}s$ is at most exponential in $6U$. □

We will also need the following:

Lemma 2.10. *If $\oint \mu(t) \mathrm{d}s$ is bounded and $\mu(t)$ satisfies affine curvature evolution equation (2.34) for $t \in [0, T)$, then $\mu(t)$ is bounded in the interval.*

Proof: Because $\oint \mu \mathrm{d}s$ is bounded, for every $\delta > 0$ we can find a constant $M(\delta)$ such that $\mu \leq M$, except on intervals $[a, b]$ of length less than δ. On such an interval, we obtain

$$\begin{aligned}
\mu(x) &= \mu(a) + \int_a^x \frac{\partial \mu}{\partial s} \mathrm{d}s \\
&\leq M + \sqrt{\delta} \left[\oint \left(\frac{\partial \mu}{\partial s}\right)^2 \mathrm{d}s\right]^{1/2} \\
&\leq M + \sqrt{\delta} B,
\end{aligned}$$

where B is the bound given by Lemma 2.9. □

Lemma 2.11. *If $\mu(t)$ satisfies PDE problem (2.54) in the time interval $[0, T)$ and the area $A^{(r)}(t)$ enclosed by the corresponding curve $\mathcal{C}(t)$ is bounded away from zero for all $t \in [0, T)$, then the affine curvature $\mu(t)$ is bounded in the interval $[0, T)$.*

Proof: We use the affine isoperimetric inequality. Indeed, because $L(t) \leq L(0)$ and $A^{(r)}(t)$ is bounded away from zero, i.e., $A^{(r)}(t) \geq A^{(r)}_{\min} > 0$, we obtain

$$\oint \mu \mathrm{d}s \leq \frac{L^2(0)}{2A^{(r)}_{\min}}.$$

Therefore, from Lemma 2.10, we see that $\mu(t)$ is bounded in the interval. □

We are now ready to prove the following:

Theorem 2.7. *The solution to PDE problem (2.54) (and so to the affine curve evolution process) continues until the areas $A^{(r)}(t)$ enclosed by the curves $\mathcal{C}(t)$ go to zero.*

Proof: As long as the area is bounded away from zero, μ is bounded (Lemma 2.11), and bounds on all its derivatives $\partial^n \mu / \partial s^n$ can be obtained by means of Lemma 2.7. By using Eq. (2.34), we can also obtain bounds on the time derivatives. Suppose the solution exists for $t \in [0, T)$ (this assumption is valid from the short-term results; for example, choose T as in Theorem 2.6). Assume that $\lim_{t \to T} A^{(r)}(t) > 0$. Then, $\mu(t)$ has a limit as $t \to T$, and the limiting curve, say $\mathcal{C}(T)$, is also smooth (see also Ref. [369]).

From the short-term time results (Theorems 2.5 and 2.6), the solutions $\mu(t)$ and the corresponding $\mathcal{C}(t)$ exist also for some interval $[T, T + \epsilon)$, $\epsilon > 0$. This process can be continued until the area $A^{(r)}(t)$ goes to zero. □

Convergence of Elliptically Curved Ovals. Bounds on the convergence time of elliptically curved curves are now presented. Assume that $\mu(p, 0)$, the affine curvature of the initial curve $\mathcal{C}(p, 0)$, is positive everywhere, i.e., $\mu(p, 0) > 0$ for all p. From Proposition 2.1, we know that also $\mu(p, t) > 0$ for $t > 0$.

Let μ_m and μ_M be the minimal and the maximal affine curvature of $\mathcal{C}(p, 0)$, respectively (i.e., the minimal and the maximal values of the function $\mu(p, 0)$). μ_m determines the greatest osculating ellipse of $\mathcal{C}(p, 0)$, $\mathcal{E}_m(p, 0)$, which entirely contains the curve. μ_M determines the smallest osculating ellipse of $\mathcal{C}(p, 0)$, $\mathcal{E}_M(p, 0)$, which is contained entirely in the curve. In particular, $\mathcal{E}_M(p, 0) \subseteq \mathcal{C}(p, 0) \subseteq \mathcal{E}_m(p, 0)$. If these three initial curves evolve according to the affine evolution equation, it is easy to show that also $\mathcal{E}_M(p, t) \subseteq \mathcal{C}(p, t) \subseteq \mathcal{E}_m(p, t)$ [328, 411] ($\mathcal{E}_M(p, t)$ and $\mathcal{E}_m(p, t)$ are not necessarily the smallest and the greatest osculating ellipses of $\mathcal{C}(p, t)$).

Then the convergence time of $\mathcal{C}(p, 0)$ is bounded by the convergence times of $\mathcal{E}_M(p, 0)$ and $\mathcal{E}_m(p, 0)$.

The convergence time of an ellipse with affine curvature μ_0 was found to be equal to $[3/(4\mu_0)]$. Therefore the convergence times of $\mathcal{E}_M(p, 0)$ and $\mathcal{E}_m(p, 0)$ are $[3/(4\mu_M)]$ and $[3/(4\mu_m)]$, respectively. If $T_\mathcal{C}$ stands for the convergence time of $\mathcal{C}(p, 0)$, we conclude that

$$\frac{3}{4\mu_M} \leq T_\mathcal{C} \leq \frac{3}{4\mu_m}.$$

Remark: If $\mu(p, 0) = 0$ on some intervals, then $\mu(p, t)$ (for $t > 0$) becomes strictly positive immediately. Thus the bounds presented above are also valid for initial curves with nonnegative affine curvature.

Comments and Affine Level-Set Motion. This concludes the analysis of the affine heat flow for the convex case. The full analysis of the nonconvex case, extending the results in Ref. [162], is presented in Ref. [15]. Once again, when dealing with the nonconvex case we use the projected flow

$$\mathcal{C}_t = \kappa^{1/3}\vec{\mathcal{N}},$$

which in level set form is

$$u_t = \kappa^{1/3}\|\nabla u\| = \left(u_{xx}u_y^2 - 2u_x u_y u_{xy} u_{yy} u_x^2\right)^{1/3},$$

which is again well defined from the theory of viscosity solutions. Note also that, in comparison with the corresponding Euclidean flow, the denominator is gone, making the numerical implementation of this flow more stable.

2.5.3. Length- and Area Preserving Flows

When we are smoothing a planar shape or computing a multiscale representation of it, we do not want its area, length, or other basic geometric properties to change; we just want the noise and the details removed. The problem of smoothing without shrinkage was addressed in the literature as well [185, 244, 275]. This problem has been addressed for Gaussian filtering, and it tends to reduce the shrinkage but not totally eliminate it. In many cases also, the solution violates the semigroup property of scale spaces or it works only for convex curves. We now show how to preserve the length or area for the geometric heat flows while keeping the scale-space characteristics of the flow.

When a curve evolves with the flow

$$\mathcal{C}_t = \beta\vec{\mathcal{N}},$$

the area A changes according to

$$A_t = -\oint \beta dv.$$

In the case of the Euclidean heat flow then,

$$A_t = -2\pi,$$

that is,

$$A(t) = A_0 - 2\pi t,$$

where A_0 is the area enclosed by the original curve. The basic idea to preserve the area of the evolving curve is to compute a new curve

$$\tilde{\mathcal{C}}(\tau) := \psi(t)\mathcal{C}(t),$$

where $\psi(t)$ stands for a normalization factor. Let us select $\psi^2(t) = (\partial\tau/\partial t)$. The new time scale τ must be chosen to obtain a fixed area, that is, $\tilde{A}_t = 0$. Then, having $T = A_0/2\pi$, let

$$\tau(t) = -T \ln(T - t).$$

Because the ratio between the areas of $\tilde{\mathcal{C}}$ and $\mathcal{C}$ is $\psi(t) = (\partial\tau/\partial t)$, we obtain the desired result of constant area with this selection. We need now to find the evolution for the dilated curve, which is easily obtained:

$$\begin{aligned}\tilde{\mathcal{C}}_\tau &= t_\tau \tilde{\mathcal{C}}_t = \psi^{-2}(\psi_t \mathcal{C} + \psi \mathcal{C}_t)\\ &= \psi^{-2}(\psi_t \mathcal{C} + \psi\kappa\vec{\mathcal{N}}) = \psi^{-3}\psi_t\tilde{\mathcal{C}} + \tilde{\kappa}\vec{\mathcal{N}}.\end{aligned}$$

Because tangential velocities do not affect the geometry of the flow, we obtain

$$\tilde{\mathcal{C}}_\tau = \psi^{-3}\psi_t\langle\tilde{\mathcal{C}}, \vec{\mathcal{N}}\rangle\vec{\mathcal{N}} + \tilde{\kappa}\vec{\mathcal{N}}.$$

Considering the support function

$$\rho := -\langle\mathcal{C}, \vec{\mathcal{N}}\rangle,$$

it is easy to show that

$$A = \frac{1}{2}\oint \rho dv.$$

Applying the last three equations to the general area change formula for $\tilde{\mathcal{C}}$, together with the condition $\tilde{A}_\tau = 0$, we obtain

$$\frac{\partial \tilde{\mathcal{C}}}{\partial \tau} = \left(\tilde{\kappa} - \frac{\pi \tilde{\rho}}{A_0}\right) \vec{\mathcal{N}},$$

as the area-preserving geometric scale-space (Euclidean invariant).

In the affine case, we have that

$$A_t = -\oint \kappa^{1/3} \mathrm{d}v.$$

In this case we consider

$$\psi^{4/3} = \frac{\partial \tau}{\partial t},$$

and, repeating the calculations above, we find that the area-preserving affine geometric scale space is given by

$$\frac{\partial \tilde{\mathcal{C}}}{\partial \tau} = \left(\tilde{\kappa}^{1/3} - \frac{\tilde{\rho} \oint \kappa^{1/3}}{2A_0}\right) \vec{\mathcal{N}}.$$

We can repeat the computations to obtain flows that preserve the Euclidean perimeter. From the relations

$$P_t = -\oint \beta \kappa \mathrm{d}v,$$

$$P = \oint \kappa \rho \mathrm{d}v,$$

where P stands for the Euclidean perimeter, it is easy to show that the corresponding perimeter-preserving flows are given by

$$\frac{\partial \tilde{\mathcal{C}}}{\partial t} = \left(\tilde{\beta} - \frac{\oint \tilde{\beta} \tilde{\kappa} \mathrm{d}v}{P_0} \tilde{\rho}\right) \vec{\mathcal{N}}.$$

2.6. Euclidean and Affine Surface Evolution

We want now to extend the basic concepts described above for curves to surfaces, at least to surfaces $\mathcal{S}$ in three dimensions. Unfortunately, as we will see, this is not straightforward and not always possible. For details see [20, 116, 117, 156, 164, 188–190, 277–280, 373, 395–397].

Of course, the constant-velocity flow is straightforward to extend, and this is just

$$\frac{\partial \mathcal{S}}{\partial t} = \vec{\mathcal{N}},$$

where $\vec{\mathcal{N}}$ is the inner normal to the surface. This gives 3D erosion, and when the sign is reversed it will produce the dilation of $\mathcal{S}$. As in the 2D case, these flows produce shocks (singularities) and can change the topology of the evolving surface.

Having the constant-velocity flow, the next step if to find an analogous of the Euclidean and affine geometric heat flows. Here is where problems begin. A general geometric surface flow has the form

$$\frac{\partial \mathcal{S}}{\partial t} = \beta(\mathbf{H}, \mathbf{K})\vec{\mathcal{N}},$$

that is, the velocity is a function of the mean and Gaussian curvatures (or just a function of the principal curvatures). For example, in analogy to minimizing length ($\mathcal{C}_t = \kappa\vec{\mathcal{N}}$), we want to minimize surface area

$$\iint da,$$

obtaining the gradient descent flow

$$\frac{\partial \mathcal{S}}{\partial t} = \mathbf{H}\vec{\mathcal{N}},$$

mean curvature motion. In contrast with the Euclidean geometric heat flow, this surface deformation can create singularities. Indeed, it deforms a convex surface into a round point (sphere) [94], but nonconvex surfaces can split before becoming convex. Examples of topological changes for the mean curvature flow in three dimensions can be found in the literature, being the classical example the deformation of a dumbbell. Flows deforming as functions of the Gaussian curvature suffer from the same problems. In particular, the flow

$$\frac{\partial \mathcal{S}}{\partial t} = (\mathbf{K}^+)^{1/4}\vec{\mathcal{N}},$$

where $\mathbf{K}^+ := \max\{\mathbf{K}, 0\}$, which is the simplest affine invariant surface flow (see below and Refs. [5, 74, and 286]), behaves better than the mean curvature flow, but still can create singularities and topological changes for generic 3D shapes. (All these flows, when embedded in the level-set function $u : \mathbb{R}^3 \to \mathbb{R}$, do have a unique viscosity solution.)

So, can we do better? Indeed we can, but we need either to limit the class of surfaces or increase the complexity of the flows. To obtain a flow that smooths a generic 3D surface, we need to go to more complex flows. For example, in Ref. [316], Polden studies the flow that minimizes

$$\iint \mathbf{K}^2 \mathrm{d}a,$$

that is, the tension of the surface (the analogous flows were studied in two dimensions as well, $\int \kappa^2 \mathrm{d}s$). This energy, which is in general normalized to keep a constant volume, has a high-order-corresponding gradient descent flow (second derivative of the mean and the Gaussian curvatures, a fourth-order flow then) and deforms the surface into a sphere [316].

We can also obtain positive smoothing results with just second-order flows if we limit the surface, for example, to be a function; see, e.g., Ref. [116]. This is not very useful for generic 3D shapes, but it is important for smoothing images; see Chap. 5.

2.7. Area- and Volume-Preserving 3D Flows

We now repeat the computations on shrinkage-free curve flows for surface deformations [349]. When a closed surface $\mathcal{S}$ evolves according to

$$\frac{\partial \mathcal{S}}{\partial t} = \beta \vec{\mathcal{N}}, \tag{2.55}$$

it is easy to prove that the enclosed volume $\mathbf{V}$ evolves according to

$$\frac{\partial \mathbf{V}}{\partial t} = -\iint \beta \mathrm{d}a \tag{2.56}$$

and the surface area $\mathbf{A}$ evolves according to

$$\frac{\partial \mathbf{A}}{\partial t} = -\iint \beta \mathbf{H} \mathrm{d}a, \tag{2.57}$$

where $\mathrm{d}a$ is the element of area. As we have seen for the curve flows, one of the advantages of performing operations such as smoothing with geometric flows of the form of Eq. (2.55) is exactly that the change of geometric characteristics can be computed a priori. These evolution equations, together with the simple relations

$$\mathbf{V} = \frac{1}{3} \iint \langle \mathcal{S}, \vec{\mathcal{N}} \rangle \mathrm{d}\mu,$$
$$\mathbf{A} = \frac{1}{2} \iint \mathbf{H} \langle \mathcal{S}, \vec{\mathcal{N}} \rangle \mathrm{d}\mu,$$

will help in the construction of geometric flows that preserve $\mathbf{V}$ or $\mathbf{A}$, that is, nonshrinking flows. The possibility of doing this was already pointed out, for example, in Ref. [188].

As in the curve flows, the normalization process is given by a change of the time scale, from t to τ, such that a new surface is obtained by means of

$$\tilde{\mathcal{S}}(\tau) := \psi(t)\,\mathcal{S}(t), \tag{2.58}$$

where $\psi(t)$ represents the normalization factor (time scaling) and is given by

$$\psi^n(t) = \frac{\partial \tau}{\partial t}. \tag{2.59}$$

The exponent n is such that $\psi^{-n+1}\beta = \tilde{\beta}$. One of the advantages of this kind of normalization is that the behavior of $\tilde{\mathcal{S}}(\tau)$ is the same as that of $\mathcal{S}(t)$, that is, both surfaces have the same geometric properties. This means that the basic geometric characteristics of the flow of Eq. (2.55) hold for $\tilde{\mathcal{S}}(\tau)$ as well.

The new time scale τ must be chosen to obtain $\tilde{\mathbf{V}}_\tau \equiv 0$ or $\tilde{\mathbf{A}}_\tau \equiv 0$. The evolution of $\tilde{\mathcal{S}}$ is obtained from the evolution of $\mathcal{S}$ and the time scaling given by Eq. (2.59). Taking partial derivatives in Eq. (2.58) we have

$$\begin{aligned}\frac{\partial \tilde{\mathcal{S}}}{\partial \tau} &= \frac{\partial t}{\partial \tau}\frac{\partial \tilde{\mathcal{S}}}{\partial t} = \psi^{-n}(\psi_t \mathcal{S} + \psi \mathcal{S}_t)\\ &= \psi^{-n}\psi_t\mathcal{S} + \psi^{-n+1}\beta\vec{\mathcal{N}} = \psi^{-n}\psi_t\mathcal{S} + \tilde{\beta}\vec{\mathcal{N}} = \psi^{-n-1}\psi_t\tilde{\mathcal{S}} + \tilde{\beta}\vec{\mathcal{N}}.\end{aligned}$$

Because the geometry of the evolution is affected by only the normal component of the velocity, the flow above is geometrically equivalent to

$$\frac{\partial \tilde{\mathcal{S}}}{\partial \tau} = \psi^{-n-1}\psi_t\langle \tilde{\mathcal{S}}, \vec{\mathcal{N}}\rangle\vec{\mathcal{N}} + \tilde{\beta}\vec{\mathcal{N}}. \tag{2.60}$$

Define again the support function ρ as

$$\rho := -\langle \mathcal{S}, \vec{\mathcal{N}}\rangle,$$

obtaining

$$\frac{\partial \tilde{\mathcal{S}}}{\partial \tau} = (\psi^{-n-1}\psi_t\tilde{\rho} + \tilde{\beta})\vec{\mathcal{N}}. \tag{2.61}$$

Therefore the rate of change of volume $\tilde{\mathbf{V}}$ of $\tilde{\mathcal{S}}$ is given by

$$\mathbf{V}_t = -\iint (\psi^{-n-1}\psi_t\tilde{\rho} + \tilde{\beta})\mathrm{d}\mu.$$

If we require that $\mathbf{V}_t \equiv 0$, then

$$\psi^{-n-1}\psi_t\iint \tilde{\rho}\,\mathrm{d}\tilde{\mu} = -\iint \tilde{\beta}\,\mathrm{d}\tilde{\mu}.$$

Therefore, because

$$\mathbf{V}(0) = \tilde{\mathbf{V}}(t) = \frac{1}{3}\iint \tilde{\rho} d\tilde{\mu},$$

we obtain

$$\psi^{-n-1}\psi_t = -\frac{\iint \tilde{\beta} d\tilde{\mu}}{3\mathbf{V}_0},$$

and

$$\frac{\partial \tilde{\mathcal{S}}}{\partial t} = \left(\tilde{\beta} - \frac{\tilde{\rho} \iint \tilde{\beta} d\tilde{\mu}}{3\mathbf{V}_0}\right)\vec{\mathcal{N}}, \tag{2.62}$$

gives the volume-preserving alternative to Eq. (2.55). In the same way, the area-preserving flow is given by

$$\frac{\partial \tilde{\mathcal{S}}}{\partial t} = \left(\tilde{\beta} - \frac{\tilde{\rho} \iint \tilde{\beta}\tilde{\mathbf{H}} d\tilde{\mu}}{2\mathbf{A}_0}\right)\vec{\mathcal{N}}. \tag{2.63}$$

The table in Fig. 2.12 gives the corresponding nonshrinking flows for a number of velocities β frequently used in the literature.

Note that the volume-preserving flow for the Gaussian velocity is local because the rate of volume changing is constant. The last flow is the unique affine invariant well-posed flow for 3D surfaces; see below. Figure 2.13, produced in collaboration with R. Malladi, shows an example of volume-preserving smoothing by means of curvature flows.

β	Volume-Preserving Flow	Area-Preserving flow
1	$\mathcal{S}_t = \left(1 - \frac{\rho \mathbf{A}}{3\mathbf{V}_0}\right)\vec{\mathcal{N}}$	$\mathcal{S}_t = \left(1 - \frac{\rho \int\int \mathbf{H} d\mu}{2\mathbf{A}_0}\right)\vec{\mathcal{N}}$
$\mathbf{H}$	$\mathcal{S}_t = \left(\mathbf{H} - \frac{\rho \int\int \mathbf{H} d\mu}{3\mathbf{V}_0}\right)\vec{\mathcal{N}}$	$\mathcal{S}_t = \left(\mathbf{H} - \frac{\rho \int\int \mathbf{H}^2 d\mu}{2\mathbf{A}_0}\right)\vec{\mathcal{N}}$
$\mathbf{K}$	$\mathcal{S}_t = \left(\mathbf{K} - \frac{\rho 4\pi}{3\mathbf{V}_0}\right)\vec{\mathcal{N}}$	$\mathcal{S}_t = \left(\mathbf{K} - \frac{\rho \int\int \mathbf{K}\mathbf{H} d\mu}{2\mathbf{A}_0}\right)\vec{\mathcal{N}}$
$\mathbf{K}_+^{1/4}$	$\mathcal{S}_t = \left(\mathbf{K}_+^{1/4} - \frac{\rho \int\int \mathbf{K}_+^{1/4} d\mu}{3\mathbf{V}_0}\right)\vec{\mathcal{N}}$	$\mathcal{S}_t = \left(\mathbf{K}_+^{1/4} - \frac{\rho \int\int \mathbf{K}_+^{1/4}\mathbf{H} d\mu}{2\mathbf{A}_0}\right)\vec{\mathcal{N}}$

Fig. 2.12. Area- and volume-preserving flows for a number of 3D geometric evolution equations frequently used.

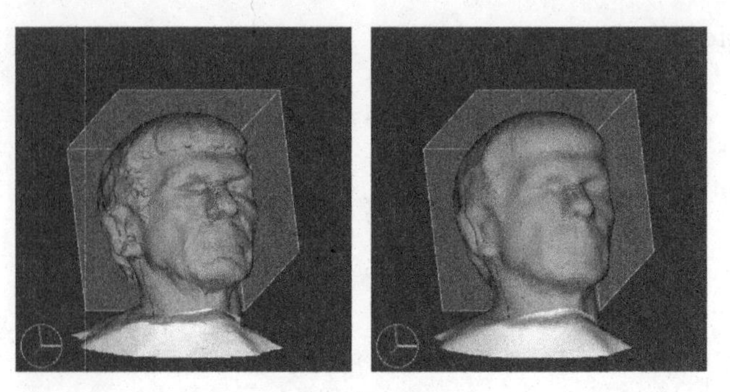

Fig. 2.13. Geometric smoothing with a volume-preserving flow.

2.8. Classification of Invariant Geometric Flows

We have seen two Euclidean invariant flows for planar curves (constant velocity and geometric heat flow) and one affine invariant flow (affine geometric heat flow). We have also seen corresponding flows for hypersurfaces. We can now ask the following questions: (1) Are these the simplest flows that achieve the desired properties? and (2) for another group, how can we find the corresponding geometric heat flow and make sure that it is the simplest possible flow for the group. These questions were answered in Refs. [284–286] based on Lie group theory. The main results are presented now.

In the same way as we have a Euclidean and an affine arc length, for every transformation group we have a parameterization that is intrinsically an invariant to the action of the group. This is the group arc length, $\mathrm{d}r$, which is a nontrivial invariant one-form of minimal order. From it, the group metric g is defined by the equality

$$\mathrm{d}r = g\mathrm{d}p$$

for any parameterization p.

With this in mind, we are ready to generalize the Euclidean and the affine geometric heat flows. Recall that $\mathcal{C}(p, t) : S^1 \times [0, \tau) \to \mathbb{R}^2$ is a family of smooth curves, where p parameterizes the curve and t is the family. (In this case, we take p to be independent of t.) Assume that this family evolves according to the following evolution equation, denoted as invariant geometric heat flow:

$$\frac{\partial \mathcal{C}(p, t)}{\partial t} = \frac{\partial^2 \mathcal{C}(p, t)}{\partial r^2}, \tag{2.64}$$

$$\mathcal{C}(p, 0) = \mathcal{C}_0(p).$$

If the group acts linearly, it is easy to see that because dr is an invariant of the group, so is $\mathcal{C}_{rr}$. $\mathcal{C}_{rr}$ is called the group normal. For nonlinear actions, the flow of Eqs. (2.64) is still invariant, because, as pointed out in Chap. 1, $\partial/\partial r$ is the invariant derivative.

We have just formulated the invariant geometric heat flow in terms of concepts intrinsic to the curve and group itself, i.e., based on the group arc length. Recall that only the normal component of $\mathcal{C}_{rr}$ affects the geometry of the flow. If v stands for the Euclidean arc length, then,

$$\frac{\partial^2 \mathcal{C}}{\partial r^2} = \frac{1}{g^2}\frac{\partial^2 \mathcal{C}}{\partial v^2} - \frac{g_v}{g^3}\frac{\partial \mathcal{C}}{\partial v}, \tag{2.65}$$

where g is the group metric defined above. (In this case, g is a function of v.) Now, using the relations

$$\mathcal{C}_{vv} = \kappa\vec{\mathcal{N}}, \quad \mathcal{C}_v = \vec{\mathcal{T}},$$

we obtain

$$\alpha = -\frac{g_v}{g^3}, \quad \beta = \frac{\kappa}{g^2}, \tag{2.66}$$

where α and β are the tangential and the normal velocities, respectively. In general, $g(v)$ in Eqs. (2.66) is written as well as a function of κ and its derivatives

Formulations (2.64) give a very intuitive formulation of the invariant geometric heat flow. The flow is nonlinear because the group arc length r is a function of time. This flow gives the invariant geometric-heat-type flow of the group and provides the invariant direction of the deformation. For subgroups of the full projective group SL($\mathbb{R}$, 3), we show below that the most general invariant evolutions are obtained if the group curvature χ and its derivatives (with respect to arc length) are incorporated into the flow:

$$\frac{\partial \mathcal{C}(p,t)}{\partial t} = \Psi\left(\chi, \frac{\partial \chi}{\partial r}, \ldots, \frac{\partial^n \chi}{\partial r^n}\right)\frac{\partial^2 \mathcal{C}(p,t)}{\partial r^2}, \tag{2.67}$$

$$\mathcal{C}(p,0) = \mathcal{C}_0(p),$$

where $\Psi(\cdot)$ is a given function. Because the group arc length and the group curvature are the basic invariants of the group transformations, it is natural to formulate Eqs. (2.67) as the most general geometric invariant flow.

Because we can expressed the flow of Eqs. (2.64) in terms of Euclidean properties, we can do the same for the general flow of Eqs. (2.67). All we do we have to do is to express χ as a function of κ and it derivatives. We do this by expressing the curve components $x(p)$ and $y(p)$ as functions of κ and then computing χ.

Let us first note that, as before, locally we may express a solution of Eqs. (2.64) as the graph of a function $y = \gamma(x, t)$.

Lemma 2.12. *Locally, the evolution of Eqs. (2.64) is equivalent to*

$$\frac{\partial \gamma}{\partial t} = \frac{1}{g^2}\frac{\partial^2 \gamma}{\partial x^2},$$

where g is the G–invariant metric ($g = \mathrm{d}r/\mathrm{d}x$).

Proof: Indeed, locally the equation

$$\mathcal{C}_t = \mathcal{C}_{rr},$$

becomes

$$x_t = x_{rr}, \quad y_t = y_{rr}.$$

Now $y(r, t) = \gamma[x(r, t), t]$, so that

$$y_t = \gamma_x x_t + \gamma_t, \quad y_{rr} = \gamma_{xx} x_r^2 + \gamma_x x_{rr}.$$

Thus

$$\gamma_t = y_t - \gamma_x x_t = y_{rr} - \gamma_x x_{rr} = x_r^2 \gamma_{xx}.$$

Therefore evolution equations (2.64) reduce to

$$\gamma_t = g^{-2}\gamma_{xx},$$

as $\mathrm{d}r = g\mathrm{d}x$. □

We can now state the following fundamental result:

Theorem 2.8. *Let G be a subgroup of the projective group SL*($\mathbb{R}$, 3). *Let* $\mathrm{d}r = g\mathrm{d}p$ *denote the G-invariant arc-length and* χ *denote the G-invariant curvature. Then*

1. *every differential invariant of G is a function*

$$I\left(\chi, \frac{\mathrm{d}\chi}{\mathrm{d}r}, \frac{\mathrm{d}^2\chi}{\mathrm{d}r^2}, \ldots, \frac{\mathrm{d}^n\chi}{\mathrm{d}r^n}\right)$$

of χ *and its derivatives with respect to arc length and*

2. *every G-invariant evolution equation has the form*

$$\frac{\partial u}{\partial t} = \frac{1}{g^2}\frac{\partial^2 u}{\partial x^2} I, \tag{2.68}$$

where I is a differential invariant for G.

We are particularly interested in the following subgroups of the full projective group: Euclidean, similarity, special affine, affine, full projective. (See the discussion below for the precise results.)

Corollary 2.2. *Let G denote the similarity, special affine, affine, or full projective group (see remark on the Euclidean group below). Then there is, up to a constant factor, a unique G-invariant evolution equation of lowest order, namely*

$$\frac{\partial u}{\partial t} = \frac{c}{g^2}\frac{\partial^2 u}{\partial x^2},$$

where c is a constant.

Part 1 of Theorem 2.8 (suitably interpreted) does not require G to be a subgroup of the projective group; however, for part 2 and the corollary this is essential (see surfaces section below for an extension). We can, of course, classify the differential invariants, invariant arc lengths, invariant evolution equations, etc., for any group of transformations in the plane, but the interconnections are more complicated. See Lie [233] and Olver [283] for the details of the complete classification of all groups in the plane and their differential invariants.

As explained before, the uniqueness of the Euclidean and affine flows was also proven in Ref. [5], in which a completely different approach was used. In contrast with the results here presented, the ones in Ref. [5] were proven independently for each group, and when a new group was considered, a new analysis had to be carried out.

Proof of Theorem 2.8: Part 1 follows immediately from the results on Lie groups in Chap. 1, and the definitions of dr and χ. (Note for a subgroup of $SL(\mathbb{R}, 3)$ acting on $\mathbb{R}^2$, we have that each differential invariant of the order of k is in fact unique.)

As for part 2, let

$$\mathbf{v} = \xi(x, u)\partial_x + \varphi(x, u)\partial_u$$

be an infinitesimal generator of G, and let pr $\mathbf{v}$ denote its prolongation to the jet space. Because dr is (by definition) an invariant one-form, we have

$$\mathbf{v}(dr) = [\,\mathrm{pr}\,\mathbf{v}(g) + g D_x \xi\,]\mathrm{d}x,$$

which vanishes if and only if

$$\mathrm{pr}\,\mathbf{v}(g) = -g D_x \xi = -g(\xi_x + u_x \xi_u). \tag{2.69}$$

Applying pr $\mathbf{v}$ to evolution equation (2.68) and using condition (2.69), we have (because ξ and φ do not depend on t)

$$\begin{aligned} \text{pr}\,\mathbf{v}[u_t - g^{-2}u_{xx}I] = (\varphi_u - u_x\xi_u)u_t - 2g^{-2}(\xi_x + u_x\xi_u)u_{xx}I \\ - g^{-2}\text{pr}\,\mathbf{v}[u_{xx}]I - g^{-2}u_{xx}\text{pr}\,\mathbf{v}[I]. \end{aligned} \tag{2.70}$$

If G is to be a symmetry group, this must vanish on solutions of the equation; thus, in the first term, we replace u_t with $g^{-2}u_{xx}I$. Now, because G was assumed to be a subgroup of the projective group, which is the symmetry group of the second-order ODE $u_{xx} = 0$, we have that pr $\mathbf{v}[u_{xx}]$ is a multiple of u_{xx}; in fact, inspection of the general prolongation formula for pr $\mathbf{v}$ (see Chap. 1) shows that in this case

$$\text{pr}\,\mathbf{v}[u_{xx}] = (\varphi_u - 2\xi_x - 3\xi_u u_x)u_{xx}. \tag{2.71}$$

(The terms in pr $\mathbf{v}[u_{xx}]$ that do not depend on u_{xx} must add up to zero, owing to our assumption on $\mathbf{v}$.) Substituting Eq. (2.71) into Eq. (2.70) and combining terms, we find

$$\text{pr}\,\mathbf{v}[u_t - g^{-2}u_{xx}I] = g^{-2}u_{xx}\text{pr}\,\mathbf{v}[I],$$

which vanishes if and only if pr $\mathbf{v}[I] = 0$, a condition that must hold for each infinitesimal generator of G. But this is just the infinitesimal condition that I be a differential invariant of G, and the theorem follows. □

The Corollary follows from the fact that, for the listed subgroups, the invariant arc length r depends on lower-order derivatives of u than those of the invariant curvature χ. (This fact holds for most (but not all) subgroups of the projective group; one exception is the group consisting of translations in x, u and isotropic scalings $(x, u) \mapsto (\lambda x, \lambda u)$.) For the Euclidean group, it is interesting to note that the simplest nontrivial flow is given by (constant motion)

$$u_t = c\sqrt{1 + u_x^2},$$

where c is a constant. (Here $g = \sqrt{1 + u_x^2}$.) In this case the curvature (the ordinary planar curvature κ) has order 2. This equation is obtained for the invariant function $I = 1/\kappa$. The Euclidean geometric heat equation is indeed given by the flow in the Corollary. The orders are indicated in the following table:

Group	Arc Length	Curvature
Euclidean	1	2
Similarity	2	3
Special affine	2	4
Affine	4	5
Projective	5	7

The explicit formulas are given in the following table:

Group	Arc Length	Curvature
Euclidean	$\sqrt{1+u_x^2}\mathrm{d}x$	$\frac{u_{xx}}{(1+u_x^2)^{3/2}}$
Similarity	$\frac{u_{xx}\mathrm{d}x}{(1+u_x^2)}$	$\frac{(1+u_x^2)u_{xxx}-3u_x u_{xx}^2}{u_{xx}^2}$
Special affine	$(u_{xx})^{1/3}\mathrm{d}x$	$\frac{P_4}{(u_{xx})^{8/3}}$
Affine	$\frac{\sqrt{P_4}}{u_{xx}}\,\mathrm{d}x$	$\frac{P_5}{(P_4)^{3/2}}$
Projective	$\frac{(P_5)^{1/3}}{u_{xx}}\,\mathrm{d}x$	$\frac{P_7}{(P_5)^{8/3}}$

Here

$$
\begin{aligned}
P_4 &= 3u_{xx}u_{xxxx} - 5u_{xxx}^2, \\
P_5 &= 9u_{xx}^2 u_{xxxxx} - 45u_{xx}u_{xxx}u_{xxxx} + 40u_{xxx}^3, \\
P_7 &= \frac{1}{3}u_{xx}^2\left[6P_5 D_x^2 P_5 - 7(D_x P_5)^2\right] + 2u_{xx}u_{xxx}P_5 D_x P_5 \\
&\quad -\left(9u_{xx}u_{xxxx} - 7u_{xxx}^2\right)P_5^2.
\end{aligned}
$$

This concludes the general results on invariant curve flows. The Euclidean and affine heat flows were described in detail before in Section 2.5. The projective flow is studied by Faugeras and Keriven [134, 135], and the similarity flow is be presented at the end this section. However, before that, the extension of the above results to surfaces is presented.

We can write the general form that any G–invariant evolution is n independent and must have one dependent variable. Thus, for $n = 1$, we get the family of all possible invariant curve evolutions in the plane under a given transformation group G and, for $n = 2$, the family of all possible invariant surface evolutions under a given transformation group G.

We let

$$\omega = g\mathrm{d}x^1 \wedge \ldots \wedge \mathrm{d}x^n$$

denote a G–invariant n–form with respect to the transformation group G acting on $\mathbb{R}^n$. Let $E(g)$ denote the variational derivative of g. We consider forms only such that $E(g) \neq 0$. We call such a g a G–invariant volume function.

Theorem 2.9. *Notation as above. Then every G–invariant evolution equation has the form*

$$u_t = \frac{g}{E(g)} I, \tag{2.72}$$

where I is a differential invariant.

Theorem 2.10. *Suppose G is a connected transformation group, and* $g\mathrm{d}x$ *a G-invariant n-form such that $E(g) \neq 0$. Then $E(g)$ is invariant if and only if G is volume preserving.*

It is now trivial to give the simplest possible invariant surface evolution. This gives for example the surface version of the affine shortening flow for curves. This equation was also derived with completely different methods by Alvarez et al. [5].

We define the (special) affine group on $\mathbb{R}^3$ as the group of transformations generated by $SL_3(\mathbb{R})$ (the group of 3×3 matrices with determinant 1) and translations.

Let $\mathcal{S}$ be a smooth strictly convex surface in $\mathbb{R}^3$, which we write locally as the graph (x, y, u). Then we can compute that the Gaussian curvature is given by

$$\mathbf{K} = \frac{u_{xx}u_{yy} - u_{xy}^2}{\left(1 + u_x^2 + u_y^2\right)^2}.$$

From Ref. [39], the affine invariant metric is given by

$$g = \mathbf{K}^{1/4}\sqrt{1 + u_x^2 + u_y^2}.$$

(We can also write

$$g = \mathbf{K}^{1/4}\sqrt{\det g_{ij}},$$

where g_{ij} are the coefficients of the first fundamental form.)

We now have the following Corollary.

Corollary 2.3. *Notation as above. Then*

$$u_t = c\mathbf{K}^{1/4}\sqrt{1 + u_x^2 + u_y^2}, \tag{2.73}$$

(for c, a constant) is the simplest affine invariant surface flow. This corresponds to the global evolution

$$\mathcal{S}_t = c\mathbf{K}^{1/4}\vec{\mathcal{N}}, \tag{2.74}$$

where $\vec{\mathcal{N}}$ denotes the inward normal.

We call the evolution

$$\mathcal{S}_t = \mathbf{K}^{1/4}\vec{\mathcal{N}}, \tag{2.75}$$

the affine surface flow. Note that it is the surface analog of the affine heat equation.

2.8.1. Similarity Heat Flow

Let us conclude this chapter with an example of these results, computing the geometric heat flow for the similarity group (Euclidean group plus dilation), considering only convex planar curves. It is easy to see that the angle θ between the tangent and the fixed coordinate system is a similarity invariant parameterization and then the similarity arc length. The heat flow is then

$$\frac{\partial \mathcal{C}}{\partial t} = \frac{\partial^2 \mathcal{C}}{\partial \theta^2}. \tag{2.76}$$

By using the relation $(\mathrm{d}\theta/\mathrm{d}v) = \kappa$ and projecting this velocity into the Euclidean normal, we obtain the geometrically equivalent flow

$$\frac{\partial \mathcal{C}}{\partial t} = \frac{1}{\kappa}\vec{\mathcal{N}}.$$

Instead of looking at this flow, which can develop singularities, we invert it and look at

$$\frac{\partial \mathcal{C}}{\partial t} = -\frac{1}{\kappa}\vec{\mathcal{N}}.$$

For this flow we obtain the following evolution equations:

$$\vec{\mathcal{T}}_t = -\frac{\kappa_v}{\kappa^2}\vec{\mathcal{N}},$$

$$\vec{\mathcal{N}}_t = \frac{\kappa_v}{\kappa^2}\vec{\mathcal{T}},$$

$$A_t = \oint \frac{1}{\kappa}\mathrm{d}v,$$

$$\kappa_t = -\left(\frac{1}{\kappa}\right)_{vv} - \kappa,$$

$$\theta_t = \frac{\kappa_v}{\kappa^2}.$$

It is also easy to prove that the convex curve remains convex and smoothly deforms toward a disk [345].

Exercises

1. Compute the Euclidean tangent, normal, and curvature of a planar graph $y = \gamma(x, t)$.
2. Repeat the exercise above for the affine tangent, normal, and curvature.
3. Prove all the evolution equations for the tangent, normal, and curvature evolutions for the Euclidean and the similarity heat flows.
4. Show that the gradient descent flow of the surface-area-minimization problem is given by mean curvature motion.
5. Using the level-set approach, implement the Euclidean and the affine geometric heat flows and the Euclidean constant velocity flow. (When the Matlab, Maple, or Mathematica programming languages and their level-set-finding capabilities are used, a simple implementation can be quickly obtained).
6. Show that the Euclidean geometric heat flow is the gradient descent of the Euclidean length.
7. Prove the formulas for the perimeter-preserving flows.
8. Which geometric flows have corresponding perimeter-preserving flows that are local?
9. Implement the 3D surface flows introduced in this chapter.
10. For the curve flow with constant Euclidean velocity, implement the fast-marching method and the narrow-band method and compare their speeds with the direct level-set implementation.
11. Consider a graph from $\mathbb{R}^2$ to $\mathbb{R}$, for example, an image, and apply to it mean and Gaussian curvature flows. Note of course that, in order to keep the graph characteristic of the shape, the normal velocity must be projected onto the vertical (z) direction.

CHAPTER THREE

Geodesic Curves and Minimal Surfaces

In this chapter we show how a number of problems in image processing and computer vision can be formulated as the computation of paths or surfaces of minimal energy. We start with the basic formulation, connecting classical work on segmentation with the computation of geodesic curves in two dimensions. We then extend this work to three dimensions and show the application of this framework to object tracking and stereo. The geodesic or minimal surface is computed by means of geometric PDEs, obtained from gradient descent flows. These flows are driven by intrinsic curvatures as well as forces that are derived from the image (data). From this point of view, with this chapter we move one step forward in the theory of curve evolution and PDEs: from equations that included only intrinsic velocities to equations that combine intrinsic with external velocities.

3.1. Basic Two-Dimensional Derivation

Since the original pioneering work by Kass et al. [198], extensive research has been done on "snakes" or active-contour models for boundary detection. The classical approach is based on deforming an initial contour $\mathcal{C}_0$ toward the boundary of the object to be detected. We obtain the deformation by trying to minimize a functional designed so that its (local) minimum is obtained at the boundary of the object. These active contours are examples of the general technique of matching deformable models to image data by means of energy minimization [38, 387]. The energy functional is basically composed of two components; one controls the smoothness of the curve and the other attracts the curve toward the boundary. This energy model is not capable of handling changes in the topology of the evolving contour when direct implementations are performed. Therefore the topology of the final curve will be as the one of $\mathcal{C}_0$ (the initial curve), unless special procedures are implemented for detecting possible splitting and merging [238, 256, 379]. This is a problem when an unknown number of objects must be simultaneously detected. This approach is also nonintrinsic, as the energy depends on the parameterization of the curve and is not directly related to the object geometry.

As we show in this chapter, a kind of reinterpretation of this model solves these problems and presents a new paradigm in image processing: the formulation of image processing problems as the search for geodesic curves or minimal surfaces. A particular case of the classical energy snake model is proved to be equivalent to finding a geodesic curve in a Riemannian space with a metric derived from the image content. This means that, in a certain framework, boundary detection can be considered equivalent to finding a curve of minimal weighted length. This interpretation gives a new approach for boundary detection by means of active contours, based on geodesic or local minimal-distance computations. We also show that the solution to the geodesic flow exists in the viscosity framework and is unique and stable. Consistency of the model is presented as well, showing that the geodesic curve converges to the right solution in the case of ideal objects (the proof is left for when we deal with the 3D case in Section 3.2). A number of examples of real images, showing the above properties, are presented.

3.1.1. Geodesic Active Contours

In this section we discuss the connection between energy-based active contours (snakes) and the computation of geodesics or minimal-distance curves in a Riemannian space derived from the image. From this geodesic model for object detection, we derive a novel geometric PDE for active contours that improves on previous curve evolution models.

Energy-Based Active Contours. Let us briefly describe the classical energy-based snakes. Let $\mathcal{C}(p) : [0, 1] \to \mathbb{R}^2$ be a parameterized planar curve and let $I : [0, a] \times [0, b] \to \mathbb{R}^+$ be a given image in which we want to detect the object's boundaries. The classical snake approach [198] associates the curve $\mathcal{C}$ with an energy given by

$$E(\mathcal{C}) = \alpha \int_0^1 |\mathcal{C}'(p)|^2 \mathrm{d}p + \beta \int_0^1 |\mathcal{C}''(p)|^2 \mathrm{d}q - \lambda \int_0^1 |\nabla I[\mathcal{C}(p)]| \mathrm{d}p, \tag{3.1}$$

where α, β, and λ are real positive constants. The first two terms control the smoothness of the contours to be detected (internal energy), and the third term is responsible for attracting the contour toward the object in the image (external energy). Solving the problem of snakes amounts to finding, for a given set of constants α, β, and λ, the curve $\mathcal{C}$ that minimizes E. Note that when considering more than one object in the image, for instance for an initial prediction of $\mathcal{C}$ surrounding all of them, it is not possible to detect all the objects. Special topology-handling procedures must be added. Actually, the solution without those special procedures will be in most cases a curve that approaches a convex-hull-type figure of the objects in the image. In other words, the classical (energy) approach of snakes cannot directly deal with changes in topology. The topology of the initial curve will be the same as the one of the, possibly wrong, final curve. The model derived below, as well as the curve evolution models in Refs. [63 and 250–252], overcomes this problem.

Another possible problem of the energy-based models is the need to select the parameters that control the trade-off between smoothness and proximity to the object. Let us consider a particular class of snake models in which the rigidity coefficient is set to zero, that is, $\beta = 0$. Two main reasons motivate this selection, which at least mathematically restricts the general model of Eq. (3.1): First, this selection will allow us to derive the relation between these energy-based active contours and geometric curve evolution ones, which is one of the goals of this chapter. This will be done in Subsection 3.1.1 through the presentation of the proposed geodesic active contours. Second, the regularization effect on the geodesic active contours comes from curvature-based curve flows, obtained only from the other terms in Eq. (3.1) (see Eq. (3.16) and its interpretation after it). This will allow us to achieve smooth curves in the proposed approach without having the high-order smoothness given by $\beta \neq 0$ in energy-based approaches. Moreover, the second-order smoothness component in Eq. (3.1), assuming an arc-length parameterization, appears in order to minimize the total squared curvature (curve known as elastica). It is easy to prove that the curvature flow used in the new approach and presented below decreases the total curvature [12]. The use of the curvature-driven curve motions introduced in Chap. 2 as smoothing terms was proved to be very efficient in previous literature [5, 63, 206, 250–252, 269, 348], and is also supported by the experiments in Subsection 3.1.3 and subsequent chapters in this book. Therefore curve smoothing will be obtained also with $\beta = 0$, having only the first regularization term. Assuming this, we can reduce Eq. (3.1) to

$$E(\mathcal{C}) = \alpha \int_0^1 |\mathcal{C}'(p)|^2 \mathrm{d}p - \lambda \int_0^1 |\nabla I[\mathcal{C}(p)]| \mathrm{d}p. \tag{3.2}$$

Observe that, by minimizing functional (3.2), we are trying to locate the curve at the points of maxima $|\nabla I|$ (acting as edge detector) while keeping a certain smoothness in the curve (object boundary). This is actually the goal in general formulation (3.1) as well. The trade-off between edge proximity and edge smoothness is played by the free parameters in the above equations.

We can extend Eq. (3.2), generalizing the edge detector part in the following way: Let $g : [0, +\infty[\to \mathbb{R}^+$ be a strictly decreasing function such that $g(r) \to 0$ as $r \to \infty$. Hence we can replace $-|\nabla I|$ with $g(|\nabla I|)^2$, obtaining a general energy functional given by

$$E(\mathcal{C}) = \alpha \int_0^1 |\mathcal{C}'(p)|^2 \mathrm{d}p + \lambda \int_0^1 g\{|\nabla I[\mathcal{C}(p)]|\}^2 \mathrm{d}p$$
$$= \int_0^1 \{E_{\mathrm{int}}[\mathcal{C}(p)] + E_{\mathrm{ext}}[\mathcal{C}(p)]\}\, \mathrm{d}p. \tag{3.3}$$

The goal now is to minimize E in Eq. (3.3) for $\mathcal{C}$ in a certain allowed space of curves. (To simplify the notation, we sometimes write $g(I)$ or $g(\mathcal{X})$ ($\mathcal{X} \in \mathbb{R}^2$) instead of $g(|\nabla I|)$.) Note that in the above energy functional, only the ratio λ/α counts. The geodesic active contours will be derived from Eq. (3.3).

The functional in Eq. (3.3) is not intrinsic as it depends on the parameterization q that until now was arbitrary. This is an undesirable property, as parameterizations are not related to the geometry of the curve (or object boundary), but only to the velocity they are traveled. Therefore it is not natural for an object-detection problem to depend on the parameterization of the representation. Actually, if we define a new parameterization of the curve by means of $p = \phi(r)$, $\phi : [c, d] \to [0, 1]$, $\phi' > 0$, we obtain

$$\int_0^1 |\mathcal{C}'(p)|^2 \mathrm{d}p = \int_c^d |(\mathcal{C} \circ \phi)'(r)|^2 [\phi'(r)]^{-1} \mathrm{d}r,$$
$$\int_0^1 g\{|\nabla I[\mathcal{C}(p)]|\} \mathrm{d}q = \int_c^d g\{|\nabla I[\mathcal{C} \circ \phi(r)]|\} \phi'(r) \mathrm{d}r,$$

and the energies can change in any arbitrary form. One of our goals will be to present a possible solution to this problem by choosing a parameterization that is intrinsic to the curve (geometric).

The Geodesic Curve Flow. We now proceed and show that the solution of the particular energy snake model of Eq. (3.3) is given by a geodesic curve in a Riemannian space induced from the image I. (A geodesic curve is a (local) minimal distance path between given points.) To show this, we use the classical Maupertuis' principle [113] from dynamical systems. Giving all the background on this principle is beyond the scope of this book, so the presentation is restricted to essential points and geometric interpretation. Let us define

$$\mathcal{U}(\mathcal{C}) := -\lambda g(|\nabla I(\mathcal{C})|)^2$$

and write $\alpha = m/2$. Therefore

$$E(\mathcal{C}) = \int_0^1 \mathcal{L}[\mathcal{C}(p)]\mathrm{d}p,$$

where $\mathcal{L}$ is the Lagrangian given by

$$\mathcal{L}(\mathcal{C}) := \frac{m}{2}|\mathcal{C}'|^2 - \mathcal{U}(\mathcal{C}).$$

The Hamiltonian [113] is then given by

$$H = \frac{q^2}{2m} + \mathcal{U}(\mathcal{C}),$$

where $q := m\mathcal{C}'$. To show the relation between energy-minimization problem (3.3) and geodesic computations, we will need the following Theorem [113].

Theorem 3.1 (Maupertuis' Principle). *Curves $\mathcal{C}(p)$ in Euclidean space that are extremal, corresponding to the Hamiltonian $H = (q^2/2m) + \mathcal{U}(\mathcal{C})$, and have a fixed energy level E_0 (law of conservation of energy), are geodesics, with a nonnatural parameter with respect to the new metric $(i, j = 1, 2)$*

$$g_{ij} = 2m[E_0 - \mathcal{U}(\mathcal{C})]\delta_{ij}.$$

This classical Theorem explains, among other things, when an energy-minimization problem is equivalent to finding a geodesic curve in a Riemannian space; that means when the solution to the energy problem is given by a curve of minimal weighted distance between given points. Distance is measured in the given Riemannian space with the first fundamental form g_{ij} (the first fundamental form defines the metric or distance measurement in the space). See the mentioned references (especially Section 3.3 in Ref. [113]) for details on the theorem and the corresponding background on Riemannian geometry. According to the above result, minimizing $E(\mathcal{C})$ as in Eq. (3.3) with $H = E_0$ (conservation of energy) is equivalent to minimizing

$$\int_0^1 \sqrt{g_{ij}\mathcal{C}'_i\mathcal{C}'_j}\mathrm{d}p, \tag{3.4}$$

$(i, j = 1, 2)$ or

$$\int_0^1 \sqrt{g_{11}\mathcal{C}'^2_1 + 2g_{12}\mathcal{C}'_1\mathcal{C}'_2 + g_{22}\mathcal{C}'^2_2}\, dp, \tag{3.5}$$

where $(\mathcal{C}_1, \mathcal{C}_2) = \mathcal{C}$ (components of $\mathcal{C}$) and $g_{ij} = 2m[E_0 - \mathcal{U}(\mathcal{C})]\delta_{ij}$.

We have just transformed the minimization of Eq. (3.3) into the energy of expression (3.5). As we see from the definition of g_{ij}, expression (3.5) has a free parameter, E_0. We deal now with this energy. From Fermat's principle, we motivate the selection of the value of E_0. We then present an intuitive approach that brings us to the same selection. In Appendix A, we extend the formulation without a priori fixing E_0.

Fixing the ratio λ/α, we may consider the search for path-minimizing equation (3.3) as a search for a path in the (x, y, p) space, indicating the nonintrinsic nature of this minimization problem. The Maupertuis principle of least action used to derive expression (3.5) presents a purely geometric principle describing the orbits of the minimizing paths [45]. In other words, it is possible to use the above theorem to find the projection of the minimizing path of Eq. (3.3) in the (x, y, p) space onto the (x, y) plane by solving an intrinsic problem. Observe that the parameterization along the path is yet to be determined after its orbit is tracked. The intrinsic problem of finding the projection of the minimizing path depends on a single free parameter E_0 that incorporates the parameterization as well as λ and α ($E_0 = E_{\text{int}} - E_{\text{ext}} = \alpha|\mathcal{C}'(p)|^2 - \lambda g[\mathcal{C}(p)]^2$).

The question to be asked is whether the problem in hand should be regarded as the behavior of springs and mass points leading to the nonintrinsic model of Eq. (3.3). We take this one step further, moving from springs to light rays, and use the following result from optics to motivate the proposed model [45, 113]:

Theorem 3.2 (Fermat's Principle). *In an isotropic medium the paths taken by light rays in passing from a point A to a point B are extrema corresponding to the traversal time (as action). Such paths are geodesics with respect to the new metric* $(i, j = 1, 2)$

$$g_{ij} = \frac{1}{c^2(\mathcal{X})}\delta_{ij}.$$

In the above equation $c(\mathcal{X})$ corresponds to the speed of light at $\mathcal{X}$. Fermat's principle defines the Riemannian metric for light waves. We define $c(\mathcal{X}) = 1/g(\mathcal{X})$, where high speed of light corresponds to the presence of an edge, whereas low speed of light corresponds to a nonedge area. The result is equivalent then to minimizing the intrinsic problem

$$\int_0^1 g\{|\nabla I[\mathcal{C}(p)]|\}|\mathcal{C}'(p)|\mathrm{d}p, \tag{3.6}$$

which is the same formulation as that of expression (3.5), with $E_0 = 0$.

We return for a while to the energy model of Eq. (3.3) to further explain the selection of $E_0 = 0$ from the point of view of object detection. As was explained above, in order to have a completely closed form for boundary detection by means of expression (3.5), we have to select E_0. It was shown that selecting E_0 is equivalent to fixing the free parameters in Eq. (3.3) (i.e., the parameterization and λ/α). Note that, by Theorem 3.1, the interpretation of the snake model of Eq. (3.3) for object detection as a geodesic computation is valid for any value of E_0. The value of E_0 is selected to be zero from now on, which means that $E_{\text{int}} = E_{\text{ext}}$ in Eq. (3.3). This selection simplifies the notation (see Appendix A in this chapter) and clarifies the relation of Theorem 3.1 and energy snakes with (geometric) curve evolution active contours that result from Theorems 3.1 and 3.2. At an ideal edge, E_{ext} in Eq. (3.3) is expected to be zero, because $|\nabla I| = \infty$ and $g(r) \to 0$ as $r \to \infty$. Then the ideal goal is to send the edges to the zeros of g. Ideally we should try as well to send the internal energy to zero. Because images are not formed by ideal edges, we choose to make equal contributions of both energy components. This choice, which coincides with the one obtained from Fermat's principle and, as stated above, allows us to show the connection with curve evolution active contours, is also consistent with the fact that when we are looking for an edge, we may travel along the curve with arbitrarily slow velocity (given by the parameterization q; see equations obtained with the above change of parameterization). More comments on different selections of E_0, as well as formulas corresponding to $E_0 \neq 0$, are given in Appendix A in this chapter.

Therefore, with $E_0 = 0$ and $g_{ij} = 2m\lambda g[|\nabla I(\mathcal{C})|]^2\delta_{ij}$, expression (3.4) becomes

$$\min \int_0^1 \sqrt{2m\lambda}\, g\{|\nabla I[\mathcal{C}(p)]|\}|\mathcal{C}'(p)|\mathrm{d}p. \tag{3.7}$$

Because the parameters above are constants, without loss of generality we can set now $2\lambda m = 1$ to obtain

$$\min \int_0^1 g\{|\nabla I[\mathcal{C}(p)]|\}|\mathcal{C}'(p)|\mathrm{d}p. \tag{3.8}$$

We have transformed the problem of minimizing Eq. (3.3) into a problem of geodesic computation in a Riemannian space, according to a new metric.

Let us, based on the above theory, give the above expression a further geodesic curve interpretation from a slightly different perspective. The Euclidean length of the curve C is given by

$$L := \oint |\mathcal{C}'(p)|\mathrm{d}p = \oint \mathrm{d}s, \tag{3.9}$$

where ds is the Euclidean arc length (or Euclidean metric). As we have seen before in Section 2.5, the flow

$$\mathcal{C}_t = \kappa \vec{\mathcal{N}}, \tag{3.10}$$

where again κ is the Euclidean curvature, gives the fastest way to reduce L, that is, it moves the curve in the direction of the gradient of the functional L. Looking now at expression (3.8), we find that a new length definition in a different Riemannian space is given:

$$L_R := \int_0^1 g\{|\nabla I[\mathcal{C}(p)]|\}|\mathcal{C}'(p)|\mathrm{d}p. \tag{3.11}$$

Because $|\mathcal{C}'(p)|\mathrm{d}p = \mathrm{d}s$, we obtain

$$L_R := \int_0^{L(C)} g\{|\nabla I[\mathcal{C}(s)]|\}\mathrm{d}s. \tag{3.12}$$

Comparing this with the classical length definition as given in Eq. (3.9), we observe that we obtain the new length by weighting the Euclidean element of length ds by $g\{|\nabla I[\mathcal{C}(s)]|\}$, which contains information regarding the boundary of the object. Therefore, when trying to detect an object, we are not just interested in finding the path of minimal classical length ($\oint \mathrm{d}s$) but we are also interested in finding the one that minimizes a new length definition that takes into account image characteristics. Note that expression (3.8) is general; besides being a positive decreasing function, no assumptions on g were made. Therefore the theory of boundary detection based on geodesic computations given above can be applied to any general edge-detector functions g. Recall that expression (3.8) was obtained from the particular case of energy-based snakes of Eq. (3.3) with Maupertuis' principle, which helps to identify variational approaches that are equivalent to computing paths of minimal length in a new metric space.

To minimize expression (3.8) (or L_R), we search for the gradient descent direction of expression (3.8), which is a way of minimizing L_R by means of the steepest-descent method. Therefore we need to compute the

Euler–Lagrange of expression (3.8). Details on this computation are given in Appendix B in this chapter. Thus, according to the steepest-descent method, to deform the initial curve $\mathcal{C}(0) = \mathcal{C}_0$ toward a (local) minima of L_R, we should follow the curve evolution equation (compare with Eq. (3.10)):

$$\frac{\partial \mathcal{C}(t)}{\partial t} = g(I)\kappa\vec{\mathcal{N}} - (\nabla g \cdot \vec{\mathcal{N}})\vec{\mathcal{N}}, \tag{3.13}$$

where κ is the Euclidean curvature, $\vec{\mathcal{N}}$ is the unit inward normal, and the right-hand side of the equation is given by the Euler–Lagrange of expression (3.8) as derived in Appendix B. This equation shows how each point in the active contour $\mathcal{C}$ should move in order to decrease the length L_R. The detected object is then given by the steady-state solution of Eq. (3.13), that is, $\mathcal{C}_t = 0$.

To summarize, Eq. (3.13) presents a curve evolution flow that minimizes the weighted length L_R, which was derived from the classical snake case of Eq. (3.3) by means of Maupertuis' principle of least action. This is the basic geodesic curve flow proposed for object detection (the full model is presented below). In the following section we embed this flow in a level-set formulation to complete the model and show its connection with previous curve evolution active contours. This embedding will also help to present theoretical results regarding the existence of the solution of Eq. (3.13), as we do in Subsection 3.1.2. We note that minimization of a normalized version of Eq. (3.12) was proposed in Ref. [145] from a different perspective, leading to a different geometric method.

The Level-Set Geodesic Flow: Derivation. To find the geodesic curve, we compute the corresponding steepest-descent flow of expression (3.8), Eq. (3.13). Equation (3.13) is represented with the level-set approach.

Assume that the curve $\mathcal{C}$ is a level set of a function $u : [0, a] \times [0, b] \to \mathbb{R}$, that is, $\mathcal{C}$ coincides with the set of points $u =$ constant (e.g., $u = 0$). Therefore u is an implicit representation of the curve $\mathcal{C}$. Recall that this representation is parameter free, then intrinsic. As we have discussed before, in Section 2.2, the representation is also topology free because different topologies of the zero level set do not imply different topologies of u. We have shown that if a curve $\mathcal{C}$ evolves according to

$$\mathcal{C}_t = \beta\vec{\mathcal{N}},$$

for a given function β, then the embedding function u should deform according to

$$u_t = \beta|\nabla u|,$$

where β is computed on the level sets. Recall that, by embedding the evolution of $\mathcal{C}$ in that of u, topological changes of $\mathcal{C}(t)$ are handled automatically and accuracy and stability are achieved with the proper numerical algorithm [294]. This level-set representation was formally analyzed in Refs. [90, 126, and 372], proving for example that, in the viscosity framework, the solution is independent of the embedding function u for a number of velocities β (see Subsection 3.1.2 as well as Theorems 5.6 and 7.1 in Ref. [90]). In our case u is initiated to be the signed distance function. Therefore, from expression (3.8) and embedding Eq. (3.13) in u, we obtain that solving the geodesic problem is equivalent to searching for the steady-state solution $(\partial u/\partial t) = 0$ of the following evolution equation ($u(0, \mathcal{C}) = u_0(\mathcal{C})$):

$$\begin{aligned}\frac{\partial u}{\partial t} &= |\nabla u| \operatorname{div}\left[g(I)\frac{\nabla u}{|\nabla u|}\right]\\ &= g(I)|\nabla u| \operatorname{div}\left(\frac{\nabla u}{|\nabla u|}\right) + \nabla g(I)\cdot\nabla u\\ &= g(I)|\nabla u|\kappa + \nabla g(I)\cdot\nabla u,\end{aligned} \tag{3.14}$$

where the right-hand side of the flow is the Euler–Lagrange of expression (3.8) with $\mathcal{C}$ represented by a level set of u, and the curvature κ is computed on the level sets of u. This means that Eq. (3.14) is obtained by embedding Eq. (3.13) into u for $\beta = g(I)\kappa - \nabla g \cdot \vec{\mathcal{N}}$. On the equation above we made use of the fact that

$$\kappa = \operatorname{div}\left(\frac{\nabla u}{|\nabla u|}\right).$$

Equation (3.14) is the main part of the proposed active-contour model.

The Level-Set Geodesic Flow: Boundary Detection. Let us proceed and explore the geometric interpretation of geodesic active-contour equation (3.14) from the point of view of object segmentation as well as its relation to other geometric curve evolution approaches to active contours. In Refs. [63 and 250–252], the authors proposed the following model for boundary detection:

$$\begin{aligned}\frac{\partial u}{\partial t} &= g(I)|\nabla u| \operatorname{div}\left(\frac{\nabla u}{|\nabla u|}\right) + cg(I)|\nabla u|\\ &= g(I)(c+\kappa)|\nabla u|,\end{aligned} \tag{3.15}$$

where c is a positive real constant. Following Refs. [63 and 250–252], we can interpret Eq. (3.15) as follows: First, the flow

$$u_t = (c + \kappa)|\nabla u|$$

means that each one of the level sets $\mathcal{C}$ of u is evolving according to

$$\mathcal{C}_t = (c + \kappa)\vec{\mathcal{N}},$$

where $\vec{\mathcal{N}}$ is the inward normal to the curve. This equation, as we have seen in Chap. 2, was first proposed in Refs. [294 and 358], in which extensive numerical research was performed on it and then studied in Ref. [206] for shape analysis. The previously presented Euclidean shortening flow

$$\mathcal{C}_t = \kappa\vec{\mathcal{N}}, \tag{3.16}$$

denoted also as Euclidean heat flow, is well known for its very satisfactory geometric smoothing properties [12, 150, 162]; see Chap. 2. The flow decreases the total curvature as well as the number of zero crossings and the value of maxima/minima curvature. Recall that this flow also moves the curve in the gradient direction of its length functional. Therefore it has the properties of shortening as well as smoothing. This shows that having only the first regularization component in Eq. (3.1), $\alpha \neq 0$ and $\beta = 0$, is enough to obtain smooth active contours, as argued in Subsection 3.1.1 when the selection $\beta = 0$ was done. The constant velocity $c\vec{\mathcal{N}}$, which as we saw in Chap. 2 is related to classical mathematical morphology [343] and shape offsetting in computer-aided design [209], is similar to the balloon force introduced in Ref. [96]. Actually this velocity pushes the curve inward (or outward) and it is crucial in the above model to allow convex initial curves to capture nonconvex shapes, that is, to detect nonconvex objects. Of course, the c parameter must be specified a priori in order to make the object-detection algorithm automatic. This is not a trivial issue, as pointed out in Ref. [63], in which possible ways of estimating this parameter are considered. Estimates of this parameter are presented when we deal with the 3D case. To summarize the force $(c + \kappa)$ acts as the internal force in the classical energy-based snake model, smoothness being provided by the curvature part of the flow. The Euclidean heat flow $\mathcal{C}_t = \kappa\vec{\mathcal{N}}$ is exactly the regularization curvature flow that replaces the high-order smoothness term in Eq. (3.1), as discussed in Section 3.1.1.

The external-image-dependent force is given by the stopping function $g(I)$. The main goal of $g(I)$ is actually to stop the evolving curve when it arrives at the object's boundaries. In Refs. [63 and 250–252], the authors chose

$$g = \frac{1}{1 + |\nabla \hat{I}|^p}, \tag{3.17}$$

where $\hat{I}$ is a smoothed version of I and $p = 1$ or 2. $\hat{I}$ was computed by Gaussian filtering, but more effective geometric smoothers, such as those in introduced in Chap. 4, can be used as well. Note that other decreasing functions of the gradient may be selected as well. For an ideal edge, $\nabla \hat{I} = \delta$, $g = 0$, and the curve stops ($u_t = 0$). The boundary is then given by the set $u = 0$.

In contrast with classical energy models of snakes, the curve evolution model given by Eq. (3.15) is topology independent, that is, there is no need to know a priori the topology of the solution. This allows it to detect any number of objects in the image, without knowing their exact number. This is achieved with the help of the mentioned level-set numerical algorithm.

Let us return to the full model. Comparing Eq. (3.14) with Eq. (3.15), we see that the term $\nabla g \cdot \nabla u$, naturally incorporated by means of the geodesic framework, is missing in the old model. This term attracts the curve to the boundaries of the objects (∇g points toward the middle of the boundaries); Fig. 3.1

Note that in the old model, the curve stops when $g = 0$. This happens at only an ideal edge. In cases in which there are different gradient values along the edge, as often happens in real images, g gets different values at different locations along the boundaries. It is necessary to restrict the g values, as well as possible gaps in the boundary, so that the propagating curve is guaranteed to stop. This makes the geometric model of Eq. (3.15)

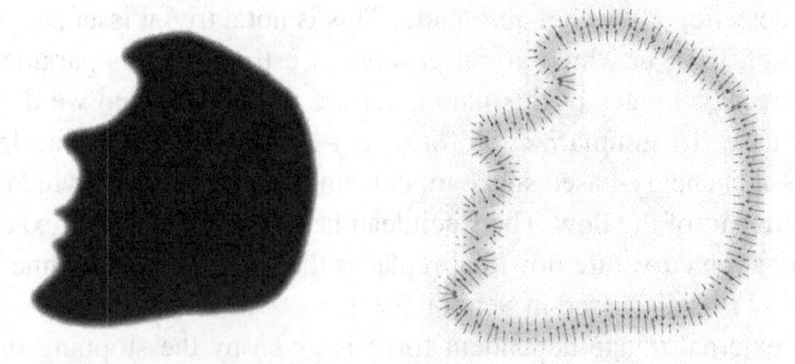

Fig. 3.1. Geometric interpretation of the new term in the proposed deformable model. The gradient vectors are all directed toward the middle of the boundary. Those vectors direct the propagating curve into the valley of the g function.

inappropriate for the detection of boundaries with (unknown) high variations of the gradients. In the proposed model, the curve is attracted toward the boundary by the new gradient term. Observe in Fig. 3.1 the way the gradient vectors are all directed toward the middle of the boundary. Those vectors direct the propagating curve into the valley of the g function. In the 2D case, $\nabla g \cdot \vec{\mathcal{N}}$ is effective in case the gradient vectors coincide with normal direction of the propagating curve. Otherwise, it will lead the propagating curve into the boundary and eventually force it to stay there. To summarize, this new force increases the attraction of the deforming contour toward the boundary, being of special help when this boundary has high variations on its gradient values. Thereby it is also possible to detect boundaries with high differences in their gradient values, as well as small gaps. The second advantage of this new term is that we partially remove the necessity of the constant velocity given by c. This constant velocity, which mainly allows the detection of nonconvex objects, introduces an extra parameter to the model that, in most cases, is an undesirable property. In the full model, the new term will allow the detection of nonconvex objects as well. This constant-motion term may help us to avoid certain local minima (as the balloon force) and is also of importance when we start from curves inside the object, as we will see in Subsection 3.1.3. In case we wish to add this constant velocity, for example to increase the speed of convergence, we can consider the term $cg(I)|\nabla u|$ as an area constraint to geodesic problem (3.8) (c being the Lagrange multiplier), obtaining

$$\frac{\partial u}{\partial t} = |\nabla u| \operatorname{div}\left[g(I)\frac{\nabla u}{|\nabla u|}\right] + cg(I)|\nabla u|. \tag{3.18}$$

Before proceeding, we note that constant velocity is derived from an energy that involves area, that is, $\mathcal{C} = c\vec{\mathcal{N}}$ minimizes the area enclosed by $\mathcal{C}$. Therefore, adding constant velocity is like solving $L_R + c$ area$(\mathcal{C})$ [67, 125, 366].

Equation (3.18) is of course equivalent to

$$\frac{\partial u}{\partial t} = g(c + \kappa)|\nabla u| + \nabla u \cdot \nabla g \tag{3.19}$$

and means that the level sets move according to

$$\mathcal{C}_t = g(I)(c + \kappa)\vec{\mathcal{N}} - (\nabla g \cdot \vec{\mathcal{N}})\vec{\mathcal{N}}. \tag{3.20}$$

Equation (3.18), which is the level-set representation of the modified solution of geodesic problem (3.8) derived from the energy of Eq. (3.3), constitutes the general geodesic active-contour model we propose. The solution to the object detection problem is then given by the zero level set of the steady state ($u_t = 0$) of this flow. As described in Subsection 3.1.3, it is possible to choose $c = 0$ (no constant velocity), and the model still converges (in a slower motion). The advantage is that we have obtained a model with fewer parameters.

An important issue of the proposed model is the selection of the stopping function g in the model. According to the results in Ref. [63] and in Subsection 3.1.2, in the case of ideal edges the described approach of object detection by means of geodesic computation is independent of the choice of g, as long as g is a positive strictly decreasing function and $g(r) \to 0$ as $r \to \infty$. Because real images do not contain ideal edges, g must be specified. In the following experimental results we use g as used in Refs. [63 and 250–252], given by Eq. (3.17). This is a very simple edge detector, similar to the ones used in the preceding active-contours models, both curve evolution and energy-based ones, and suffers from the well-known problems of gradient-based edge detectors. In spite of this, and as we can appreciate from the following examples, accurate results are obtained with this simple function. The use of better edge detectors, such as for example energy ones [143, 311], will immediately improve the results. The use of different metrics to define edges can be further investigated, and these metrics can be incorporated in the geodesic model. As pointed out before, the results here described and the described approach of object segmentation by means of geodesic computation are independent of the specific selection of g.

Remark: In the derivations above we have followed the formulations in Refs. [64 and 65]. The obtained geodesic equation, as well as its 3D extension (see next section), was independently proposed by Kichenassamy et al. [201, 202, 417, 418] and Shah [363], based on a different initial approaches. In the case of Ref. [363], g is obtained from an elaborated segmentation procedure obtained from the approach of Mumford and Shah [265]. In Ref. [202] Kichenassamy et al. give a number of important theoretical results as well. The formal mathematical connections between energy models and curve evolution models was done in Ref. [65]; before this work the two approaches were considered independent. 3D examples are given in Ref. [410], in which equations similar to those presented in Subsection 3.1.2 are proposed. We obtain the equations in Subsection 3.1.2 by extending the flows in Refs. [63 and 250]. In Ref. [383], Tek and Kimia extend the models in Refs. [63 and 250], motivated by work reported in Refs. [204 and 206]. One of the key ideas, motivated by the shape theory of shocks developed by Kimia et al., is to perform multiple initializations while using the same equations as those in Refs. [63 and 250]. The possible advantages of this are reported in Ref. [406], which uses the same equations as those in Refs. [63 and 250] and not the new ones described in this chapter (and in Refs. [201 and 202]), also without showing its connection with classical snakes. A normalized version of expression (3.8) was derived in Ref. [145] from a different point of view, giving as well different flows for 2D active contours. See also [306] for a related curve evolution technique.

3.1.2. Existence, Uniqueness, Stability, and Consistency of the Geodesic Model

Before the experimental results are given, results are presented regarding the existence and uniqueness of the solution to Eq. (3.18). Based on the theory of viscosity solutions [105], the Euclidean heat flow and the geometric model of Eq. (3.15) are well defined for nonsmooth images as well [63, 90, 126]. Similar results are now presented for our model of Eq. (3.18). Note that, besides the work in Ref. [63], there is not much formal analysis for active-contour approaches in the literature. The results presented in this subsection, together with the results on numerical analysis of viscosity solutions, ensure the existence and the uniqueness of the solution of the geodesic active contours model.

Let us first recall from Chap. 1 the notion of viscosity solutions for this specific equation; see Ref. [105] for details. We rewrite Eq. (3.18) in the form

$$\begin{cases} \dfrac{\partial u}{\partial t} - g(\mathcal{X})a_{ij}(\nabla u)\partial_{ij}u - \nabla g \cdot \nabla u - cg(\mathcal{X})|\nabla u| = 0 \\ [t, \mathcal{X}) \in [0, \infty) \times \mathbb{R}^2 \\ u(0, \mathcal{X}) = u_0(\mathcal{X}), \end{cases} \tag{3.21}$$

where $a_{ij}(q) = \delta_{ij} - [(p_i, p_j)/|p|^2]$ if $p \neq 0$. We used in Eqs. (3.21) and we use below the usual notations $\partial_i = (\partial/\partial x_i)$ and $\partial_{ij} = [\partial^2/(\partial x_i \partial x_j)]$, together with the classical Einstein summation convention. The terms $g(\mathcal{X})$ and ∇g are assumed to be continuous.

Equation (3.21) should be solved in $D = [0, 1]^2$ with Neumann boundary conditions. To simplify the notation and as is usual in the literature, we extend the images by reflection to $\mathbb{R}^2$ and we look for solutions verifying $u(\mathcal{X} + 2h) = u(\mathcal{X})$ for all $\mathcal{X} \in \mathbb{R}^2$ and $h \in \mathbb{Z}^2$. The initial condition u_0 as well as the data $g(\mathcal{X})$ are taken to be extended to $\mathbb{R}^2$ with the same periodicity.

Let $u \in \mathrm{C}([0, T] \times \mathbb{R}^2)$ for some $T \in]0, \infty[$. We say that u is a viscosity subsolution of Eq. (3.21) if for any function $\phi \in \mathrm{C}(\mathbb{R} \times \mathbb{R}^2)$ and any local maxima $(t_0, \mathcal{X}_0) \in]0, T] \times \mathbb{R}^2$ of $u - \phi$ we have if $\nabla\phi(t_0, \mathcal{X}_0) \neq 0$, then

$$\frac{\partial \phi}{\partial t}(t_0, \mathcal{X}_0) - g(\mathcal{X}_0) a_{ij}[\nabla\phi(t_0, \mathcal{X}_0)]\partial_{ij}\phi(t_0, \mathcal{X}_0) \\ - \nabla g(\mathcal{X}_0) \cdot \nabla\phi(t_0, \mathcal{X}_0) - c g(\mathcal{X}_0)|\nabla\phi(t_0, \mathcal{X}_0)| \leq 0,$$

and if $\nabla\phi(t_0, \mathcal{X}_0) = 0$, then

$$\frac{\partial \phi}{\partial t}(t_0, \mathcal{X}_0) - g(\mathcal{X}_0) \limsup_{q \to 0} a_{ij}(q)\partial_{ij}\phi(t_0, \mathcal{X}_0) \leq 0,$$

$$u(0, \mathcal{X}) \leq u_0(\mathcal{X}).$$

In the same way, we define a viscosity supersolution by changing in the expressions above "local maxima" to "local minima", "$\leq$" to "$\geq$", and "lim sup" to "lim inf." A viscosity solution is a function that is both a viscosity subsolution and a viscosity supersolution. The viscosity solution is one of the most popular frameworks for the analysis of nonsmooth solutions of PDEs, having physical relevance as well. The viscosity solution coincides with the classical one if this exists.

With the notion of viscosity solutions, we can now present the following result regarding the geodesic model:

Theorem 3.3. *Let* $\mathrm{W}^{1,\infty}$ *denote the space of bounded Lipschitz functions in* $\mathbb{R}^2$. *Assume that* $g \geq 0$ *is such that* $\sup_{\mathcal{X}\in\mathbb{R}^2} |Dg^{1/2}(\mathcal{X})| < \infty$ *and* $\sup_{\mathcal{X}\in\mathbb{R}^2} |D^2 g(\mathcal{X})| < \infty$. *Let* $u_0 \in \mathrm{BUC}(\mathbb{R}^2) \cap W^{1,\infty}(\mathbb{R}^2)$. *(In the experimental results, the initial function* u_0 *will be the distance function, with* $u_0 = 0$ *at the boundary of the image.) Then*

1. *Equation (3.21) admits a unique viscosity solution*

$$u \in \mathrm{C}([0, \infty) \times \mathbb{R}^2) \cap L^\infty[0, T; W^{1,\infty}(\mathbb{R}^2)]$$

for all $T < \infty$. *Moreover, u satisfies*

$$\inf u_0 \leq u(t, \mathcal{X}) \leq \sup u_0.$$

2. *Let* $v \in \mathrm{C}([0, \infty) \times \mathbb{R}^2)$ *be the viscosity solution of Eq. (3.21) corresponding to the initial data* $v_0 \in \mathrm{C}(\mathbb{R}^2) \cap \mathrm{W}^{1,\infty}(\mathbb{R}^2)$. *Then*

$$\|u(t, \cdot) - v(t, \cdot)\|_\infty \leq \|u_0 - v_0\|_\infty$$

for all $t \geq 0$. *This shows that the unique solution is stable.*

The assumptions of Theorem 3.3 above are just technical. They imply that the smoothness of the coefficients of Eq. (3.21) is required for proving the result with the method in Refs. [6 and 63]. In particular, Lipschitz continuity in $\mathcal{X}$ is required. This implies a well-defined trajectory of the flow $\mathcal{X}_t = \nabla g(\mathcal{X})$, going to every point $\mathcal{X}_0 \in \mathbb{R}^2$, which is reasonable in our context.

The proof of this theorem follows the same steps of the corresponding proofs for the model of Eq. (3.15); see Ref. [63], Theorem 3.1, and we shall omit the details (see also Ref. [6]).

In Theorem 3.4, we recall results of the independence of the generalized evolution with respect to the embedding function u_0. Let Γ_0 be the initial active contour, oriented such that it contains the object. In this case the initial condition u_0 is selected to be the signed distance function, such that it is negative in the interior of Γ_0 and positive in the exterior. Then we have the following

Theorem 3.4 (Theorem 7.1 of Ref. [90]). *Let* $u_0 \in W^{1,\infty}(\mathbb{R}^2) \cap \mathrm{BUC}(\mathbb{R}^2)$. *Let* $u(t,x)$ *be the solution of the proposed geodesic evolution equation as in Theorem 3.3. Let* $\Gamma(t) := \{\mathcal{X} : u(t,\mathcal{X}) = 0\}$ *and* $\mathcal{D}(t) := \{\mathcal{X} : u(t,\mathcal{X}) < 0\}$. *Then* $[\Gamma(t), \mathcal{D}(t)]$ *are uniquely determined by* $[\Gamma(0), \mathcal{D}(0)]$.

This Theorem is adapted from Ref. [90], in which a slightly different formulation is given. The techniques there can be applied to the present model.

Let us make some further remarks on the proposed geodesic flows of Eqs. (3.14) and (3.18), as well as the previous geometric model of Eq. (3.15). First, note that these equations are invariant under increasing rearrangements of contrast (morphology invariant [5]). This means that $\Theta(u)$ is a viscosity solution of the flow if u is and $\Theta : \mathbb{R} \to \mathbb{R}$ is an increasing function. On the other hand, although Eq. (3.14) is also contrast invariant, i.e., invariant to the transformation $u \leftarrow -u$ (remember that u is the embedding function used by the level-set approach), Eqs. (3.15) and (3.18) are not due to the presence of the constant-velocity component $cg(I)|\nabla u|$. This has a double effect. First, for Eq. (3.14), it can be shown that the generalized evolution of the level sets $\Gamma(t)$ depends on only Γ_0 (see Ref. [126], Theorem 2.8), whereas for Eq. (3.18), the result in Theorem 3.3 is given. Second, for Eq. (3.14) we can show that if a smooth classical solution of the curve flow of Eq. (3.13) exists and is unique, then it coincides with the generalized solution obtained by means of the level-set representation of Eq. (3.14) during the lifetime of the classical solution (see Ref. [126], Theorem 6.1). The same result can then be proved for the general curve flow of Eq. (3.20) and its level-set representation in Eq. (3.18), although a more delicate proof, on the lines of Corollary 11.2 in Ref. [372], is required.

We have just presented results concerning the existence, uniqueness, and stability of the solution of the geodesic active contours. Moreover, we have observed that the evolution of the curve is independent of the embedding function, at least as long as we are precise in giving its interior and exterior regions. These results are presented within the viscosity framework. To conclude this section, in the case of a smooth ideal edge $\hat{\Gamma}$, we can prove that the generalized motion $\Gamma(t)$ converges to $\hat{\Gamma}$ as $t \to \infty$, making the proposed approach consistent.

Theorem 3.5. *Let* $\hat{\Gamma} = \{\mathcal{X} \in \mathbb{R}^2 : g(\mathcal{X}) = 0\}$ *be a simple Jordan curve of class* C^2 *and* $Dg(\mathcal{X}) = 0$ *in* $\hat{\Gamma}$. *Furthermore, assume that* $u_0 \in W^{1,\infty}(\mathbb{R}^2) \cap \mathrm{BUC}(\mathbb{R}^2)$ *is of class* C^2 *and such that the set* $\{\mathcal{X} \in \mathbb{R}^2 : u_0(\mathcal{X}) \leq 0\}$ *contains* $\hat{\Gamma}$ *and its interior. Let* $u(t, \mathcal{X})$ *be the solution of Eq. (3.18) and* $\Gamma(t) = \{\mathcal{X} \in \mathbb{R}^2 : u(t, \mathcal{X}) = 0\}$. *Then, if c, the constant component of the velocity, is sufficiently large,* $\Gamma(t) \to \hat{\Gamma}$ *as* $t \to \infty$ *in the Hausdorff distance.*

This theorem is proved below for the extension of the geodesic model for 3D object segmentation. In this theorem, we assumed c to be sufficiently large. A similar result can be proved for the basic geodesic model, that is, for $c = 0$, assuming that the maximal distance between $\hat{\Gamma}$ and the initial curve $\Gamma(0)$ is given and bounded (to avoid local minima).

3.1.3. Experimental Results

Some examples are presented of the proposed geodesic active-contour model of Eq. (3.18). In the numerical implementation of Eq. (3.18) we choose the central-difference approximation in space and the forward-difference approximation in time. This simple selection is possible because of the stable nature of the equation; however, when the coefficient c is taken to be of high value, more sophisticated approximations are required [294]. See the mentioned references for details on the numerics.

In the following figures, the original image is presented on the left and the one with the deforming contours is presented on the right. The deforming contour ($u = 0$) is represented by a green contour and the final one (the geodesic) by a red one. In the case of inward motion, the original curve surrounds all the objects. In the case of outward motion, it is any given curve in the interior of the object.

Figure 3.2 presents two wrenches with inward flow. Note that this is a difficult image, not only for the existence of two objects, separated by only a few pixels, but also for the existence of many artifacts, such as the shadows, which can lead the edge detection to a wrong solution. The geodesic model of Eq. (3.18) was applied to the image and, indeed, both objects are detected. The original connected curve splits in order to detect both objects. The geodesic contours also managed not to be stopped by the shadows (false contours), because of the stronger attraction force provided

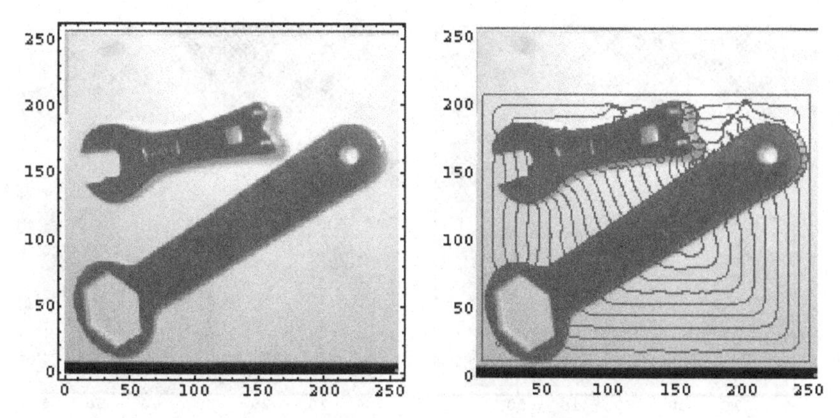

Fig. 3.2. Detecting two wrenches with the geodesic flow moving inward.

by the term $\nabla g \cdot \nabla u$ toward the real boundaries. Observe that the process of preferring the real edge over the shadow edge starts at their connection points, and the contour is pulled to the real edge, "like closing a zipper." The model was also run with $c = 0$, and practically the same results were obtained, with slower convergence.

Figure 3.3 presents an outward flow. The original curves are the two small circles, one inside each of the objects. Note that both curves deform simultaneously and independently. In the case of energy approaches, disjoint curves must be tracked to ensure that they contribute to different energy functionals. This tracking is not necessary in the model; it is not necessary to know how many disjoint deforming contours there are in the image. Note also here that the interior and the exterior boundaries are both detected. The

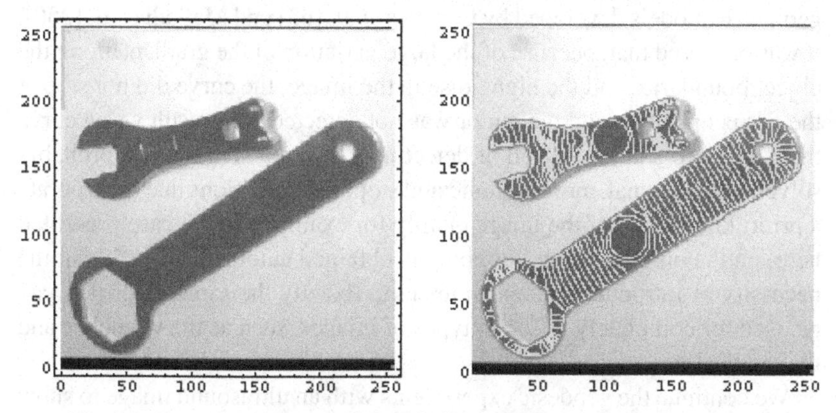

Fig. 3.3. Detecting two wrenches with the geodesic flow moving outward.

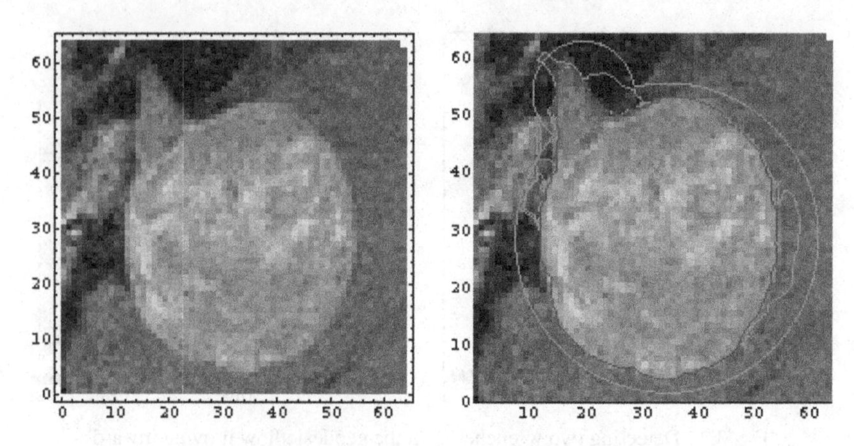

Fig. 3.4. Inward geodesic flow for tumor detection.

initial curves manage to split and detect all the contours in both objects. This splitting is automatic. We can also appreciate, in the lower left-hand corner, that the geodesic active contours split in a very narrow band (only a few pixels' width), managing to enter in very small regions.

Figure 3.4 presents another example of a medical image. The tumor in the image is an acoustics neurinoma and includes the triangular-shaped portion in the top left part. For this image, an inward deforming contour was used. The results are presented on the right, where the tumor portion is shown after zoom out for better presentation. Note that, because of the intrinsic subpixel accuracy of the algorithm, very accurate measurements, such as of the tumor area, can be computed. For comparison, the same image was also applied to the model without the new gradient term ($\nabla g \cdot \nabla u$), that is, the geometric models developed by Caselles et al. [63] and Malladi et al. [250]. It was observed that, because of the large variation of the gradient along the object boundaries and the high noise in the image, the curve did not stop at the correct position and the tumor was not detected. The result was a curve that shrunk to a point instead of detecting the tumor. This can be probably solved by additional, more complicated stopping conditions that incorporate a priori knowledge of the image quality for example. In the case presented here, on the other hand, the stopping is obtained automatically without the necessity of introducing new parameters. Exactly the same algorithm can be used for completely different types of images, such as the wrenches and the medical one.

We continue the geodesic experiments with an ultrasound image to show the flexibility of the approach. This is shown in Fig. 3.5, in which a human

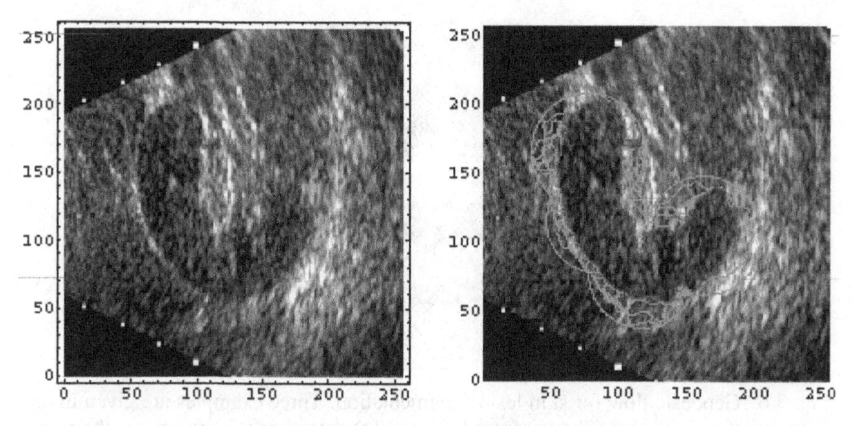

Fig. 3.5. Inward geodesic flow for ultrasound detection.

fetus is detected. In this case, the image was smoothed with a Gaussian-type kernel (two-three iterations of a 3×3 window filter are usually applied) before the detection was performed. This avoids possible local minima, and, together with the attraction force provided by the new term, allowed the detection of an object with gaps in its boundary. In general, gaps in the boundary (flat gradient) can be detected if they are of the order of magnitude of $1/(2c)$ (after smoothing). Note also that the initial curve is closer to the object to be detected to avoid further possible detection of false contours (local minima). Although this problem is significantly reduced by the new term incorporated into the geodesic model, is not completely solved. In many applications, such as interactive segmentation of medical data, this is not a problem, as the user can provide a rough initial contour like the one in Fig. 3.5 (or remove false contours). This problem might be automatically solved if better stopping function g is used, as explained in the previous sections, or by higher values of c, the constant velocity, imitating the balloon force of Cohen et al. [96]. Another classical technique for avoiding some local minima is to solve the geodesic flow in a multiscale fashion. Starting from a contour surrounding the entire image, and a low resolution of it, the algorithm is applied. Then, its result (steady state) is used as the initial contour for the next-higher resolution, and the process continues up to the original resolution. Multiresolution can help as well to reduce the computational complexity [151].

Figure 3.6 shows results for the segmentation of skin lesions. Three examples are given in the first row. The second row shows the combination of the geodesic active contours with the continuous-scale morphology introduced in Chap. 2. The hairy image on the left is first preprocessed with

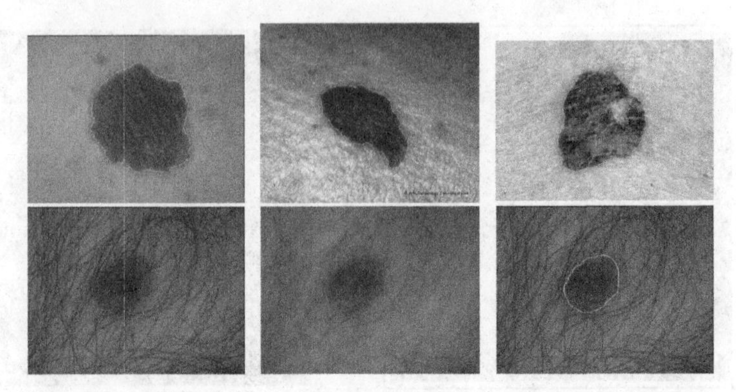

Fig. 3.6. Geodesic flow for skin lesion segmentation. Three examples are given in the first row; the second row shows (from left to right) the original hairy image, the result of hair removal by means of continuous-scale morphological PDEs, and the segmented lesion.

a directional morphology PDE to remove the hairs (middle figure), and the segmentation is performed on this filtered image, right.

3.1.4. Intermezzo

A geodesic formulation for active contours was presented. It was shown that a particular case of the classical energy snake or active-contour approach for boundary detection leads to finding a geodesic curve in a Riemannian space derived from the image content. This proposes a scheme for object boundary detection based on geodesic or minimal-path computations. This approach also gives possible connections between classical energy-based deformable contours and geometric curve evolution ones, improving over both of them. The result is an active-contour approach that is intrinsic (geometric) and topology independent. Results were also presented regarding existence, uniqueness, stability, and consistency of the solution obtained by the proposed active contours.

Experiments for different kinds of images were presented. These experiments demonstrate the ability to detect several objects, as well as the ability to detect interior and exterior boundaries at the same time. The subpixel accuracy intrinsic to the algorithm allows performing accurate measurements after the object has been detected [342].

Beyond the particular result of object detection, we have just formulated a fundamental problem in image processing as the computation of a geodesic curve. In the rest of this chapter we will see how this geometric approach is useful in addressing additional image processing and computer vision problems.

3.2. Three-Dimensional Derivation

In this section we extend the results for 2D geodesic active contours to 3D object detection. The obtained geometric flow is based on geometric deformable surfaces. We show that the desired boundary is given by a minimal surface in a space defined by the image. Therefore segmentation is achieved by means of the computation of surfaces of minimal area, in which the area is defined in an image-dependent space. The obtained flow has the same advantages over other 3D deformable models such as the advantages of the geodesic active contours over previous 2D approaches. Although the formal mathematical connection between the classical 3D snakes and the novel minimal-area ones (both models are given below) cannot be obtained by following the same technique we used for the 2D case, the two models are deeply connected, being almost equivalent as well, as shown by G. Aubert and colleagues.

3.2.1. Classical Three-Dimensional Formulations

It is clear that the classical 2D snake energy-based method can be generalized to 3D data images, for which the boundaries of the objects are surfaces. This extension is known as the deformable surface model and was introduced by Terzopoulos et al. [387] for a 3D representation of objects and extended and used for a 3D segmentation by many others (see, for example, Refs. [96–98 and 238]). In the 3D case, a parameterized surface

$$\mathcal{S}(r, s) = [x(r, s), y(r, s), z(r, s)], \quad (r, s) \in [0, 1] \times [0, 1]$$

is considered, and the energy functional is given by

$$E(v) = \int_\Omega \left[\omega_{10} \left| \frac{\partial \mathcal{S}}{\partial r} \right|^2 + \omega_{01} \left| \frac{\partial \mathcal{S}}{\partial s} \right|^2 + 2\omega_{11} \left| \frac{\partial^2 \mathcal{S}}{\partial r \partial s} \right|^2 \right] drds + \int_\Omega \left[\omega_{20} \left| \frac{\partial^2 \mathcal{S}}{\partial r^2} \right|^2 + \omega_{02} \left| \frac{\partial^2 \mathcal{S}}{\partial s^2} \right|^2 + P[v(r, s)] \right] drds, \quad (3.22)$$

where $P := -\|\nabla I\|^2$ or any related decreasing function of the gradient. As in the 2D case, the algorithm starts with an initial surface $\mathcal{S}_0$, generally near the desired 3D boundary $\mathcal{O}$, and tries to move $\mathcal{S}_0$ toward a local minimum of E.

The model of Eq. (3.15) can easily be extended to 3D object detection as well. Let us consider for each $t \geq 0$ a 3D function $u(t, .) : \mathbb{R}^3 \rightarrow \mathbb{R}$ and denote by $\mathcal{S}$ its level sets (3D surfaces). Then the 3D geometric deformable model is given by

$$\begin{aligned} \frac{\partial u}{\partial t} &= g(I)|\nabla u| \operatorname{div}\left(\frac{\nabla u}{|\nabla u|}\right) + \nu g(I)|\nabla u| \\ &= g(I)(\nu + \mathbf{H})|\nabla u|, \end{aligned} \tag{3.23}$$

where now $\mathbf{H}$ is the sum of the two principal curvatures of the level sets $\mathcal{S}$, that is, twice its mean curvature. This model has the same concepts as those of the 2D one, and is composed of the following three elements:

1. A smoothing term. In the case of Eq. (3.23), this smoothing term $\mathbf{H}$ is twice the mean curvature, but other more efficient smoothing velocities as those proposed in Refs. [5, 74, and 286] can be used. (Recall that although curvature flows smooth curves in two dimensions, in order to find a similar flow in three dimensions, we must go to higher-order flows, including second-order derivatives of the Gaussian and mean curvatures.)
2. A constant balloon-type force ($\nu|\nabla u|$).
3. A stopping factor [$g(I)$]. This is a function of the gradient or other 3D edge detectors [430].

3.2.2. Three-Dimensional Deformable Models as Minimal Surfaces

In Subsection 3.2.1 a model for 2D object detection was presented based on the computation of geodesics in a given Riemannian space. This means that we are computing paths or curves of minimal (weighted) length. This is extended to 3D surfaces by the computation of surfaces of minimal area, where area is defined in an image-dependent space. In the 2D case, length is given by Eq. (3.9) and the new length that allows us to perform object detection is given by Eq. (3.11). In the case of surfaces, Eq. (3.9) is replaced with area

$$A := \iint da, \tag{3.24}$$

and Eq. (3.11) is replaced with weighted area

$$A_R := \iint g(I)\mathrm{d}a, \tag{3.25}$$

where $\mathrm{d}a$ is the (Euclidean) element of area. The weighted area A_R is a natural 3D extension to the weighted 2D arc length L_R and is thereby the natural analog to the nonintrinsic energy minimization (3.22). Surfaces minimizing Eq. (3.24) are denoted as minimal surfaces [296]. In the same way, we denote as minimal surfaces those surfaces that minimize Eq. (3.25). The difference between A and A_R is like the difference between L and L_R. In A, the element of area is given by the classical element $\mathrm{d}a$ in Euclidean space, whereas in A_R, the area element $\mathrm{d}a_r$ is given by $g(I)\mathrm{d}a$. The basic element of the deformable model is given by minimizing Eq. (3.25) by means of an evolution equation obtained from its Euler–Lagrange. Given the definition of A_R above, the computations are straightforward and are a direct extension of the geodesic active-contour computation as presented in Subsection 3.2.1 and in Ref. [65]. Only the basic characteristics of this flow are pointed out below.

The Euler–Lagrange of A is given by the mean curvature $\mathbf{H}$, and we obtain a curvature flow

$$\frac{\partial \mathcal{S}}{\partial t} = \mathbf{H}\vec{\mathcal{N}}, \tag{3.26}$$

where $\mathcal{S}$ is the 3D surface and $\vec{\mathcal{N}}$ is its inner unit normal. In level-set notation, if for each $t \geq 0$, the evolving surface $\mathcal{S}$ is the zero level set of a function $u(t, .) : \mathbb{R}^3 \rightarrow \mathbb{R}$, which we take as negative inside $\mathcal{S}$ and positive outside, we obtain

$$u_t = |\nabla u| \mathrm{div}\left(\frac{\nabla u}{|\nabla u|}\right) = \mathbf{H}|\nabla u|. \tag{3.27}$$

Therefore the mean curvature motion provides a flow that computes (local) minimal surfaces [93].

Computing now the Euler–Lagrange of A_R, we get

$$\mathcal{S}_t = (g\mathbf{H} - \nabla g \cdot \vec{\mathcal{N}})\vec{\mathcal{N}}. \tag{3.28}$$

This is the basic weighted minimal-surface flow. Taking a level-set representation in analogy with Eq. (3.14), we find that the steepest-descent method to minimize Eq. (3.25) gives

$$\frac{\partial u}{\partial t} = |\nabla u| \mathrm{div}\left[g(I)\frac{\nabla u}{|\nabla u|}\right] = g(I)|\nabla u| \mathrm{div}\left(\frac{\nabla u}{|\nabla u|}\right) + \nabla g(I) \cdot \nabla u. \tag{3.29}$$

However, now u is a four-dimensional (4D) function with $\mathcal{S}$ as its 3D zero level set. We note again that, compared with previous surface evolution approaches for 3D object detection, the minimal-surface model includes a new term, $\nabla g \cdot \nabla u$.

As in the 2D case, we can add a constant force to the minimization problem (it is easy to show that this force minimizes the enclosed weighted volume $\int g dx dy dz$), obtaining the general minimal-surface model for object detection:

$$\frac{\partial u}{\partial t} = |\nabla u| \mathrm{div}\left[g(I)\frac{\nabla u}{|\nabla u|}\right] + \nu g(I)|\nabla u|. \tag{3.30}$$

This is the flow we will further analyze and use for 3D object detection. It has the same properties and geometric characteristics as the geodesic active contours, leading to accurate numerical implementations and topology-free object segmentation.

Estimation of the Constant Velocity ν. One of the critical issues of the model presented above is to estimate ν (see Subsection 3.2.3 for the related theoretical results). A possible way of doing this is now presented. (Another technique for estimating ν can be obtained from the results in Ref. [428].)

In Ref. [100] it was shown that the maximum curvature magnitude along the geodesics that minimize L_R is given by

$$\max\{|\kappa|\} = \sup_{\tau\in[0,1]} \left\{ \frac{|\nabla g[\mathcal{C}(\tau)] \cdot \vec{\mathcal{N}}|}{g[\mathcal{C}(\tau)]} \right\}.$$

This result is obtained directly from Euler–Lagrange equation (3.11). It leads to an upper bound over the maximum curvature magnitude along the geodesics, given by

$$|\kappa| \leq \sup_{p\in[0,a]\times[0,b]} \left\{ \frac{|\nabla g[I(p)]|}{g[I(p)]} \right\},$$

which does not require the geodesic itself for limiting the curvature values. In Ref. [100] this bound helped in the construction of different potential functions.

A straightforward generalization of this result to the 3D model yields the bound over the mean curvature $\mathbf{H}$. From the equations above, it is clear that for a steady state (i.e., $\mathcal{S}_t = 0$) the mean curvature along the surface $\mathcal{S}$ is given by

$$\mathbf{H} = \frac{\nabla g \cdot \vec{\mathcal{N}}}{g} - \nu.$$

We readily obtain the following upper bound for the mean curvature magnitude along the final surface

$$|\mathbf{H}| \leq \sup\left\{\frac{|\nabla g|}{g}\right\} + |\nu|,$$

where the sup operation is taken over all the 3D domain. The above bound gives an estimation of the allowed gaps in the edges of the object to be detected as a function of ν. A pure gap is defined as a part of the object boundary at which, for some reason, g =constant$\neq 0$ at a large enough neighborhood. At these locations $|\mathbf{H}| = |\nu|$. Therefore pure gaps of radius larger than $1/\nu$ will cause the propagating surface to penetrate into the segmented object. It is also clear that $\nu = 0$ allows the detection of gaps of any given size, and the boundary at such places will be detected as the minimal surface gluing the gaps' boundaries.

3.2.3. Existence and Uniqueness Results for the Minimal-Surface Model

As shown in Subsection 3.2.2, the 3D object-detection model is given by

$$u_t = g(I)|\nabla u|\left[\operatorname{div}\left(\frac{\nabla u}{|\nabla u|}\right) + \nu\right] + \nabla g \cdot \nabla u \;\; (t, x) \in [0, \infty) \times \mathbb{R}^3, \tag{3.31}$$

with initial condition $u(0, x) = u_0(x)$, and (a specific g function is selected for the analysis, whereas, as explained before, the model is general)

$$g(I) = \frac{1}{1 + |\nabla G_\sigma * I|^2}$$

where $\nu > 0$ represents a constant force in the normal direction to the level sets of u, I is the original image where we are looking for the boundary of an object O, and $G_\sigma * I$ is the regularized version of it by convolution with a Gaussian G_σ of variance σ (once again, the Gaussian is just an example of smoothing operator used for the analysis). The initial condition is $u_0(x)$, which is usually taken as a regularized version of $1 - \chi_C$, where χ_C is the characteristic function of a set C containing O in the case of outer deforming models (surfaces evolving toward the objects' boundary ∂O from the exterior of O) or a regularized version of χ_C where C is a set in the interior of O in the case of inner deforming models (surfaces evolving toward ∂O, starting from the inner side of O). Although only 2D outer snakes were considered in Ref. [63], here we also consider the inner snakes, as it seems natural for some applications [65].

Model (3.31) should be solved in $R = [0, 1]^3$ with Neumann boundary conditions. To simplify the presentation and as is usually done in the literature, we extend the images by reflection to $\mathbb{R}^3$ and we look for solutions of model (3.31) that are periodic, i.e., that satisfy $u(t, x + 2h) = u(t, x)$ for all $x \in \mathbb{R}^3$ and $h \in \mathbb{Z}$. The initial condition $u_0(x)$ and $g(x)$ are extended to $\mathbb{R}^3$ with the same periodicity as u.

We can prove existence and uniqueness results for Eq. (3.31) by using once again the theory of viscosity solutions. First we rewrite Eq. (3.31) in the form

$$\frac{\partial u}{\partial t} - g(x)a_{ij}(\nabla u)\partial_{ij}u - \nu g(x)|\nabla u| - \nabla g \cdot \nabla u = 0,$$

$$(t, x) \in [0, \infty) \times \mathbb{R}^3, \qquad (3.32)$$

where $a_{ij}(p) = \delta_{ij} - [(p_i p_j)/|p|^2]$ if $p \neq 0$. We use the usual notations $\partial_i u = (\partial u/\partial x_i)$, $\partial_{ij} u = [\partial^2 u/(\partial x_i \partial x_j)]$, and the classical Einstein summation convention in Eq. (3.32) and in all that follows.

Let us recall the definition of viscosity solutions. Let $u \in C([0, T] \times \mathbb{R}^3)$ for some $T \in (0, \infty)$. We say that u is a viscosity subsolution of Eq. (3.31) if for any function $\phi \in C^2(\mathbb{R} \times \mathbb{R}^3)$ and any local maximum $(t_0, x_0) \in (0, T] \times \mathbb{R}^3$ of $u - \phi$ we have the following: if $\nabla\phi(t_0, x_0) \neq 0$, then

$$\frac{\partial \phi}{\partial t}(t_0, x_0) - g(x_0)a_{ij}[\nabla\phi(t_0, x_0)]\partial_{ij}\phi(t_0, x_0) - \nu g(x)|\nabla\phi(t_0, x_0)|$$
$$- \nabla g(x_0) \cdot \nabla\phi(t_0, x_0) \leq 0$$

and if $\nabla\phi(t_0, x_0) = 0$, then

$$\frac{\partial \phi}{\partial t}(t_0, x_0) - g(x_0) \limsup_{p\to 0} a_{ij}(p)\partial_{ij}\phi(t_0, x_0) \leq 0$$

and $u(0, x) \leq u_0(x)$ for all $x \in \mathbb{R}^3$. In the same way we define the notion of the viscosity supersolution's changing "local maximum" to "local minimum", "≤ 0" to "≥ 0" and "lim sup" to "lim inf" in the expressions above. A viscosity solution is a function that is a viscosity subsolution and a viscosity supersolution.

The existence result in Refs. [63 and 65] can be easily adapted to the 3D case and we recall them without proof.

Theorem 3.6. *Let $W^{1,\infty}$ denote the space of bounded Lipschitz functions in $\mathbb{R}^3$. Assume that $g \geq 0$ is such that $\sup\{|\nabla g^{\frac{1}{2}}(x)| : x \in \mathbb{R}^3\} < \infty$ and $\sup\{|\partial_{ij} g(x)| : x \in \mathbb{R}^3, i, j = \{1, 2, 3\}\} < \infty$. Let $u_0, v_0 \in C(\mathbb{R}^3) \cap W^{1,\infty}(\mathbb{R}^3)$. Then*

1. *Equation (3.31) admits a unique viscosity solution $u \in C([0, \infty) \times \mathbb{R}^3) \cap L^\infty(0, T; W^{1,\infty}(\mathbb{R}^3))$ for all $T < \infty$. Moreover, it satisfies*
$$\inf_{\mathbb{R}^3} u_0 \leq u(t, x) \leq \sup_{\mathbb{R}^3} u_0.$$
2. *Let $v \in C([0, \infty) \times \mathbb{R}^3)$ be the viscosity solution of (3.31) with initial data v_0. Then for all $T \in [0, \infty)$ we have*
$$\sup_{0 \leq t \leq T} \|u(t, x) - v(t, x)\|_{L^\infty(\mathbb{R}^3)} \leq \| u_0(x) - v_0(x)\|_{L^\infty(\mathbb{R}^3)},$$
which means that the solution is stable.

The assumptions of Theorem 3.6 are just technical. They imply the smoothness of the coefficients of Eq. (3.32) required for proving the result with the method given in Refs. [6 and 63]. In particular, Lipschitz continuity in x is required. This implies a well-defined trajectory of the flow $X_t = \nabla g(X)$, passing through any point $X_0 \in \mathbb{R}^3$, which is a reasonable assumption in our context. The proof of this theorem follows the same steps of the corresponding proofs for the model of Eq. (3.15) (see Ref. [63], Theorem 3.1), and we shall omit the details (see also Ref. [6]).

In Theorem 3.7 we recall a result on the independence of the generalized evolution with respect to the embedding function u_0.

Theorem 3.7. *Let $u_0 \in W^{1,\infty}(\mathbb{R}^3) \cap \mathrm{BUC}(\mathbb{R}^3)$. Let $u(t, x)$ be the solution of Eq. (3.32) as in Theorem 3.6. Let $\Gamma_t := \{x : u(t, x) = 0\}$ and $D_t := \{x : u(t, x) \geq 0\}$. Then (Γ_t, D_t) are uniquely determined by (Γ_0, D_0).*

This theorem is adapted from Ref. [90], in which a slightly different formulation is given. The techniques there can be applied to the present model.

As in the 2D case, let us present again some remarks on the proposed flows of Eqs. (3.29) and (3.30), as well as the previous geometric model of Eq. (3.15). First note that these equations are invariant under increasing rearrangements of contrast (morphology invariance [5]). This means that

$\Theta(u)$ is a viscosity solution of the flow if u and $\Theta : \mathbb{R} \to \mathbb{R}$ are increasing functions. On the other hand, although Eq. (3.29) is also contrast invariant, i.e., invariant to the transformation $u \leftarrow -u$, Eqs. (3.15) and (3.30) are not, because of the presence of the constant-velocity term $\nu g(I)|\nabla u|$. This has a double effect. First, for Eq. (3.29), it can be shown that the generalized evolution of the level sets Γ_t depends on only Γ_0 (Ref. [126], Theorem 2.8), whereas for Eq. (3.30), the result in Theorem 3.7 is given. Second, for Eq. (3.29) we can show that if a smooth classical solution of the weighted minimal surface flow with $\nu = 0$ exists and is unique, then it coincides with the generalized solution obtained by means of the level-set representation of Eq. (3.29) during the lifetime of existence of the classical solution (Ref. [126], Theorem 6.1). The same result can then be proved for the general minimal-surface flow ($\nu \neq 0$) and its level-set representation of Eq. (3.30), although a more delicate proof, on the lines of Corollary 11.2 in Ref. [372], is required.

The next general result (see also Lemma 3.2 below) will be needed in Subsection 3.2.4 to study the asymptotic behavior of the minimal-surface model of Eq. (3.31). Because the proof is an easy adaptation of the one in Ref. [90] (Theorem 3.2), we shall omit it as well.

Lemma 3.1. *Assume that* $S = \{x \in [0, 1]^3 : g(x) = 0\}$ *is a smooth compact surface,* $g \geq 0$*, and* $\nabla g(x) = 0$ *for all* $x \in S$*, and assume that* $u_0(x) \in W^{1,\infty}(\mathbb{R}^3)$ *is periodic with the fundamental domain* $[0, 1]^3$ *vanishing in an open neighborhood of* S*. Let* $u(t, x)$ *be the viscosity solution of the minimal surfaces flow. Then*

$$u(t, x) = 0 \qquad \forall x \in S, \ \ \forall t \geq 0.$$

3.2.4. Correctness of the Geometric Minimal-Surface Model

By correctness we mean the consistency of the results with the initial purpose of object detection in an ideal case. A smooth surface in an ideal image with no noise should be recovered by the model. In this section we deal with this point.

To study the asymptotic behavior of the equation

$$u_t = g(x)|\nabla u| \left[\operatorname{div}\left(\frac{\nabla u}{|\nabla u|}\right) + \nu \right] + \nabla g \cdot \nabla u \ (t, x) \in [0, \infty) \times \mathbb{R}^3 \tag{3.33}$$

with initial condition

$$u(0, x) = u_0(x), \qquad \forall x \in \mathbb{R}^3,$$

we assume that $S = \{x \in [0, 1]^3 : g(x) = 0\}$ is a compact surface of class C^2. S divides the cube $[0, 1]^3$ into two regions: the interior region and the exterior region of S. Denote these regions by $I(S)$ and $E(S)$, respectively. Observe that the interior $I(S)$ may have several connected components and $I(S)$ describes all of them together. The initial datum u_0 will be always taken in $C^2(\mathbb{R}^3)$ periodic with fundamental domain $[0, 1]^3$ and vanishing in an open neighborhood of $S \cup I(S)$. Moreover, we take $u_0(x)$ such that its level sets have uniformly bounded curvatures. Let $u(t, x)$ be the unique viscosity solution of Eq. (3.33) given by Theorem 3.6 above. We follow the evolution of the set $G(t) = \{x \in [0, 1]^3 : u(t, x) = 0\}$ whose boundary $S(t)$ we are interested in.

Before going into the details, let us recall some elementary notions of differential geometry required below. A surface of genus p is a surface we obtain by removing the interiors of $2p$ disjoint disks from the sphere S^2 and by attaching p disjoint cylinders to their boundaries. We define the Euler–Poincaré characteristic of a surface S as $\chi(S) = 2 - 2p$, where p is its genus. We say that a surface S such that $\chi(S) = 2 - 2p$ is unknotted if every diffeomorphism from S to a standard p torus can be extended to a diffeomorphism of $\mathbb{R}^3$.

We can assume that $g \geq 0, \ \ \forall x \in [0, 1]^3$ and $\nabla g = 0, \ \ \forall x \in S$. Then, for some function $h \geq 0$, we have $g(x) = h(x)^2$ and Eq. (3.33) in the form

$$u_t = h(x)\left\{h(x)|\nabla u|\left[\operatorname{div}\left(\frac{\nabla u}{|\nabla u|}\right) + \nu\right] + 2\nabla h \cdot \nabla u\right\}. \tag{3.34}$$

With this formulation, $S = \{x \in [0, 1]^3 : h(x) = 0\}$.

In this subsection we are going to prove the following theorems.

Theorem 3.8. *Assume that $S = \{x \in [0, 1]^3 : g(x) = 0\}$ is diffeomorphic to a sphere, i.e. $\chi(S) = 2$. If the constant ν is sufficiently large, then $S(t)$ converges to S in the Hausdorff distance as $t \to \infty$.*

Theorem 3.9. *Assume that $S = \{x \in [0, 1]^3 : g(x) = 0\}$ is diffeomorphic to a p torus, i.e., $\chi(S) = 2 - 2p$ and is unknotted. If ν is sufficiently large, then $S(t)$ converges to S in the Hausdorff distance as $t \to \infty$.*

Theorem 3.10. *Assume that $S = \{x \in [0, 1]^3 : g(x) = 0\}$ is the knotted surface in Subsection 3.2.5. If ν is sufficiently large, then $S(t)$ converges to S in the Hausdorff distance as $t \to \infty$.*

Proof of Theorem 3.8: The proof is based on Lemmas 3.2 and 3.3 below.

Lemma 3.2. $I(S) \subset \{x \in [0, 1]^3 : u(t, x) = 0\}$ *for all* $t > 0$.

Proof: In Theorem 3.1 we proved that $u(t, x) = 0$ for all $x \in S$ and all $t > 0$. Consider the problem

$$z_t = g(x)|\nabla z|\left[\operatorname{div}\left(\frac{\nabla z}{|\nabla z|}\right) + \nu\right] + \nabla g \cdot \nabla u, \quad (t, x) \in [0, \infty) \times I(S),$$
$$z(0, x) = u_0(x), \quad x \in I(S), \tag{3.35}$$
$$z(t, x) = 0, \quad x \in S, \quad t \geq 0.$$

We know that $z(t, x) = 0$ and $z(t, x) = u(t, x)$ are two viscosity solutions of Eqs. (3.35). Because there is uniqueness of viscosity solutions of this problem, it follows that $u(t, x) = 0$ for all $(t, x) \in [0, \infty) \times I(S)$. □

Remark: Lemma 3.2 is also true if S is as in Theorem 3.9.

Lemma 3.3. *Assume that* $S = \{x \in [0, 1]^3 : g(x) = 0\}$ *is diffeomorphic to a sphere, i.e.,* $\chi(S) = 2$. *If the constant* ν *is sufficiently large, for any* $\eta > 0$, *there exists some* $T_\eta > 0$ *such that*

$$G(t) \subset \{x \in [0, 1]^3 : \mathrm{d}[x, I(S)] < 2\eta\}$$

for all $t > T_\eta$.

Proof: Essentially the proof of this lemma consists of constructing a subsolution of Eq. (3.33) that becomes strictly positive as $t \to \infty$. If S is a convex C^2 surface, this is not a difficult task. In this case the distance function to S, $d(x) = d(x, S), x \in E(S)$, is of class C^2 in $E(S)$. This function is the tool to construct the desired subsolution. In the general case S need not be convex, the proof is a bit more technical, and we need a geometric construction.

Let S_1 be a compact surface of class C^2 contained in $\{x \in [0, 1]^3 : u_0(x) > 0\} \cap E(S)$. Consider $E = \{x \in [0, 1]^3 : x \in I(S_1) \cap E(S)\}$. E is diffeomorphic to the closed annulus $A = \{x \in \mathbb{R}^3 : r' \leq |x| \leq r''\}$. There is a C^2 diffeomorphism ϕ between A and E. Consider the family ζ_r of surfaces $\zeta_r(\theta, \varphi) = (r \cos\theta \cos\varphi, r \cos\theta \sin\varphi, r \sin\theta)$ with $r' \leq r \leq r''$, $-\frac{\pi}{2} < \theta \leq \frac{\pi}{2}$, $0 \leq \varphi < 2\pi$. This family is mapped by ϕ into a family of surfaces $S(r)$ in E of class C^2, i.e., $S(r) = \phi \circ \zeta_r$. Without loss of generality we may suppose that $\Gamma(r') = S$ and $\Gamma(r'') = S_1$. Because the surfaces $\zeta_r, r' \leq r \leq r''$, have uniformly bounded curvatures it follows that the family of surfaces $S(r)$, $r' \leq r \leq r''$, have uniformly bounded curvatures.

We may choose $\rho > 0$ such that we have $\nabla h \cdot \nabla d(x, S) > 0$ on $E(S) \cap [S + B(0, \rho)]$.

For any $\eta > 0$, we can take $n = n(\eta)$ and $r'' = r_1 > r_2 > \cdots > r_n$ as sufficiently close to each other and r_n near to r' such that the family of surfaces $S_i = \phi \circ \zeta_{r_i}$ satisfies the following:

1. $S_i \in C^2$ with interior region $I(S_i)$ and exterior region $E(S_i)$, $i = 1, \ldots, n$, and with $S_i \subset I(S_{i-1})$, $i = 2, \ldots, n$.
2. $S_1 \subset \{x \in [0, 1]^3 : u_0(x) > 0\}$ and $S_n \subset \{x \in E(S) : 0 < d(x, S) < \eta\}$.
3. For each $x \in S_i$, let $k_i^j(x)$, $j = 1, 2$, be the principal curvatures of S_i at the point x. Let $K_i = \max\{|k_i^j(x)| : x \in S_i,\ j = 1, 2\}$. We suppose (to ensure regularity) that $S_i \subset \{x \in E(S_{i+1}) : d(x, S_{i+1}) < [1/(2K_{i+1})]\}$
4. $\sup\{K_i, i = 1, \ldots, n\} \leq M$ with a constant M independent of η.

From these surfaces we construct another family S_i^* such that $S_1^* = S_1$ and for each $i = 2, \ldots, n-1$, we let S_i^* be a regular surface contained in $\{x \in E(S_i) \cap I(S_{i-1}) : \mathrm{d}(x, S_i) < [1/(4K_{i+1})]\}$. Finally, let S_n^* be a smooth surface contained in $\{x \in E(S_n) \cap I(S_{n-1}) : \mathrm{d}(x, S) < 2\eta\}$. Each surface S_i^* is in a neighborhood of S_{i+1} of radius $[3/(4K_{i+1})]$.

Let R_i be the region between the surfaces S_{i-1}^* and S_i, $i = 2, \ldots, n$ (see Fig. 3.7). Let $d_i(x) = \mathrm{d}(x, S_i)$ for $x \in R_i$, and let $C_i > 0$ be a constant such that $d_i(x) \leq C_i h(x)$ for $x \in R_i$, $i = 2, \ldots, n$. Note also that we are working in the unit cube and $d_i(x) \leq \sqrt{3}$. We may also suppose that there exists $n' < n$ such that for $i \geq n'$ we have $R_i \subset E(S) \cap [S + B(0, \rho)]$. Because we assume that $\nabla h \cdot \nabla d(x, S) > 0$, by continuity we may choose ρ small enough such that when $i \geq n'$ we have $\nabla h \cdot \nabla d_i \geq 0$.

Our purpose is to construct a family of subsolutions that becomes asymptotically positive as $t \to \infty$ on each R_i. By the last of our assumptions on u_0,

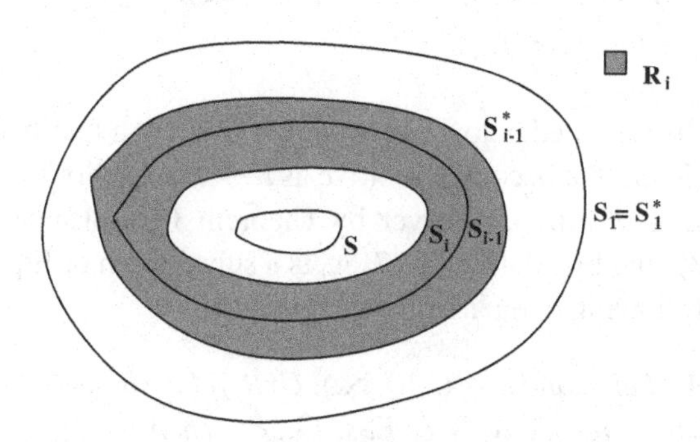

Fig. 3.7. Construction for Lemma 3.3.

we may choose a constant ν in Eq. (3.34) sufficiently largely independent of the geometric construction such that for some $\delta > 0$

$$\Delta d_i + \nu \geq \delta > 0 \tag{3.36}$$

in R_i $i = 2, \dots, n$ and u_0 is a subsolution of Eq. (3.34), i.e.,

$$|\nabla u_0| h(x) \left[\operatorname{div} \left(\frac{\nabla u_0}{|\nabla u_0|} \right) + \nu \right] + 2\nabla h \cdot \nabla u_0 \geq 0. \tag{3.37}$$

□

Remark: If $i \leq n' - 1$ we take ν large enough such that

$$\delta > \frac{2C}{\inf_{E(S)\cap[S+B(0,\rho)]} h}$$

where C is an upper bound for $|\nabla h \cdot \nabla d_i|$, $i = 2, \dots, n' - 1$. With this we obtain that, for some $\delta_1 > 0$,

$$h(x)(\Delta d_i + \nu) + 2\nabla h \cdot \nabla d_i \geq \delta_1 > 0, \qquad x \in R_i, \;\; i = 2, \cdots n' - 1.$$

If $i \geq n'$ we have $R_i \subset E(S) \cap [S + B(0, \rho)]$ and $\mathrm{D}h \cdot \mathrm{D}d_i > 0$. In this case, we also obtain that, for some $\delta_2 > 0$,

$$h(x)(\Delta d_i + \nu) + 2\nabla h \cdot \nabla d_i \geq \delta_2 > 0, \quad i \geq n'.$$

On each region R_i we consider the problem:

$$\begin{aligned}
& z_t = g(x)|\nabla z| \left[\operatorname{div} \left(\frac{\nabla z}{|\nabla z|} \right) + \nu \right] + \nabla g \cdot \nabla z \; (t, x) \in [T_{i-1}, \infty) \times R_i, \\
& z(t, x) = 0, \quad x \in S_i, \quad t \in [T_{i-1}, \infty), \\
& z(t, x) = u(t, x), \quad x \in S^*_{i-1}, \quad t \in [T_{i-1}, \infty), \\
& z(T_{i-1}, x) = 0, \quad x \in R_i,
\end{aligned} \tag{3.38}$$

where T_i will be specified below. We want to construct a family of subsolutions of Eqs. (3.38) that becomes positive as $t \to \infty$. Obviously $u(t, x)$, the viscosity solution of Eq. (3.33) given by Theorem 3.6, is a supersolution of problem (3.38), and by relation (3.37) u_0 is a subsolution of Eq. (3.33). We shall use the following comparison principle [105].

Theorem 3.11. *Let w and $v \in C\{[0, \infty), C(R_i)\}$ be a bounded subsolution and supersolution, respectively of Eqs. (3.38). Then $w(t, x) \leq v(t, x)$ for all $(t, x) \in [0, \infty) \times R_i$. The same comparison holds for Eq. (3.33).*

The preceding result implies that $u(t,x) \geq u_0(x)$ for all $(t,x) \in [0,\infty) \times \mathbb{R}^3$. In particular we have

$$\inf\{u(t,x) : t \in [0,\infty), x \in E(S_1^*)\} \geq \inf\{u_0(x), x \in E(S_1^*)\} > 0. \tag{3.39}$$

Assume we have shown that for all $j < i$ there exists some T_j such that

$$\inf\{u(t,x) : t \in [T_j,\infty),\ \ x \in E(S_j^*)\} \geq \beta_j > 0. \tag{3.40}$$

By relation (3.39) this is true when $i = 2$ with $T_1 = 0$.

For constructing the subsolution of Eqs. (3.38) in $[T_{i-1},\infty) \times R_i$, $i = 2,\ldots,n$, we change variables and take $T_{i-1} = 0$. Let

$$w_m(t,x) = f_m(t)d_i(x) + g_m(t), \tag{3.41}$$

where $(t,x) \in [0,\infty) \times R_i,\ \ m > 0$,

$$f_m(t) = \lambda\left[1 - \frac{1}{(1+t)^m}\right], \qquad \lambda > 0, \tag{3.42}$$

$$g_m(t) = g_m t, \quad t \in [0,t_m]; \quad g_m(t) = g_m t_m, \quad t > t_m, \tag{3.43}$$

where $g_m = -2m\lambda$, $t_m = (1 + \frac{mC_i}{\delta_0})^{\frac{1}{m}} - 1$ $[\delta_0 = \min(\delta_1,\delta_2)]$. With these functions we have Lemma 3.4.

Lemma 3.4. *For $\lambda > 0$ small enough and for all $m > 0$, w_m is a subsolution of Eqs. (3.38).*

Proof: It is clear by construction that $w_m(t,x) \leq 0$ for $(t,x) \in [T_{i-1},\infty) \times S_i$ and $w_m(T_{i-1},x) \leq 0$ for $x \in R_i$.

Using relation (3.40) and taking $\lambda > 0$ sufficiently small, we have $w_m(t,x) \leq u(t,x)$ for $(t,x) \in [T_{i-1},\infty) \times S_{i-1}^*$.

The function w_m has been chosen such that

$$\frac{\partial w_m}{\partial t} - h(x)\left\{h(x)|\nabla w_m|\left[\operatorname{div}\left(\frac{\nabla w_m}{|\nabla w_m|}\right) + \nu\right] + 2Dh \cdot \nabla w_m\right\} \leq 0 \tag{3.44}$$

in $[0,\infty) \times R_i$. Indeed, if $t < t_m$ we have

$$\begin{aligned} &f_m'(t)d_i(x) + g_m'(t) - h(x)f_m(t)[h(x)(\Delta d_i + \nu] + 2\nabla h \cdot \nabla d_i \\ &\quad \leq f_m'(t)d_i(x) + g_m \leq f_m'(0)d_i(x) + g_m \leq \sqrt{3}\lambda m + g_m \leq q0. \end{aligned}$$

If $t > t_m$ we have

$$\begin{aligned} &f_m'(t)d_i(x) - h(x)f_m(t)[h(x)(\Delta d_i + \nu] + 2\nabla h \cdot \nabla d_i \\ &\quad \leq f_m'(t)C_i h(x) - h(x)f_m'(t)\delta_0 \leq f_m'(t_m)C_i h(x) - h(x)f_m(t_m)\delta_0. \end{aligned}$$

Using the expressions for t_m we immediately see that the last expression is ≤ 0. This completes the proof of Lemma 3.4. □

Lemma 3.5. *$u(t, x) \geq w_m(t, x)$ for all $(t, x) \in [T_{i-1}, \infty) \times R_i$. Hence there exist $m_0 > 0$ and $T_i > 0$ such that*

$$u(t, x) \geq \inf\{w_m(t, x) : t \in [T_{i-1}, \infty),\ x \in R_i \cap E(S_i^*)\} > 0$$

for all $t \in [T_{i-1}, \infty)$, $x \in R_i \cap E(S_i^*)$ and all $m < m_0$.

Proof: The first inequality follows from Lemma 3.4 and Theorem 3.11. On the other hand, we observe that

$$w_m(t, x) \to \lambda d_i(x) + g_m t_m \qquad as \qquad t \to \infty. \tag{3.45}$$

Because t_m is bounded and $g_m \to 0$ as $m \to 0$, there exists some $m_0 > 0$ such that

$$\inf\{\lambda d_i(x) + g_m t_m : x \in R_i \cap E(S_i^*)\} > 0 \tag{3.46}$$

for all $0 < m < m_0$. The lemma follows from relations (3.45) and (3.46). □

Lemma 3.5 is a consequence of the last two lemmas.

Extension for $\nu = 0$: If we consider the model of Eq. (3.34) with $\nu = 0$, i.e.,

$$u_t = g(x)|\nabla u| \operatorname{div}\left(\frac{\nabla u}{|\nabla u|}\right) + \nabla g(x) \cdot \nabla u \tag{3.47}$$

with $g(x) = h^2(x)$ as above, a theorem similar to Theorem 3.8 holds if we take our initial surface sufficiently close to S. For that, we consider Eq. (3.47) on $[0, +\infty) \times V$, where V is a neighborhood of $I(S)$, together with initial and boundary conditions

$$\begin{aligned} u(0, x) &= u_0(x), \quad x \in V, \\ u(t, x) &= u_0(x), \quad t > 0, \quad x \in \partial V. \end{aligned}$$

Moreover, V should be taken sufficiently near to S, i.e., $V \subset I(S) + B(0, \rho)$ for ρ small enough so that

$$h(x)\Delta d(x) + 2\nabla h(x) \cdot \nabla d(x) \geq 0, \qquad x \in V - \overline{I(S)},$$

where $d(x) = d(x, S)$. In that case, we can adapt the ideas above to prove that $S(t) = \partial\{x : u(t, x) = 0\} \to S$ as $t \to \infty$.

Proof of Theorem 3.9:

Lemma 3.6. *Suppose that* $S = \{x \in [0, 1]^3 : g(x) = 0\}$ *is diffeomorphic to a* p *torus, i.e.,* $\chi(S) = 2 - 2p$ *and is unknotted. If* ν *is sufficiently large, for any* $\eta > 0$ *there exist some* $T_\eta > 0$ *such that*

$$G(t) \subset \{x \in [0, 1]^3 : \mathrm{d}[x, I(S)] < \eta\}, \qquad t > T_\eta.$$

Proof: The idea of the proof is to construct two surfaces Γ_1 and Γ_2 of class C^2 and diffeomorphic to a sphere, i.e., $\chi(\Gamma_i) = 2,\ i = 1, 2$, such that the boundary of $I(\Gamma_1) \cap I(\Gamma_2)$ is diffeomorphic to a p torus and near S. We prove that for a large t, $S(t)$ is near the boundary of $I(\Gamma_1) \cap I(\Gamma_2)$; therefore $S(t)$ will be near S. Let us first construct the surfaces Γ_1 and Γ_2.

Given $\eta > 0$, we choose Γ_1 and Γ_2 that satisfy the following.

1. Γ_i is a compact surface of class C^2, $\chi(\Gamma_i) = 2$ and $I(S) \subset I(\Gamma_i),\ i = 1, 2$.
2. $U_\eta := [I(\Gamma_1) + B(0, \frac{\eta}{2})] \cap [I(\Gamma_2) + B(0, \frac{\eta}{2})]$ is an open set whose boundary is a regular surface diffeomorphic to a p torus.
3. $U_\eta \subset I(S) + B(0, \eta)$.

Recall that the sum of two sets A, B is defined by $A + B = \{a + b : a \in A, b \in B\}$.

If S is a standard p torus, which we denote by T_p, it is easy to construct surfaces Γ_1 and Γ_2 that satisfy the conditions above. Because this is technically delicate, but easy to see graphically, it is illustrated in Fig. 3.8.

If S is not a standard p torus (see Ref. [56], Chap. 4), then it is diffeomorphic to a standard p torus and there exists a diffeomorphism $f : S \to T_p$. As S is unknotted this diffeomorphism can be extended to all $\mathbb{R}^3$ $F : \mathbb{R}^3 \to \mathbb{R}^3$ such that $F|_S = f$. In this case we construct the surfaces $\widetilde{\Gamma}_1$ and $\widetilde{\Gamma}_2$ as $\widetilde{\Gamma}_i = F^{-1}(\Gamma_i),\ i = 1, 2$. They also satisfy the above conditions.

By Theorem 3.8, we know that for ν sufficiently large there exists $T_\eta^i > 0$ such that $G(t) \subset \{x \in [0, 1]^3 : d[x, I(\Gamma_i)] < \frac{\eta}{2}\}$ for all $t > T_\eta^i$, $i = 1, 2$.

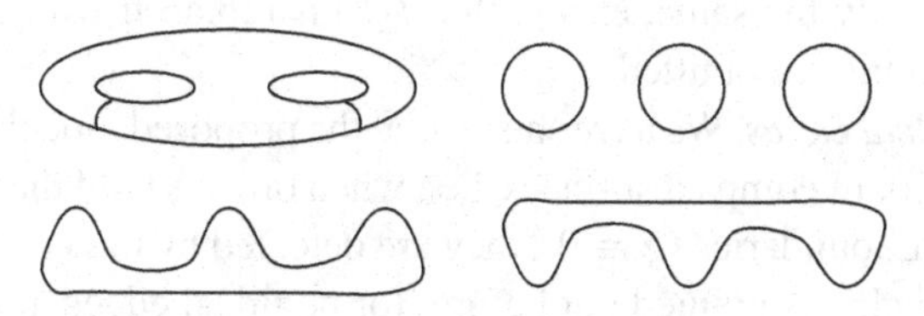

Fig. 3.8. Construction of Γ_1 and Γ_2.

Then we choose $T_\eta = \max(T_\eta^1, T_\eta^2)$ and we have

$$G(t) \subset \left[I(\Gamma_1) + B\left(0, \frac{\eta}{2}\right) \right] \cap \left[I(\Gamma_2) + B\left(0, \frac{\eta}{2}\right) \right], \quad t > T_\eta.$$

Hence $G(t) \subset U_\eta \subset I(S) + B(0, \eta)$ for all $t > T_\eta$. □

Remark: From the proof above, it is easy to see that if S_1 are S_2 are two surfaces for which the condition "for any $\eta > 0$ there exist some $T_\eta > 0$ such that $G(t) \subset \{x \in [0, 1]^3 : d[x, I(S_i)] < \eta\}, \quad i = 1, 2$, for $t > T_\eta$" is true, then it is also true for $S_1 \cap S_2$.

Using this remark, we can prove that if $S = \{x \in [0, 1]^3 : g(x) = 0\}$ is a knotted surface as in the examples, then for any $\eta > 0$ there exist some $T_\eta > 0$ such that

$$G(t) \subset \{x \in [0, 1]^3 : d[x, I(S)] < \eta\}, \quad t > T_\eta.$$

For the proof, we consider two unknotted surfaces diffeomorphic to a p torus such that their intersection is the knotted surface, and we use the remark above to conclude the proof. How do we construct such surfaces? First, observe that by attaching a finite number of thin cylinders (which can be done in a smooth way) we can get a smooth surface that is diffeomorphic to a p torus for some $p \in \mathbb{N}$. This can be done in two different ways Γ_1, Γ_2 such that the boundary of $I(\Gamma_1) \cap I(\Gamma_2)$ is diffeomorphic to the knotted surface of the examples and near to it. Of course such a result can be proven for all knotted surfaces for which the above strategy can be used.

Remark:

1. *Expanding motions.* If $u_0(x)$ is taken as

$$u_0(x) = \begin{cases} >0, & \text{if } x \in \overline{B(y, r)} \subset I(S) \\ 0, & \text{if } x \text{ in a neigborhood of } S \cup E(S) \end{cases},$$

i.e., a function vanishing in a neighborhood of $S \cup E(S)$ and such that its level sets have uniformly bounded curvatures, then we may recover a surface S by starting from its inner region. The proof of this result is similar to the proof of Theorem 3.8. Lemmas 3.2 and 3.3 are essentially the same, except that $I(S)$ is replaced with $E(S)$. Hence the details are omitted.

2. *Nonideal edges.* We have shown that the proposed model is consistent for smooth compact surfaces, i.e., when objects hold the basic definition of boundaries ($g \equiv 0$), they are detected by the minimal-surface approach. As pointed out before, for nonideal edges, previous algorithms [63, 250, 383] will fail, as $g \neq 0$ and the surface will not stop.

The new term ∇g creates a potential valley that attracts the surface, forcing the surface to stay there even if $g \neq 0$.

3. *Zero constant velocity.* Similar results can be proved when the constant velocity is equal to zero. In this case, the initial surface should be closer to the final one to avoid local minima. Again, the existence of the new term ∇g allows also the detection of nonconvex objects, a task that cannot be achieved without the constant velocity in previous models.

3.2.5. Experimental Results

Now some examples are presented of the minimal-surface deformable model. In the examples, the initialization is in general given by a surface (curve) surrounding all the possible objects in the scene. In the case of outward flows [65], a surface (curve) is initialized inside each object. Multiple initializations are performed in Refs. [250–252 and 383]. Although multiple initializations help in many cases, they may lead to false contours in noisy images. Therefore multiple initializations should in general be controlled (by rough detections of points inside the objects for example) or they should be followed by a validation step.

The first example of the minimal-surface deformable model is presented in Fig. 3.9. This object is composed of two tori, one inside the other (knotted

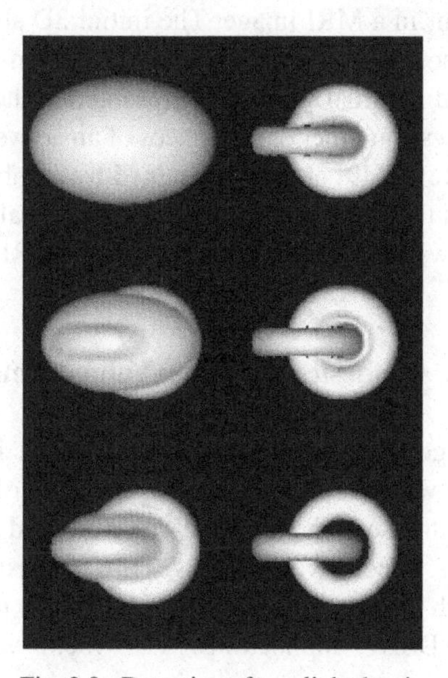

Fig. 3.9. Detection of two linked tori.

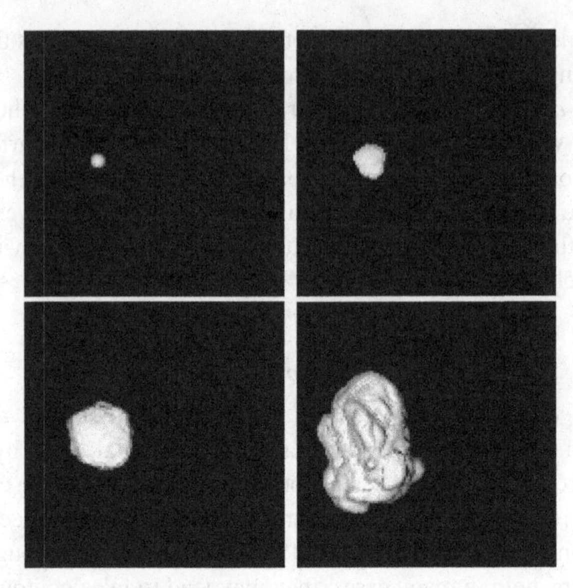

Fig. 3.10. Surface evolution toward the detection of a tumor in MRI.

surface). The initial surface is an ellipsoid surrounding the two tori (top left). Note how the model manages to split and detect this very different topology (bottom right).

A medical example is given in Fig. 3.10. Figure 3.10 presents the 3D detection of a tumor in a MRI image. The initial 3D shape is presented in the first row. The second row presents three evolution steps, and the final shape, the weighted minimal surface, is presented in the bottom.

Figure 3.11 shows a number of sequences of an active hart. This figure is reproduced from Ref. [249] and was obtained by combining the minimal-surface model with the fast numerics developed by Malladi and Sethian.

Figure 3.12 shows the detection of a head from MRI data.

3.3. Geodesics in Vector-Valued Images

We now extend the geodesic formulation to object detection in vector-valued images, presenting what we denote as color snakes (color active contours) [53, 54, 55]. (In this section the word color is used to refer to general multivalued images.) Vector-valued images are not just obtained in image modalities in which the data are recorded in a vector fashion, as in color, medical, and LANDSAT applications. The vector-valued data can be

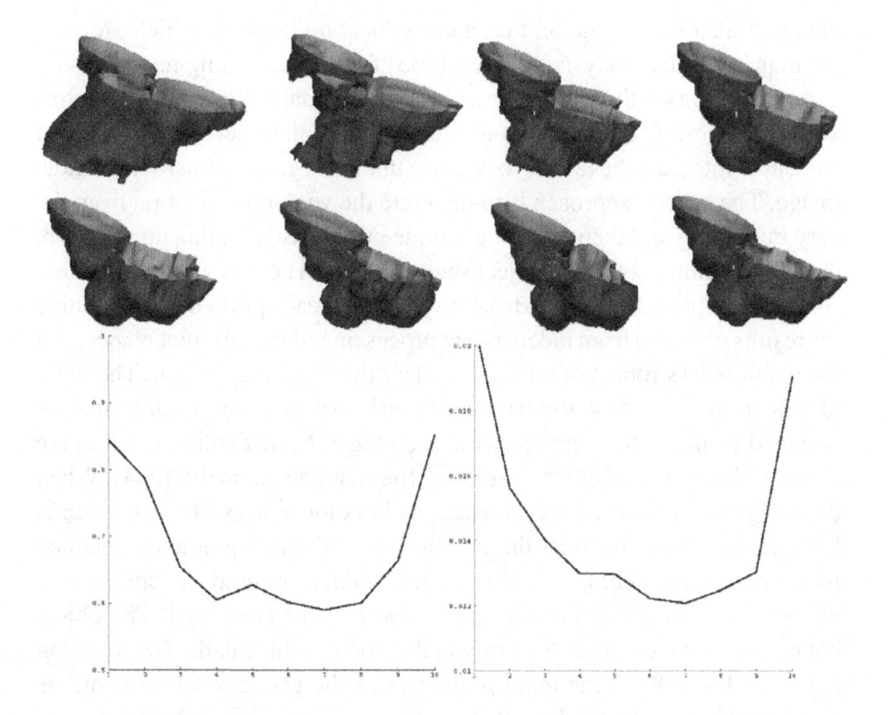

Fig. 3.11. Detection of a heart with the minimal-surface model. Several consecutive time stages are shown, together with the plot of the area and the volume (vertical axes) against time (horizontal axes).

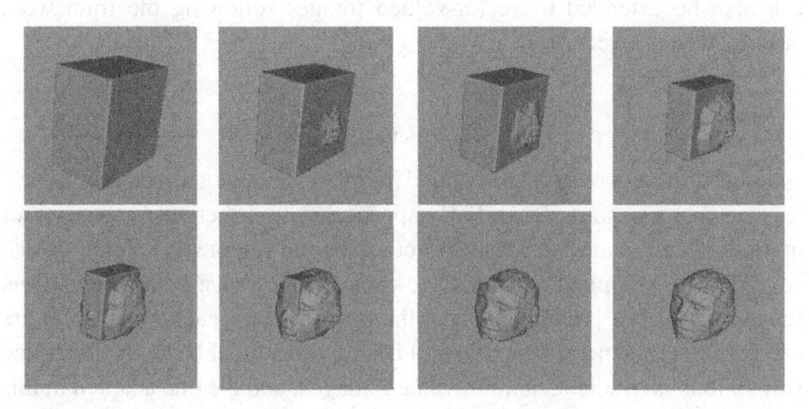

Fig. 3.12. Detecting a head from 3D MRI. Eight steps of the evolution of the minimal surface are shown.

obtained also from scale and orientation decompositions, which are very popular in texture analysis; see Ref. [333] for corresponding references.

In general, two different approaches can be adapted to work on vector-valued images. The first one is to process each plane separately and then somehow integrate the results to obtain a unique segmentation for the whole image. The second approach is to integrate the vector information from the very beginning and then deform a unique curve based on this information, directly obtaining a unique object segmentation. The first approach suffers from several problems [333]. Basically, it is not clear a priori how to combine the results obtained from independent processing of the different planes, and the result is less robust to noise. We adapt the second approach. The main idea is to define a new Riemannian (metric) space based on information obtained from all the components in the image. More explicitly, edges are computed based on classical results on Riemannian geometry [224]. When the image components are correlated, as in color images, this approach is less sensitive to noise than the combination of scalar gradients obtained from each component [231]. These vector edges are used to define a new metric space on which the geodesic curve is to be computed. The object boundaries are then given by a minimal color-weighted path. The resulting approach holds the same main properties as the geodesic active contours developed in Subsection 3.1.

A number of results on vector-valued segmentation were reported in the literature, e.g., Ref. [428]. See Ref. [333] for the relevant references and further comparisons. Here we address the geodesic active-contour approach with vector-image metrics, a simple and general approach. Other algorithms can also be extended to vector-valued images following the framework described in this section.

3.3.1. Vector-Valued Edges

A definition of edges in vector-valued images is presented based on classical Riemannian geometry [224]. Early approaches to detecting discontinuities in multivalued images attempted to combine the response of single-valued edge detectors applied separately to each of the image components (see, for example, Ref. [267]). The way the responses for each component are combined is in general heuristic and has no theoretical basis. A principled way to look at gradients in multivalued images, and the one adapted in this book, has been described in Ref. [110].

The idea is the following. Let $I(x_1, x_2) : \mathbb{R}^2 \to \mathbb{R}^m$ be a multivalued image with components $I_i(x_1, x_2) : \mathbb{R}^2 \to \mathbb{R}$, $i = 1, 2, \ldots, m$. For color images, for example, we have $m = 3$ components. The value of the image at a given point (x_1^0, x_2^0) is a vector in $\mathbb{R}^m$, and the difference of image values at two points $P = (x_1^0, x_2^0)$ and $Q = (x_1^1, x_2^1)$ is given by $\Delta I = I(P) - I(Q)$. When the (Euclidean) distance $d(P, Q)$ between P and Q tends to zero, the difference becomes the arc element,

$$\mathrm{d}I = \sum_{i=1}^{2} \frac{\partial I}{\partial x_i} \mathrm{d}x_i, \tag{3.48}$$

and its squared norm is given by

$$\mathrm{d}I^2 = \sum_{i=1}^{2} \sum_{j=1}^{2} \frac{\partial I}{\partial x_i} \frac{\partial I}{\partial x_j} \mathrm{d}x_i \mathrm{d}x_j. \tag{3.49}$$

This quadratic form is called the first fundamental form [224]. Let us denote $g_{ij} := [(\partial I/\partial x_i) \cdot (\partial I/\partial x_j)]$, then

$$\mathrm{d}I^2 = \sum_{i=1}^{2} \sum_{j=1}^{2} g_{ij} \mathrm{d}x_i \mathrm{d}x_j = \begin{bmatrix} \mathrm{d}x_1 \\ \mathrm{d}x_2 \end{bmatrix}^T \begin{bmatrix} g_{11} & g_{12} \\ g_{21} & g_{22} \end{bmatrix} \begin{bmatrix} \mathrm{d}x_1 \\ \mathrm{d}x_2 \end{bmatrix}. \tag{3.50}$$

The first fundamental form allows the measurement of changes in the image. For a unit vector $\hat{v} = (v_1, v_2) = (\cos\theta, \sin\theta)$, $\mathrm{d}I^2(\hat{v})$ is a measure of the rate of change of the image in the $\hat{v}$ direction. The extrema of the quadratic form of Eq. (3.50) are obtained in the directions of the eigenvectors of the matrix $[g_{ij}]$, and the values attained there are the corresponding eigenvalues. Simple algebra shows that the eigenvalues are

$$\lambda_\pm = \frac{g_{11} + g_{22} \pm \sqrt{(g_{11} - g_{22})^2 + 4g_{12}^2}}{2}, \tag{3.51}$$

and the eigenvectors are $(\cos\theta_\pm, \sin\theta_\pm)$, where the angles $\theta_\pm$ are given (modulo π) by

$$\theta_+ = \frac{1}{2} \arctan \frac{2g_{12}}{g_{11} - g_{22}}, \quad \theta_- = \theta_+ + \pi/2. \tag{3.52}$$

Thus the eigenvectors provide the direction of maximal and minimal changes at a given point in the image, and the eigenvalues are the corresponding rates of change. We call θ_+ the direction of maximal change and λ_+ the maximal rate of change. Similarly, θ_- and λ_- are the direction of minimal change and the minimal rate of change, respectively. Note that for $m = 1$, $\lambda_+ \equiv \|\nabla I\|^2$, $\lambda_- \equiv 0$, and $(\cos\theta_+, \sin\theta_+) = \nabla I/\|\nabla I\|$.

In contrast to gray-level images ($m = 1$), the minimal rate of change λ_- may be different from zero. In the single-valued case, the gradient is always perpendicular to the level sets and $\lambda_- \equiv 0$. As a consequence, the strength of an edge in the multivalued case is not simply given by the rate of maximal change λ_+, but by how λ_+ compares with λ_-. For example, if $\lambda_+ = \lambda_-$, we know that the image changes at an equal rate in all directions. We can detect image discontinuities by defining a function $f = f(\lambda_+, \lambda_-)$ that measures the dissimilarity between λ_+ and λ_-. A possible choice is $f = f(\lambda_+ - \lambda_-)$, which has the nice property of reducing to $f = f(\|\nabla I\|^2)$ for the 1D case, $m = 1$.

Cumani [107] extended some of the above ideas. He analyzed the directional derivative of λ_+ in the direction of its corresponding eigenvector and looked for edges by localizing the zero crossings of this function. Cumani's work attempts to generalize the ideas of Torre and Poggio [390] to multivalued images. Note that this approach entirely neglects the behavior of λ_-. As already mentioned, it is the relationship between λ_- and λ_+ that is important.

A noise analysis for the above vector-valued edge detector has been presented in Ref. [231]. It was shown that, for correlated data, this scheme is more robust to noise than the simple combination of the gradient components.

Before concluding this section it should be pointed out that, from the theory above, improved edge detectors for vector-valued images can be obtained following, for example, the developments on energy-based edge detectors [143]. To present the color snakes, the theory developed above is sufficient.

Metrics of Color Space

Given a number of image attributes, we can measure psychophysically the degree to which the human visual system is sensitive to small changes in each of the components. Psychophysical studies therefore can help in defining a metric in the "attribute" space that is relevant to human vision.

In color vision, for example, we experimentally measure psychophysical human thresholds (or isoperformance surfaces) from a reference point in different directions in color space [415]. This defines the local Riemannian metric at the reference point. This kind of measurement has to be repeated at different positions in color space. The method was pioneered by MacAdam [248]. An alternative approach has been to develop theoretical models of human detection of color differences based on our knowledge of the physiology and psychophysics of the visual system [415]. The predictions of these line-element models also provide a metric of color space. In general, all the line-element models derived from simple theoretical assumptions fail to represent many of the important features present in empirical data. Our algorithm can easily incorporate any of the empirical or theoretical line-element models mentioned above (empirical data, however, have to be interpolated to cover the entire space).

The approach followed here is to adapt one of the CIE standards (CIE stands for Commission Internationale de P'Eclairage, the International Commission on Illumination) that attempts to achieve an approximate uniform color space (in which color threshold surfaces are roughly spherical) under the viewing conditions usually found in practice. We use the CIE 1976 $L^*a^*b^*$ space [415] with its associated color-difference formula (which is simply the Euclidean distance in the space). The white reference point (X_w, Y_w, Z_w) was taken as the one obtained when the red, green, and blue guns were driven to half of their maximum amplitude. The $L^*a^*b^*$ space is a first approximation to perceptually uniform color spaces. More accurate approximations can be obtained, for example, by taking into account spatial frequencies [401]. In this case, a new vector-valued image is obtained with $m > 3$, and the same theory presented above can be applied.

3.3.2. Color Snakes

Let $f_{\text{color}} = f(\lambda_+, \lambda_-)$ be the edge detector defined in Subsection 3.3.1. The edge-stopping function g_{color} is then defined such that $g_{\text{color}} \to 0$ when $f \to$ max (∞), as in the gray-scale case. For example, $f_{\text{color}} := (\lambda_+ - \lambda_-)^{1/p}$, $p > 0$, or $f_{\text{color}} := \sqrt{\lambda_+}$ and $g_{\text{color}} := \frac{1}{1+f}$, or $g_{\text{color}} := \exp\{-f\}$. The function (metric) g_{color} defines the space on which we compute the geodesic curve. Defining

$$L_{\text{color}} := \int_0^{\text{length}} g_{\text{color}} dv, \tag{3.53}$$

we find that the object-detection problem in vector-valued images is then associated with minimizing L_{color}. We have formulated the problem of object segmentation in vector-valued images as a problem in finding a geodesic curve in a space defined by a metric induced from the whole vector image.

To minimize L_{color}, that is, the color length, we compute, as before, the gradient descent flow. The equations developed for the geodesic active contours are independent of the specific selection of the function g. Replacing g_{gray} with g_{color} and embedding the evolving curve C in the function $u : \mathbb{R}^2 \to \mathbb{R}$, we obtain the general flow, with additional unit speed, for the color snakes:

$$\frac{\partial u}{\partial t} = g_{\text{color}}(\nu + \kappa)|\nabla u| + \nabla u \cdot \nabla g_{\text{color}}. \tag{3.54}$$

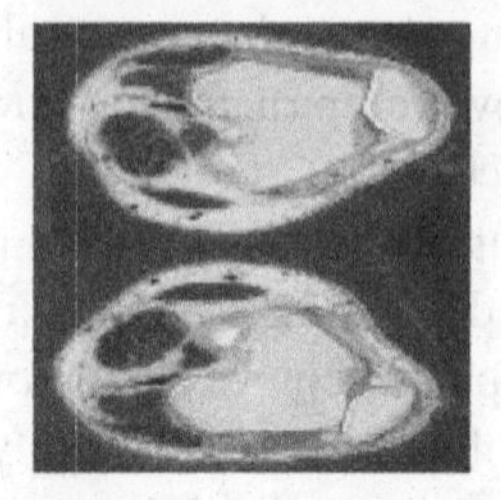
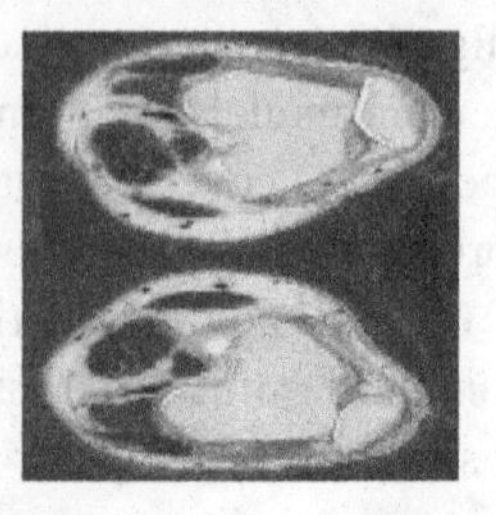

Fig. 3.13. Example of the color snakes. The original image is on the left, and the one with the segmented objects (red lines) is on the right. The original curve contained both objects. The computations were done on the $L^*a^*b^*$ space.

To recap, Eq. (3.54) is the modified level-set flow corresponding to the gradient descent of L_{color}. Its solution (steady state) is a geodesic curve in the space defined by the metric $g_{\text{color}}(\lambda_\pm)$ of the vector-valued image. This solution gives the boundaries of objects in the scene. (Note that $\lambda_\pm$ can be computed on a smooth image obtained from vector-valued anisotropic diffusion [344].) Following work presented previously in this chapter, theoretical results regarding the existence, uniqueness, stability, and correctness of the solutions to the color active contours can be obtained.

Figure 3.13 presents an example of the vector snake model for a medical image. Figure 3.14 shows an example for texture segmentation. The original image is filtered with Gabor filters tuned to frequency and orientation, as proposed in Ref. [232] for texture segmentation (see Ref. [333] for additional

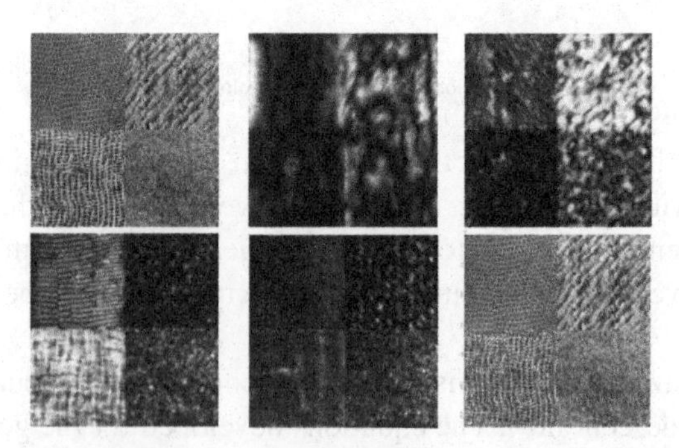

Fig. 3.14. Example of the vector snakes for a texture image. The original texture (top left) is decomposed into frequency/orientation (four frequencies and four orientations) components by means of Gabor filters and this collection of images is used to compute the metric g_{color} for the snake flow. A subset of the different components is shown, followed (bottom right) by the result of the evolving vector snakes (green), segmenting one of the texture boundaries (red).

related references). From this set of frequency/orientation images, g_{color} is computed according to the formulas in Subsection 3.3.1, and the vector-valued snake flow is applied. Four frequencies and four orientations are used, and 16 images are obtained. More examples can be found in Ref. [333].

3.3.3. Level Lines of Vector-Valued Images

As we have seen, and we will continue to develop later in this book, level sets provide a fundamental concept and representation for scalar images. The basic idea is that a scalar image $I(x, y) : \mathbb{R}^2 \to \mathbb{R}$ is represented as a collection of sets

$$\Lambda_h := \{(x, y) : I(x, y) = h\}$$

or

$$\hat{\Lambda}_h := \{(x, y) : I(x, y) \leq h\}.$$

This representation not only brings state-of-the-art image processing algorithms, it is also the source of the connected-components concept that we will develop later in this chapter and that has given light to important applications such as contrast enhancement and image registration.

One of the fundamental questions then is whether we can extend the concept of level sets (and, after that, the concept of connected components) to multivalued data, that is, images of the form $I(x, y) : \mathbb{R}^2 \to \mathbb{R}^n$, $n > 1$. As we have seen, these data include color images, multispectral images, and video.

One straightforward possibility is to consider the collection of classical level sets for each one of the image components $I_i(x, y) : \mathbb{R}^2 \to \mathbb{R}$, $1 \leq i \leq n$. Although this is an interesting approach, this has a number of caveats and is not entirely analogous to the scalar level sets. For example, it is not clear how to combine the level sets from the different components. In contrast with the scalar case, a point on the plane belongs to more than one level set $\Lambda_i(x, y)$, and therefore it might belong to more than one connected component. A possible solution to this is to consider lexicographic orders. That means that some arbitrary decisions need to be taken to combine the multiple level sets.

Let us now pursue a different approach. Basically, we redefine the level-set lines of a scalar image $I(x, y) : \mathbb{R}^2 \to \mathbb{R}$ as the integral curves of the directions of minimal change θ_- as defined above. In other words, we select a given pixel (x, y) and travel the image plane $\mathbb{R}^2$ always in the direction of minimal change. Scalar images have the particular property that the minimal

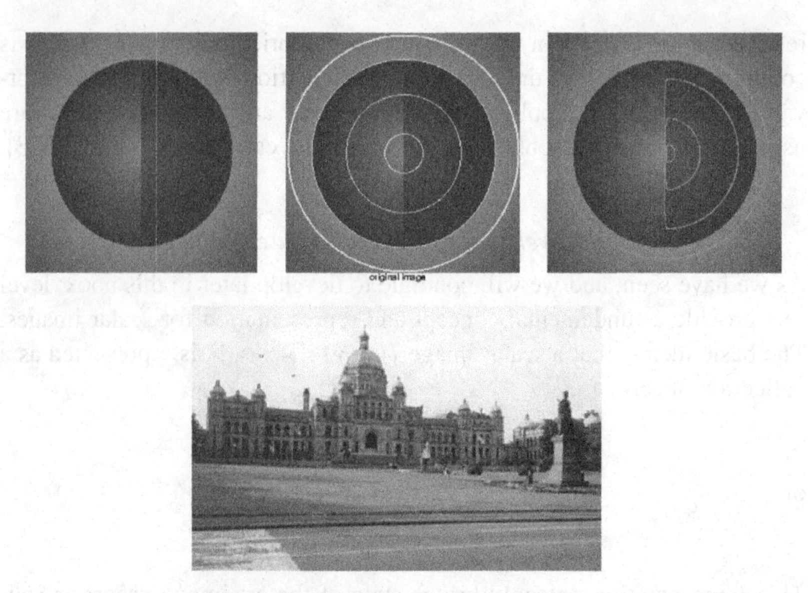

Fig. 3.15. Level sets for multivalued images. The first row shows a toy example, with the original image on the left, followed by the level lines for the corresponding gray-valued image and the level lines for the vector-valued image. The second row shows a segment of the level lines for the vector-valued image.

change is zero, and therefore the integral curves of the directions of minimal change are exactly the classical level-set lines defined above. The advantage of this definition is that it is dimension independent, and, as we have seen before, minimal change and direction of minimal change can also be defined for multivalued images following classical Riemannian geometry; we have obtained a definition of level sets for multivalued images as well. As in the scalar case, there will be a unique set of level sets, in contrast with the case in which we treat each component separately. Figure 3.15 shows segments of these multivalued level sets for color images. Note how the level set follows our intuition, and for example, it is located at the color boundary of the castle.

What is very interesting is to see if we can reconstruct a single-valued image whose level sets coincide with the multivalued level sets of a color image (or of any multivalued image in general). We can formulate this as a variational problem, that is, the multivalued level sets give a direction, θ_- (and θ_+), and a value, $f(\lambda_+, \lambda_-)$, for the level sets and gradient of the unknown single-valued image. In other words, we search for a single-valued image $\tilde{I} : \mathbb{R}^2 \to \mathbb{R}$ such that $\nabla \tilde{I}$ is as close as possible to $f(\lambda_+, \lambda_-)\theta_+$ (or $\nabla \tilde{I}/\|\nabla \tilde{I}\|$ is as close as possible to θ_+). This can be formulated as a variational problem, and we can for example search for the minimizer of

$$\int \|\nabla \tilde{I} - f(\lambda_+, \lambda_-)\theta_+\|^2 .$$

Recall that the unknown is $\tilde{I}$, the single-valued image, and θ_+, λ_+, and λ_- are obtained from the given color or multivalued image. The gradient descent flow equation corresponding to this energy is a Poisson equation, and existence and uniqueness results can be obtained (similar results can be obtained if we choose to work with only the direction of the gradient θ_+, ignoring its magnitude). There is still a technical problem, as θ_+ is defined only between $-\pi/2$ and $+\pi/2$, although this can be solved in a number of different ways. Figure 3.16 gives examples of color images and their corresponding single-valued ones. The basic idea is that the single-valued image captures the basic geometry of the color image given by the multivalued level-sets. Details on this can be found in Ref. [95], as well as in Refs. [19 and 199]. Other techniques for addressing this problem can be found in Ref. [174].

3.4. Finding the Minimal Geodesic

Figure 3.17 shows an image of a neuron from the central nervous system. This image was obtained by means of electronic microscopy (EM). After the neuron is identified, it is marked by means of the injection of a color fluid. Then a portion of the tissue is extracted, and, after some processing, it is cut into thin slices and observed and captured by the EM system. Figure 3.17 shows the output of the EM after some simple postprocessing, mainly composed by contrast enhancement. The goal of the biologist is to obtain a 3D reconstruction of this neuron. As we observe from the example in Fig. 3.17, the image is very noisy and the boundaries of the neuron are difficult to identify. Segmenting the neuron is then a difficult task.

One of the most commonly used approaches to segment objects as the neuron in Fig. 3.17 is the use of active contours such as those described in Section 3.1. As shown before, this reduces to the minimization of a weighted length given by

$$\int_C g(\|\nabla(I)\|)\mathrm{d}s, \tag{3.55}$$

where $C : \mathbb{R} \to \mathbb{R}^2$ is the deforming curve, $I : \mathbb{R}^2 \to \mathbb{R}$ is the image, $\mathrm{d}s$ stands for the curve arc length ($\|\partial C/\partial s\| = 1$), $\nabla(\cdot)$ stands for the gradient, and $g(\cdot)$ is such that $g(r) \to 0$ while $r \to \infty$ (the edge detector).

Fig. 3.16. Single-valued representatives of vector-valued images. The first three rows from the top show examples for color images (vector data on the left and scalar data on the right), and the last row shows several components of LANDSAT data followed by their scalar representation.

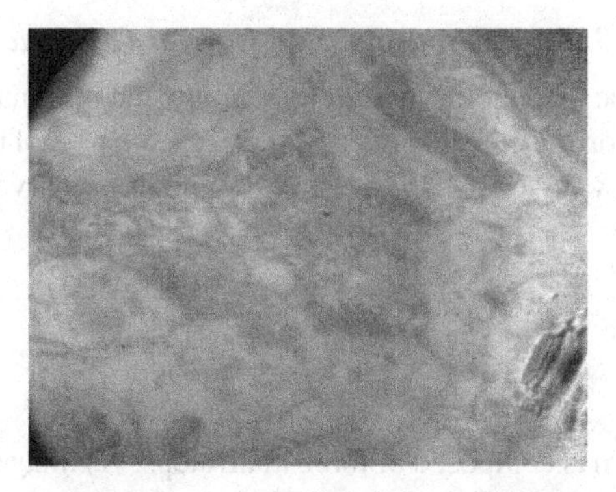

Fig. 3.17. Example of an EM image of a neuron (one slice).

There are two main techniques to find the geodesic curve, that is, the minimizer of Eq. (3.55):

1. Compute the gradient descent of Eq. (3.55) and, starting from a closed curve either inside or outside the object, deform it toward the (possibly local) minima, finding a geodesic curve. This approach gives a curve evolution flow, based on curvature motion, leading to very efficient solutions for a large number of applications. This was the approach followed in the preceding subsections in this chapter. This model, which computes a geodesic, gives a completely automatic segmentation procedure. When tested with images such as the one in Figure 3.17, two major drawbacks to this model are found [26]. First, because of the large amount of noise, spurious objects are detected, and it is left to the user to manually eliminate them. Second, because the boundary of the real neuron is very weak, this is not always detected.
2. Connect between a few points marked by the user on the neuron's boundary, while keeping the weighted length of Eq. (3.55) to a minimum. This was developed in Ref. [100], and it is the approach presented now. In contrast with the technique described above, this approach always needs user intervention to mark the initial points. On the other hand, for images such as the one in Fig. 3.17, it permits a better handling of the noise. In the rest of this section this technique is briefly described and the additions are incorporated to address our specific problem.

3.4.1. Computing the Minimal Geodesic

We now describe the algorithm used to compute the minimal weighted path between points on the object boundary, that is, given a set of boundary points $\{\mathcal{P}\}_{i=1}^{N}$ and following Eq. (3.55), we have to find the N curves that minimize $(\mathcal{P}_{N+1} \equiv \mathcal{P}_1)$:

$$d[I(\mathcal{P}_i), I(\mathcal{P}_{i+1})] := \int_{\mathcal{P}_i}^{\mathcal{P}_{i+1}} g(\|\nabla I\|)\mathrm{d}s. \tag{3.56}$$

The algorithm is composed of three main steps: (1) image regularization, (2) computation of equal-distance contours, and (3) backpropagation. Each one of these steps is briefly described now. For details on the first step, see Ref. [36]. For details on the other steps, see Ref. [100].

Image Regularization. To reduce the noise on the images obtained from EM, we perform the following two steps (these are just examples of common approaches used to reduce noise before segmenting).

1. Subsampling: We use a four-tap filter, approximating a Gaussian function, to smooth the image before a 2×2 subsampling is performed. This not only removes noise but also gives a smaller image to work with, thereby accelerating the algorithm by a factor of 4; that is, we will work on the subsampled image (although the user marks the end points on the original image), and only after the segmentation is computed is the result is extrapolated to the original-size image. Further subsampling was found to already produce inaccurate results. The result from this step is then an image $I_{2\times 2}$ that is one quarter of the original image I.
2. Smoothing: To further reduce noise in the image, we smooth the image either with a Gaussian filter or with one of the anisotropic diffusion flows presented in Chap. 4.

At the end of the preprocessing stage we then obtain an image $\hat{I}_{2\times 2}$ that is the result of the subsampling of I followed by noise removal. Although the user marks the points $\{\mathcal{P}\}_{i=1}^{N}$ on the original image I, the algorithm makes all the computations on $\hat{I}_{2\times 2}$ and then extrapolates and displays them on I.

Equal-Distance-Contour Computation. After the image $\hat{I}_{2\times 2}$ is computed, we have to compute, for every point $\hat{\mathcal{P}}_i$, where $\hat{\mathcal{P}}_i$ is the point in $\hat{I}_{2\times 2}$ corresponding to the point $\mathcal{P}_i$ in I (coordinates divided by two), the weighted-distance map according to the weighted distance d, that is, we have to compute the function

$$\mathcal{D}_i(x, y) := d[\hat{I}_{2\times 2}(\hat{\mathcal{P}}_i), \hat{I}_{2\times 2}(x, y)],$$

or in words, the weighted distance between the pair of image points $\hat{\mathcal{P}}_i$ and (x, y).

There are basically two ways of making this computation, computing equal-distance contours or directly computing $\mathcal{D}_i$. Each one is now described.

Equal-distance contours $\mathcal{C}_i$ are curves such that every point on them has the same distance d to $\hat{\mathcal{P}}_i$, that is the curves $\mathcal{C}_i$ are the level sets or isophotes of $\mathcal{D}_i$. It is easy to see [100] that, following the definition of d, these contours are obtained as the solution of the curve evolution flow:

$$\frac{\partial \mathcal{C}_i(x, y, t)}{\partial t} = \frac{1}{g(\|\nabla \hat{I}_{2\times 2}\|)} \vec{\mathcal{N}},$$

where $\vec{\mathcal{N}}$ is the outer unit normal to $\mathcal{C}_i(x, y, t)$. This type of flow should be implemented with the standard level-set method [294].

A different approach is the one presented in Section 2.2, which is based on the fact that the distance function $\mathcal{D}_i$ holds the following Hamilton–Jacobi equation:

$$\frac{1}{g(\|\nabla \hat{I}_{2\times 2}\|)} \|\nabla \mathcal{D}_i\| = 1.$$

As detailed in Section 2.2, optimal numerical techniques have been proposed to solve this static Hamilton–Jacobi equation. Because of this optimality, this is the approach we follow in the examples below. At the end of this step, we have $\mathcal{D}_i$ for each point $\hat{\mathcal{P}}_i$. We should note that we do not need to compute $\mathcal{D}_i$ for all the image plane. It is actually enough to stop the computations when the value at $\hat{\mathcal{P}}_{i+1}$ is obtained.

Backpropagation. After the distance functions $\mathcal{D}_i$ are computed, we have to trace the actual minimal path between $\hat{\mathcal{P}}_i$ and $\hat{\mathcal{P}}_{i+1}$ that minimizes d. Once again it is easy to show (see, for example, Refs. [214 and 361]) that this path should be perpendicular to the level curves $\mathcal{C}_i$ of $\mathcal{D}_i$ and therefore tangent to $\nabla \mathcal{D}_i$. The path is then computed backward from $\hat{\mathcal{P}}_{i+1}$, in the gradient direction, until we return to the point $\hat{\mathcal{P}}_i$. This backpropagation is of course guaranteed to converge to the point $\hat{\mathcal{P}}_i$, and then gives the path of minimal weighted distance.

3.4.2. Examples

A number of examples of the algorithm described above are now presented. We compare the results with those obtained with *PictureIt*, a commercially available general-purpose image processing package developed by Microsoft (to the best of my knowledge, the exact algorithm used by this product was not published). As in the tracing algorithm, this software allows the user to click a few points on the object's boundary while the program automatically completes the rest of it. Three to five points are used for each one of the examples. The points are usually marked at extrema of curvature or at areas where the user, after some experience, predicts possible segmentation difficulties. The same points were marked in the tracing algorithm and in *PictureIt*. The results are shown in Fig. 3.18. [398]. We observe that the tracing technique outperforms *PictureIt*. Moreover, *PictureIt* is extremely sensible to the exact position of the marked points; a difference of one or two pixels can cause a very large difference in the segmentation results. Our algorithm is very robust to the exact position of the points marked by the user (see also [60].)

Additional examples for natural images are given in Fig. 3.19. The first row show the edge tracing on a portrait of Cartan. The second row shows

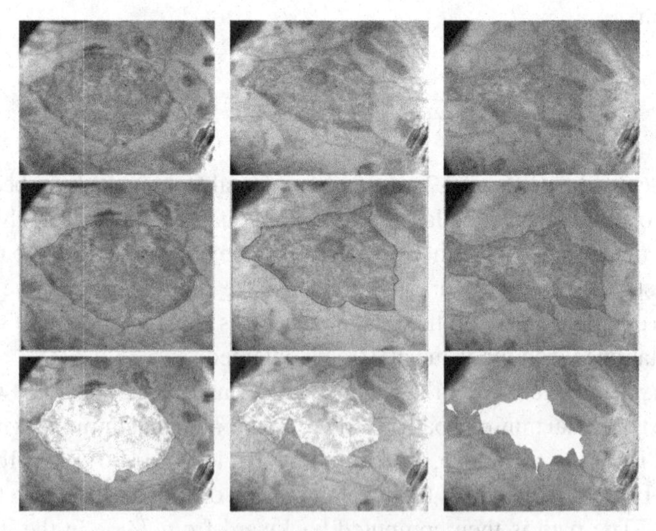

Fig. 3.18. Comparison of the minimal geodesic results with those obtained with the commercial software *PictureIt*. For each column, the original image is shown on the top, the result of the tracing algorithm (green line) in the middle, and the result of *PictureIt* on the bottom. For this last, the area segmented by the algorithm is shown brighter.

Fig. 3.19. Boundary tracing with global minimal geodesics.

an example for a color image. In this case, g, the weighting function, is computed with the color edge detector described in Section 3.3. For each row, the traced image is on the left, the edge map is in the middle (with the traced contour on it), and the weighted distance corresponding to the last marked pair of points is on the right.

3.5. Affine Invariant Active Contours

The geodesic formulation allows us to perform affine invariant segmentation as well. Affine invariant active contours are now described. To obtain them, we must replace the classical gradient-based edge detector with an affine invariant one, and we should also compute affine invariant gradient descent flows. Both extensions are given now [287, 288].

3.5.1. Affine Invariant Gradient

Let $I : \mathbb{R}^2 \to \mathbb{R}^+$ be a given image in the continuous domain. To detect edges in an affine invariant form, a possible approach is to replace the classical gradient magnitude $\|\nabla I\| = \sqrt{I_x^2 + I_y^2}$, which is only Euclidean invariant, with an affine invariant gradient. By this we mean that we search for an affine invariant function from $\mathbb{R}^2$ to $\mathbb{R}$ that has, at image edges, values significantly different from those at flat areas and such that this values are preserved, at corresponding image points, under affine transformations. To accomplish this, we have to verify that we can use basic affine invariant descriptors that can be computed from I in order to find an expression that (qualitatively) behaves like $\|\nabla I\|$. With the classification developed in Refs. [281 and 283], it was found that the two basic independent affine invariant descriptors are (note that the simplest Euclidean invariant differential descriptor is exactly $\|\nabla I\|$, which is enough to formulate a basic Euclidean invariant edge detector)

$$H := I_{xx}I_{yy} - I_{xy}^2, \qquad J := I_{xx}I_y^2 - 2I_xI_yI_{xy} + I_x^2I_{yy}.$$

There is no (nontrivial) first-order affine invariant descriptor, and all other second-order differential invariants are functions of H and J. Therefore the simplest possible affine gradient must be expressible as a function $\mathcal{F} = \mathcal{F}(H, J)$ of these two invariant descriptors.

The differential invariant J is related to the Euclidean curvature of the level sets of the image. Indeed, if a curve $\mathcal{C}$ is defined as the level set of I, then the curvature of $\mathcal{C}$ is given by $\kappa = (J/\|\nabla I\|^3)$. Lindeberg [242] used J to compute corners and edges in an affine invariant form, that is,

$$\mathcal{F} := J = \kappa\ \|\nabla I\|^3 .$$

This singles out image structures with a combination of high gradient (edges) and high curvature of the level sets (corners). Note that, in general, edges and corners do not have to lie on a unique level set. Here, by combining both H and J, we present a more general affine gradient approach. Because both H and J are second-order derivatives of the image, the order of the affine gradient is not increased while both invariants are being used.

Definition 3.1. *The (basic) affine invariant gradient of a function I is defined by the equation*

$$\widehat{\nabla}_{\text{aff}} I := \left|\frac{H}{J}\right|. \tag{3.57}$$

Technically, because $\widehat{\nabla}_{\text{aff}} I$ is a scalar (a map from $\mathbb{R}^2$ to $\mathbb{R}$), it measures just the magnitude of the affine gradient, so our definition may be slightly misleading. However, an affine invariant gradient direction does not exist, as directions (angles) are not affine invariant, and so we are justified in omitting magnitude for simplicity.

Note also that if photometric transformations are allowed, then $\widehat{\nabla}_{\text{aff}} I$ becomes only a relative invariant. To obtain an absolute invariant, we can use, for example, the combination $H^{3/2}/J$. Because in this case going from relative to absolute invariants is straightforward, we proceed with the development of the simpler function $\widehat{\nabla}_{\text{aff}} I$.

The justification for our definition is based on a (simplified) analysis of the behavior of $\widehat{\nabla}_{\text{aff}} I$ near edges in the image defined by I. Near the edge of an object, the gray-level values of the image can be (ideally) represented by $I(x, y) = f[y - h(x)]$, where $y = h(x)$ is the edge and $f(t)$ is a slightly smoothed step function with a jump near $t = 0$. Straightforward computations show that, in this case,

$$H = -h'' f' f'', \qquad J = -h'' f'^3.$$

Therefore

$$H/J = f''/f'^2 = (-1/f')'.$$

Clearly H/J is large (positive or negative) on either side of the object $y = f(x)$, creating an approximation of a zero crossing at the edge. (Note that the Euclidean gradient is the opposite, high at the ideal edge and zero everywhere else. Of course, this does not make any fundamental difference, as the important part is to differentiate between edges and flat regions. In the affine case, edges are given by doublets.) This is because $f(x) = \text{step}(x)$, $f'(x) = \delta(x)$, and $f''(x) = \delta'(x)$. (We are omitting the points where $f' = 0$.) Therefore $\widehat{\nabla}_{\text{aff}} I$ behaves like the classical Euclidean gradient magnitude.

To avoid possible difficulties when the affine invariants H or J are zero, we replace $\widehat{\nabla}_{\text{aff}}$ with a slight modification. Indeed, other combinations of H and J can provide similar behavior and hence can be used to define affine gradients. Here the general technique is presented, as well as a few examples.

We will be more interested in edge-stopping functions than in edge maps. These are functions that are as close as possible to zero at edges and close to the maximal possible value at flat regions. We then proceed to make modifications on $\widehat{\nabla}_{\text{aff}}$ that allow us to compute well-defined stopping functions instead of just edge maps.

In Euclidean invariant edge-detection algorithms based on active contours as well as in anisotropic diffusion, the stopping term is usually taken in the form $(1+ \|\nabla I\|^2)^{-1}$, the extra 1 being taken to avoid singularities where the Euclidean gradient vanishes. Thus, in analogy, the corresponding affine invariant stopping term should have the form

$$\frac{1}{1 + (\widehat{\nabla}_{\text{aff}} I)^2} = \frac{J^2}{H^2 + J^2}.$$

However, this can still present difficulties when both H and J vanish, so a second modification is proposed.

Definition 3.2. *The normalized affine invariant gradient is given by*

$$\nabla_{\text{aff}} I = \sqrt{\frac{H^2}{J^2 + 1}}. \tag{3.58}$$

The motivation comes from the form of the affine invariant stopping term, which is now given by

$$\frac{1}{1 + (\nabla_{\text{aff}} I)^2} = \frac{J^2 + 1}{H^2 + J^2 + 1}. \tag{3.59}$$

Formula (3.59) avoids all difficulties where either H or J vanishes, and hence is a proper candidate for affine invariant edge detection. Indeed, in the neighborhood of an edge we obtain

$$\frac{J^2 + 1}{H^2 + J^2 + 1} = \frac{f'^6 h''^2 + 1}{h''^2 f'^2 (f'^4 + f''^2) + 1},$$

which, assuming that h'' is moderate, gives an explanation of why it serves as a barrier for the edge. Barriers, that is, functions that go to zero at (salient) edges, will be important for the affine active contours presented in the following sections.

Examples of the affine invariant edge detector of Eq. (3.59) are given in Fig. 3.20. As with the affine invariant edge-detection scheme introduced in the previous section, this algorithm might produce gaps in the object boundaries as a result of the existence of perfectly straight segments with the same gray value. In this case, an edge-integration algorithm is needed to complete the object boundary. The affine invariant active contour presented in Section 3.5.2 below is a possible remedy of this problem.

3.5.2. Affine Invariant Gradient Snakes

From the gradient active contours and affine invariant edge detectors above, it is almost straightforward to define affine invariant gradient active contours. To carry this program out, we will first have to define the proper norm. Because affine geometry is defined for only convex curves [39], we will initially have to restrict ourselves to the (Fréchet) space of thrice-differentiable convex closed curves in the plane, i.e.,

$$\mathbf{C}_0 := \{C : [0, 1] \to \mathbb{R}^2 : C \text{ is convex, closed and } C^3\}.$$

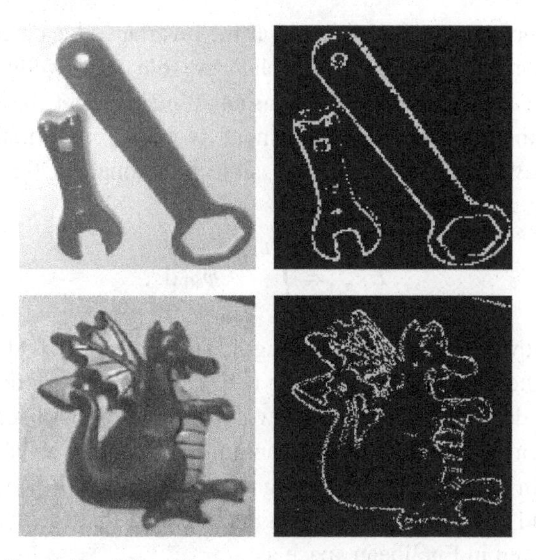

Fig. 3.20. Examples of the affine invariant edge detector (after thresholding).

As in Section 1.2, let $\mathrm{d}s$ denote the affine arc length. Then, letting $L_{\text{aff}} := \oint \mathrm{d}s$ be the affine length defined in Section 1.2, we proceed to define the affine norm on the space $\mathbf{C}_0$:

$$\|\mathcal{C}\|_{\text{aff}} := \int_0^1 \|\mathcal{C}(p)\|_a \ \mathrm{d}p = \int_0^{L_{\text{aff}}} \|\mathcal{C}(s)\|_a \ \mathrm{d}s,$$

where

$$\|\mathcal{C}(p)\|_a := [\mathcal{C}(p), \mathcal{C}_p(p)].$$

Note that the area enclosed by $\mathcal{C}$ is just

$$A = \frac{1}{2}\int_0^1 \|\mathcal{C}(p)\|_a \ \mathrm{d}p = \frac{1}{2}\int_0^1 [\mathcal{C}, \mathcal{C}_p]\mathrm{d}p = \frac{1}{2}\ \|\mathcal{C}\|_{\text{aff}}. \quad (3.60)$$

Observe that

$$\|\mathcal{C}_s\|_a = [\mathcal{C}_s, \mathcal{C}_{ss}] = 1, \qquad \|\mathcal{C}_{ss}\|_a = [\mathcal{C}_{ss}, \mathcal{C}_{sss}] = \mu,$$

where μ is the affine curvature, i.e., the simplest nontrivial differential affine invariant. This makes the affine norm $\|\cdot\|_{\text{aff}}$ consistent with the properties of the Euclidean norm on curves relative to the Euclidean arc length $\mathrm{d}v$. (Here we have that $\|\mathcal{C}_v\| = 1$, $\|\mathcal{C}_{vv}\| = \kappa$.)

We can now formulate the functionals that will be used to define the affine invariant snakes. Accordingly, assume that $\phi_{\text{aff}} = \phi(w_{\text{aff}})$ is an affine invariant stopping term, based on the affine invariant edge detectors considered in Section 3.5.1. Therefore $\phi_{\rm aff}$ plays the role of the weight ϕ in L_ϕ. As in the Euclidean case, we regard $\phi_{\rm aff}$ as an affine invariant conformal factor, and replace the affine arc-length element $\mathrm{d}s$ with a conformal counterpart $\mathrm{d}s_{\phi_{\rm aff}} = \phi_{\rm aff}\,\mathrm{d}s$ to obtain the first possible functional for the affine active contours:

$$L_{\phi_{\rm aff}} := \int_0^{L_{\rm aff}(t)} \phi_{\rm aff}\,\mathrm{d}s, \tag{3.61}$$

where, as above, $L_{\rm aff}$ is the affine length. The obvious next step is to compute the gradient flow corresponding to $L_{\phi_{\rm aff}}$ in order to produce the affine invariant model. Unfortunately, as we will see, this will lead to an impractically complicated geometric contour model that involves four spatial derivatives. In the meantime, by using the connection of Eq. (1.11) between the affine and the Euclidean arc lengths, we note that the above equation can be rewritten in Euclidean space as

$$L_{\phi_{\rm aff}} = \int_0^{L(t)} \phi_{\rm aff}\kappa^{1/3}\mathrm{d}v, \tag{3.62}$$

where $L(t)$ denotes the ordinary Euclidean length of the curve $\mathcal{C}(t)$ and $\mathrm{d}v$ stands for the Euclidean arc-length.

The snake model that we will use comes from another (special) affine invariant, namely area; see Eq. (3.60). Let $\mathcal{C}(p, t)$ be a family of curves in $\mathbf{C}_0$. A straightforward computation reveals that the first variation of the area functional

$$A(t) = \frac{1}{2}\int_0^1 [\mathcal{C}, \mathcal{C}_p]\mathrm{d}p$$

is

$$A'(t) = -\int_0^{L_{\rm aff}(t)} [\mathcal{C}_t, \mathcal{C}_s]\mathrm{d}s.$$

Therefore the gradient flow that will decrease the area as quickly as possible relative to $\|\cdot\|_{\rm aff}$ is exactly

$$\mathcal{C}_t = \mathcal{C}_{ss},$$

which, in modulo tangential terms, is equivalent to

$$\mathcal{C}_t = \kappa^{1/3}\vec{\mathcal{N}},$$

which is precisely the affine invariant heat equation described in Section 2.5. It is this functional that we will proceed to modify with the conformal factor ϕ_{aff}. Therefore we define the conformal area functional as

$$A_{\phi_{\text{aff}}} := \int_0^1 [\mathcal{C}, \mathcal{C}_p]\phi_{\text{aff}}\,\mathrm{d}p = \int_0^{L_{\text{aff}}(t)} [\mathcal{C}, \mathcal{C}_s]\phi_{\text{aff}}\,\mathrm{d}s.$$

(An alternative definition could be to use ϕ_{aff} to define the affine analog of the weighted area that produces the Euclidean weighted constant motion $\mathcal{C}_t = \phi\vec{\mathcal{N}}$.) The first variation of $A_{\phi_{\text{aff}}}$ will turn out to be much simpler than that of $L_{\phi_{\text{aff}}}$ and will lead to an implementable geometric snake model.

The precise formulas for the variations of these two functionals are given in the following result. They use the definition of $Y^{\perp}$, which is the unit vector perpendicular to Y. The proof follows by an integration-by-parts argument and some manipulations as in Appendix B [65, 66, 201, 202].

Lemma 3.7. *Let $L_{\phi_{\text{aff}}}$ and $A_{\phi_{\text{aff}}}$ denote the conformal affine length and area functionals, respectively.*

1. *The first variation of $L_{\phi_{\text{aff}}}$ is given by*

$$\frac{\mathrm{d}L_{\phi_{\text{aff}}}(t)}{\mathrm{d}t} = -\int_0^{L_{\text{aff}}(t)} [\mathcal{C}_t, (\nabla\phi_{\text{aff}})^{\perp}]\mathrm{d}s + \int_0^{L_a(t)} \phi_{\text{aff}}\mu[\mathcal{C}_t, \mathcal{C}_s]\mathrm{d}s. \tag{3.63}$$

2. *The first variation of $A_{\phi_{\text{aff}}}$ is given by*

$$\frac{\mathrm{d}A_{\phi_{\text{aff}}}(t)}{\mathrm{d}t} = -\int_0^{L_{\text{aff}}(t)} \left[\mathcal{C}_t, (\phi_{\text{aff}}\mathcal{C}_s + \frac{1}{2}[\mathcal{C}, (\nabla\phi)^{\perp}\mathcal{C}_s])\right]\mathrm{d}s. \tag{3.64}$$

The affine invariance of the resulting variational derivatives follows from a general result governing invariant variational problems that have volume-preserving symmetry groups [286]:

Theorem 3.12. *Suppose G is a connected transformation group and $\mathcal{I}[\mathcal{C}]$ is a G-invariant variational problem. Then the variational derivative (or gradient) $\delta\mathcal{I}$ of $\mathcal{I}$ is a differential invariant if and only if G is a group of volume-preserving transformations.*

We now consider the corresponding gradient flows computed with respect to $\|\cdot\|_{\text{aff}}$. First, the flow corresponding to the functional $L_{\phi_{\text{aff}}}$ is

$$\mathcal{C}_t = \{(\nabla\phi_{\text{aff}})^{\perp} + \phi_{\text{aff}}\mu\,\mathcal{C}_s\}_s = [(\nabla\phi_{\text{aff}})^{\perp}]_s + (\phi_{\text{aff}}\mu)_s\mathcal{C}_s + \phi_{\text{aff}}\mu\,\mathcal{C}_{ss}.$$

We ignore the tangential components, which do not affect the geometry of the evolving curve, and so we obtain the following possible model for geometric affine invariant active contours:

$$\mathcal{C}_t = \phi_{\text{aff}}\, \mu\kappa^{1/3}\vec{\mathcal{N}} + \langle [(\nabla\phi_{\text{aff}})^{\perp}]_s, \vec{\mathcal{N}}\rangle\vec{\mathcal{N}}. \tag{3.65}$$

The geometric interpretation of the affine gradient flow of Eq. (3.65) that minimizes $L_{\phi_{\text{aff}}}$ is analogous to that of the corresponding Euclidean geodesic active contours. The term $\phi_{\text{aff}}\, \mu\kappa^{1/3}$ minimizes the affine length L_{aff} while smoothing the curve according to the results in Section 2.5, being stopped by the affine invariant stopping function ϕ_{aff}. The term associated with $[(\nabla\phi_{\text{aff}})^{\perp}]_s$ creates a potential valley, attracting the evolving curve to the affine edges. Unfortunately, this flow involves μ, which makes it difficult to implement. (Possible techniques to compute μ numerically were recently reported in Refs. [57, 58, and 136].)

The gradient flow coming from the first variation of the modified area functional, on the other hand, is much simpler:

$$\mathcal{C}_t = \left\{\phi_{\text{aff}}\, \mathcal{C}_s + \frac{1}{2}\, [\mathcal{C}, (\nabla\phi_{\text{aff}})^{\perp}]\, \mathcal{C}_s\right\}_s. \tag{3.66}$$

Ignoring tangential terms (those involving $\mathcal{C}_s$), we find that this flow is equivalent to

$$\mathcal{C}_t = \phi_{\text{aff}}\, \mathcal{C}_{ss} + \frac{1}{2}\, [\mathcal{C}, (\nabla\phi_{\text{aff}})^{\perp}]\, \mathcal{C}_{ss}, \tag{3.67}$$

which in Euclidean form gives the second possible affine contour snake model:

$$\mathcal{C}_t = \phi_{\text{aff}}\, \kappa^{1/3}\vec{\mathcal{N}} + \frac{1}{2}\, \langle \mathcal{C}, \nabla\phi_{\text{aff}}\rangle \kappa^{1/3}\vec{\mathcal{N}}. \tag{3.68}$$

Note that although both models (3.65) and (3.68) were derived for convex curves, the flow of Eq. (3.68) makes sense in the nonconvex case as well, which makes this the only candidate for a practical affine invariant geometric-contour method. Thus we will concentrate on Eq. (3.68) from now on, and just consider Eq. (3.65) as a model with some theoretical interest.

To better capture concavities, to speed up the evolution, as well as to be able to define outward evolutions, a constant inflationary force of the type $\nu\phi\vec{\mathcal{N}}$ may be added to Eq. (3.68). This can be obtained from the affine gradient descent of $A_{\phi_{\text{aff}}} := \iint \phi_{\text{aff}}\, dx dy$. Note that the inflationary force $\mathcal{C}_t = \phi\vec{\mathcal{N}}$ in the Euclidean case of active contours if obtained from the (Euclidean) gradient descent of $A := \iint \phi dx dy$, where ϕ is a regular edge detector. We should note that although this constant-speed term is not always needed to capture a given contour, for real-world images it is certainly very helpful. We have not found an affine invariant inflationary-type term, and given the fact that the affine normal involves higher derivatives, it is doubtful that there is such an expression. Formal results regarding existence and uniqueness of solutions to Eq. (3.68) can be derived by following the same techniques used for the Euclidean case.

3.6. Additional Extensions and Modifications

A number of fundamental extensions to the theory of geodesic active contours exist. A few of them are briefly presented now in order to encourage and motivate the reader to refer to those very important contributions.

In Ref. [234], Lorigo et al. showed how to combine the geodesic active-contour framework with the high codimension level-set theory of Ambrosio and Soner [9] (see Chap. 2) in order to segment thin tubes in three dimensions.

One of the fundamental assumptions of the geodesic active contour in all its variants described so far is the presence of significant edges at the boundary of the objects of interests. Although this is significantly alleviated because of the attraction term $\nabla g \cdot \vec{\mathcal{N}}$ and can be further alleviated with the use of advanced edge-detection functions g, the existence of edges is an intrinsic assumption of the schemes (note that, as explained before, in Section 3.2.2 those edges do not have to have the same gradient value and can have gaps). A number of works have been proposed in the literature to further address this problem. Paragios and Deriche [302, 303] proposed the geodesic active regions, in which the basic idea is to have the geodesic contours driven not only by edges but also by regions. In other words, the goal is not only to have the geodesic contours converge to regions of high gradients, as in the original models described above, but also to have them separate into uniform regions. The segmentation is then driven both by the search of uniform regions and the search of jumps in uniformity (edges). Uniform regions are defined with statistical measurements and texture analysis. Figure 3.21 shows an example of their algorithm. The reader is encouraged to read their work in order to obtain details and to see how the geodesic active contours can be combined with statistical analysis, texture analysis, and unsupervised learning capabilities to produce a state-of-the-art framework for image segmentation.

A related approach was developed by Chan and Colleagues [82,83]. (Another related approach was developed by Yezzi, Tsai, and Willsky, personal communication.) The authors have connected active contours, level sets,

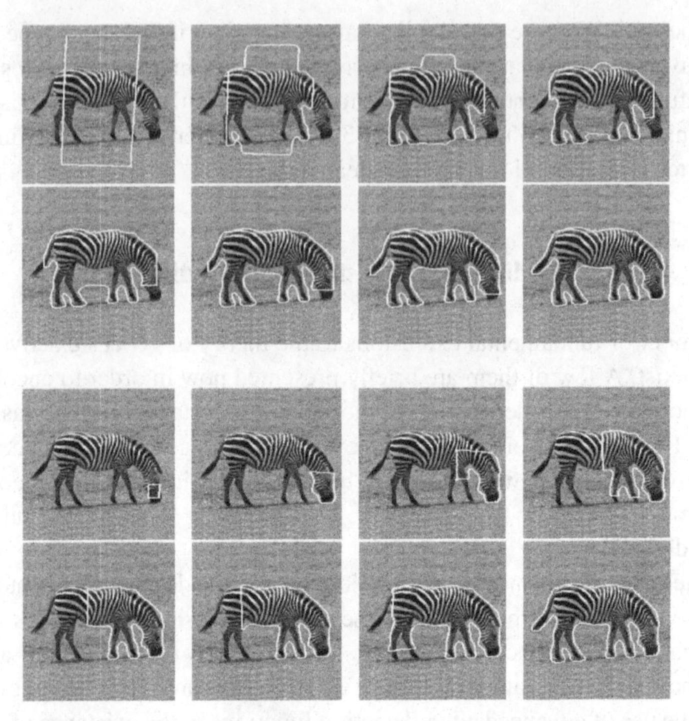

Fig. 3.21. Region active-contour example. The evolution of the contour is presented, for two different initial conditions. Note how for the geodesic active regions there is no need to have the initial contour completely inside or completely outside of the object of interest. (Figure courtesy of N. Paragios and R. Deriche.)

variational level sets, and the Mumford–Shah segmentation algorithm, and developed a framework for image segmentation without edges. The basic idea is to formulate a variational problem, related to Mumford–Shah, that searches for uniform regions while penalizing for the number of distinct regions. This energy is then embedded in the variational level set framework described in Chap. 2 and then solved with the appropriate numerical analysis machinery. A bit more detailed, the basic idea is to find two regions defined by a closed curve $\mathcal{C}$, each region with (unknown) gray-value averages A_1 and A_2 such that the set $(\mathcal{C}, A_1, A_2)$ minimizes

$$
\begin{aligned}
E(A_1, A_2, \mathcal{C}) = {} & \mu_1 \, \text{length}(\mathcal{C}) + \mu_2 \, \text{area inside}(\mathcal{C}) \\
& + \mu_3 \int_{\text{inside}(\mathcal{C})} [I(x, y) - A_1]^2 \mathrm{d}x\mathrm{d}y \\
& + \mu_4 \int_{\text{outside}(\mathcal{C})} [I(x, y) - A_2]^2 \mathrm{d}x\mathrm{d}y,
\end{aligned}
$$

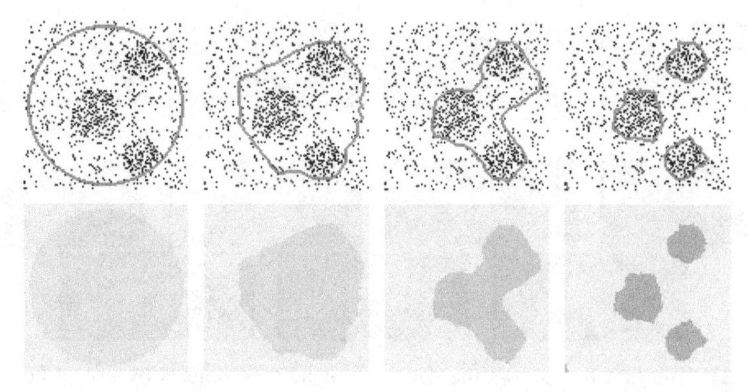

Fig. 3.22. Examples of segmentation without edges. Top: evolving curve, bottom: average of the approximation function inside and outside the evolving curve. (Figure courtesy of T. Chan and L. Vese.)

where μ_i are parameters and $I(x, y)$ is the image. As mentioned above, this is embedded in the variational level-set framework described before. Figure 3.22 presents an example of this algorithm. Again, state-of-the-art results are obtained with this scheme, especially for images with weak edges.

We can obtain, even without any basic modifications in the geodesic active contours, a number of improvements and extensions by playing with the g function, that is, with the edge-detection component of the framework. An example of this was presented in Section 3.3, in which it was shown how to redefine g for vector-valued images, permitting the segmentation of color and texture data. Paragios and Deriche pioneered the use of the geodesic active contours framework for tracking and the detection of moving objects, e.g., [300, 304]. The basic idea is to add to g a temporal component, that is, edges are defined not only by spatial gradients (as in still images), but also by temporal gradients. An object that is not moving will not be detected because its temporal gradient is zero. Figure 3.23 presents an example of this algorithm. An alternative to this approach is presented below.

3.7. Tracking and Morphing Active Contours

In the previous sections, the metric g used to detect the objects in the scene was primarily based on the spatial gradient or spatial vector gradient, that is, the metric favors areas of high gradient. In Ref. [298], Paragios and Deriche propose a different metric that allows for the detection of not just the scene objects, but of the scene objects that are moving. In other words, given two consecutive frames of a video sequence, the goal is now to detect and track objects that are moving in the scene. The idea is then to define a new

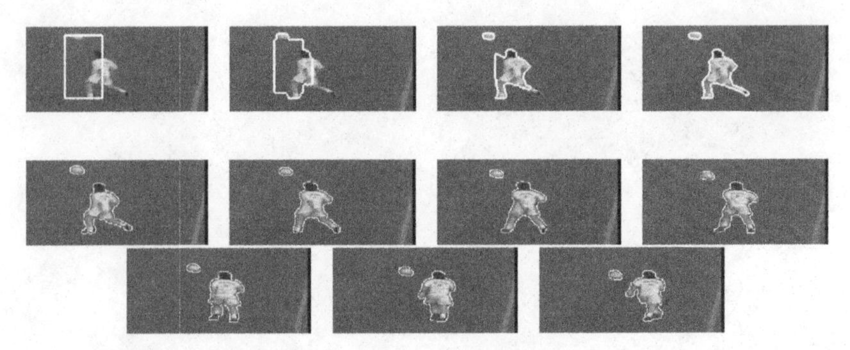

Fig. 3.23. The first row shows the detection of moving objects by use of both spatial and temporal gradients. The second and the third rows show the tracking with the active-region scheme of Paragios and Deriche. The algorithm combines optical-flow computation and moving-object detection. (Figure courtesy of N. Paragios and R. Deriche.)

function g that includes not only spatial gradient information, which will direct the geodesic curve to all the objects in the scene, but also temporal gradient, driving the geodesic to only the objects that are moving. With this concept in mind, many different g functions can be proposed.

A related approach for tracking is now described. The basic idea, inspired in part by the work on geodesic active contours and the work of Paragios and Deriche mentioned above, is to use information from one or more images to perform some operation on an additional image. Examples of this are given in Fig. 3.24. On the top row we have two consecutive slices of a 3D image obtained from EM. The image on the left has, superimposed, the contour of an object (a slice of a neuron). We can use this information to drive the segmentation of the next slice, the image on the right. On the bottom row we see two consecutive frames of a video sequence. The image on the left shows a marked object that we want to track. Once again, we can use the image on the left to perform the tracking operation in the image on the right. These are the types of problems that are addressed in this section. See [386] for a early use of the snakes framework for tracking.

Our approach is based on deforming the contours of interest from the first image toward the desired place in the second one. More specifically, we use a system of coupled PDEs to achieve this (coupled PDEs have already been used in the past to address other image processing tasks; see Refs. [324 and 322] and references therein). The first PDE deforms the first image, or features of it, toward the second one. The additional PDE is driven by the deformation velocity of the first one, and it deforms the curves of interest in the first image toward the desired position in the second one. This last deformation is implemented by the level-set numerical scheme, allowing

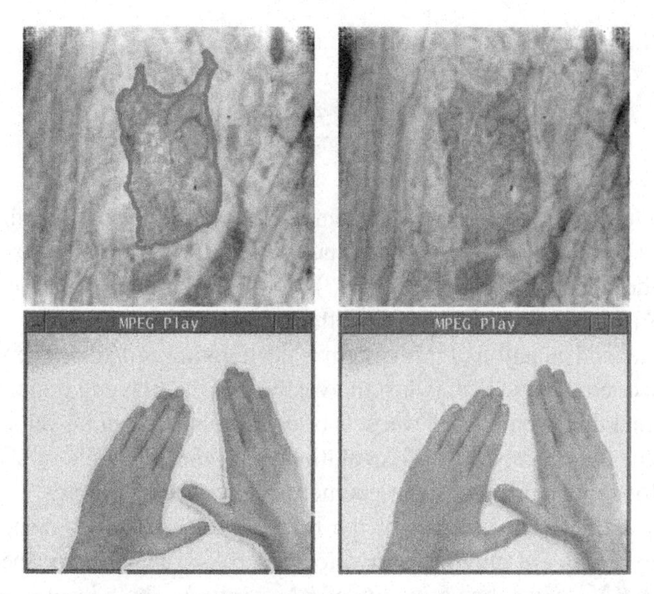

Fig. 3.24. Examples of the problems addressed in this section. See text.

for changes in the topology of the deforming curve; that is, if the objects of interest split or merge from the first image to the second one, these topology changes are automatically handled by the algorithm. This means that we will be able to track scenes with dynamic occlusions and to segment 3D medical data in which the slices contain cuts with different topologies.

Let $I_1(x, y, 0) : \mathbb{R}^2 \to \mathbb{R}$ be the current frame (or slice), where we have already segmented the object of interest. The boundary of this object is given by $\mathcal{C}_{I_1}(p, 0) : \mathbb{R} \to \mathbb{R}^2$. Let $I_2(x, y) : \mathbb{R}^2 \to \mathbb{R}$ be the image of the next frame, where we have to detect the new position of the object originally given by $\mathcal{C}_{I_1}(p, 0)$ in $I_1(x, y, 0)$. Let us define a continuous and Lipschitz function $u(x, y, 0) : \mathbb{R}^2 \to \mathbb{R}$, such that its zero level set is the curve $\mathcal{C}_{I_1}(p, 0)$. This function can be, for example, the signed distance function from $\mathcal{C}_{I_1}(p, 0)$. Finally, let us also define $\mathcal{F}_1(x, y, 0) : \mathbb{R}^2 \to \mathbb{R}$ and $\mathcal{F}_2(x, y) : \mathbb{R}^2 \to \mathbb{R}$ to be images representing features of $I_1(x, y, 0)$ and $I_2(x, y)$, respectively (e.g., $\mathcal{F}_i \equiv I_i$, or $\mathcal{F}_i$ equals the edge maps of I_i, $i = 1, 2$). With these functions as initial conditions, we define the following system of coupled evolution equations (t stands for the marching variable):

$$\frac{\partial \mathcal{F}_1(x, y, t)}{\partial t} = \beta(x, y, t)\ \|\nabla \mathcal{F}_1(x, y, t)\|, \tag{3.69}$$

$$\frac{\partial u(x, y, t)}{\partial t} = \hat{\beta}(x, y, t)\ \|\nabla u(x, y, t)\|$$

where the velocity $\hat{\beta}(x, y, t)$ is given by

$$\hat{\beta}(x, y, t) := \beta(x, y, t)\frac{\nabla \mathcal{F}_1(x, y, t)}{\|\nabla \mathcal{F}_1(x, y, t)\|} \cdot \frac{\nabla u(x, y, t)}{\|\nabla u(x, y, t)\|}. \tag{3.70}$$

The first equation of this system is the morphing equation, where $\beta(x, y, t) : \mathbb{R}^2 \times [0, \tau) \to \mathbb{R}$ is a function measuring the discrepancy between the selected features $\mathcal{F}_1(x, y, t)$ and $\mathcal{F}_2(x, y)$. This equation is morphing $\mathcal{F}_1(x, y, t)$ into $\mathcal{F}_2(x, y, t)$ so that $\beta(x, y, \infty) = 0$.

The second equation of this system is the tracking equation. The velocity in the second equation, $\hat{\beta}$, is just the velocity of the first one projected into the normal direction of the level sets of u. Because tangential velocities do not affect the geometry of the evolution, both the level sets of $\mathcal{F}_1$ and u are following exactly the same geometric flow. In other words, with $\vec{\mathcal{N}}_{\mathcal{F}_1}$ and $\vec{\mathcal{N}}_u$ as the inner normals of the level sets of $\mathcal{F}_1$ and u, respectively, these level sets are moving with velocities $\beta\vec{\mathcal{N}}_{\mathcal{F}_1}$ and $\hat{\beta}\vec{\mathcal{N}}_u$, respectively. Because $\hat{\beta}\vec{\mathcal{N}}_u$ is just the projection of $\beta\vec{\mathcal{N}}_{\mathcal{F}_1}$ into $\vec{\mathcal{N}}_u$, both level sets follow the same geometric deformation. In particular, the zero level set of u is following the deformation of $\mathcal{C}_{I_1}$, the curves of interest (detected boundaries in $I_1(x, y, 0)$). It is important to note that because $\mathcal{C}_{I_1}$ is not necessarily a level set of $I_1(x, y, 0)$ or $\mathcal{F}_1(x, y, 0)$, u is needed to track the deformation of this curve.

Because the curves of interest in $\mathcal{F}_1$ and the zero level set of u have the same initial conditions and they move with the same geometric velocity, they will then deform in the same way. Therefore, when the morphing of $\mathcal{F}_1$ into $\mathcal{F}_2$ has been completed, the zero level set of u should be the curves of interest in the subsequent frame $I_2(x, y)$.

We could argue that the steady state of Eqs. (3.69) is not necessarily given by the condition $\beta = 0$, as it can also be achieved with $\|\nabla \mathcal{F}_1(x, y, t)\| = 0$. This is correct, but it should not affect the tracking because we are assuming that the boundaries to track are not placed over regions where there is no information and then the gradient is flat. Therefore, for a certain band around the boundaries, the evolution will stop only when $\beta = 0$, thus allowing for the tracking operation.

For the examples below, we have opted for a very simple selection of the functions in the tracking system, namely

$$\mathcal{F}_i = \mathcal{L}(I_i), \quad i = 1, 2, \tag{3.71}$$

$$\beta(x, y, t) = \mathcal{F}_2(x, y) - \mathcal{F}_1(x, y, t), \tag{3.72}$$

where $\mathcal{L}(\cdot)$ indicates a band around $\mathcal{C}_{I_1}$, that is, for the evolving curve $\mathcal{C}_{I_1}$ we have an evolving band B of width w around it, and $\mathcal{L}[f(x, y, t)] = f(x, y, t)$ if (x, y) is in B, and it is zero otherwise. This particular morphing term is a local measure of the difference between $I_1(t)$ and I_2. It works to increase the gray value of $I_1(x_0, y_0, t)$ if it is smaller than $I_2(x_0, y_0)$, and to decrease it otherwise. Therefore the steady state is obtained when both values are equal to $\forall x_0, y_0$ in B, with $\|\nabla I_1\| \neq 0$. Note that this is a local measure, and that no hypothesis concerning the shape of the object to be tracked has been made. Having no model of the boundaries to track, the algorithm becomes very flexible. The main drawback of this particular selection, being so simple, is that it requires an important degree of similarity among the images for the algorithm to track the curves of interest and not to detect spurious objects. If the set of curves $\mathcal{C}_{I_1}$ isolates an almost uniform interior from an almost uniform exterior, then there is no need for a high similarity among consecutive images. On the other hand, when working with more cluttered images, if $\mathcal{C}_{I_1}(0)$ is too far away from the expected limit $\lim_{t\to\infty} \mathcal{C}_{I_1}(t)$, then the above-mentioned errors in the tracking procedure may occur. This similarity requirement concerns not only the shapes of the objects depicted in the image but especially their gray levels, as this β function measures gray-level differences. Therefore histogram equalization is always performed as a preprocessing operation.

We should also note that this particular selection of β involves information of the two present images. Better results are expected if information from additional images in the sequence is taken into account to perform the morphing between these two.

The first example of the tracking algorithm is presented in Fig. 3.25. This figure shows nine consecutive slices of neural tissue obtained by EM. The goal of the biologist is to obtain a 3D reconstruction of this neuron. As we observe from these examples, the EM images are very noisy, and the boundaries of the neuron are not easy to identify or to tell apart from other similar objects. Segmenting the neuron is then a difficult task. Before processing for segmentation, the images are regularized by anisotropic diffusion (see Chap. 4). Active-contour techniques such as those in described in Section 3.1 will normally fail with these types of images. Because the variation between consecutive slices is not too large, we can use the segmentation obtained for the first slice (segmentation obtained either manually or with the edge-tracing technique described in Section 3.4) to drive the segmentation of the next one, and then automatically proceed to find the segmentation in the following images. In this figure, the top left image shows the manual or semiautomatic segmentation superimposed, and the following ones show

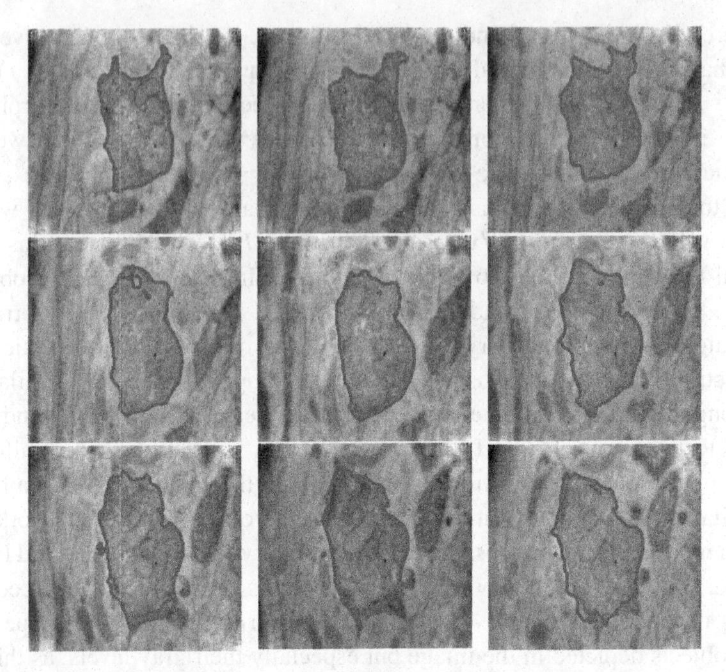

Fig. 3.25. Nine consecutive slices of neural tissue. The first image has been segmented manually. The segmentation over the sequence has been performed with the algorithm described in this section.

the boundaries found by the algorithm. Because of the particular choice of the β function, dissimilarities among the images cause the algorithm to mark, as part of the boundary, small objects that are too close to the object of interest. These can be removed by simple morphological operations. Cumulative errors might cause the algorithm to lose track of the boundaries after several slices, and reinitialization would be required.

We could argue that we could also use the segmentation of the first frame to initialize the active-contours technique mentioned above for the next frame. We still encounter a number of difficulties with this approach: (1) The deforming curve gets attracted to local minima and often fails to detect the neuron and (2) Those algorithms normally deform either inward or outward (mainly because of the presence of balloon-type forces), and the boundary curve corresponding to the first image is in general neither inside nor outside the object in the second image. To solve this, more elaborate techniques, e.g., as in Ref. [301], have to be used. Therefore, even if the image is not noisy, special techniques need to be developed and implemented to direct different points of the curve in different directions.

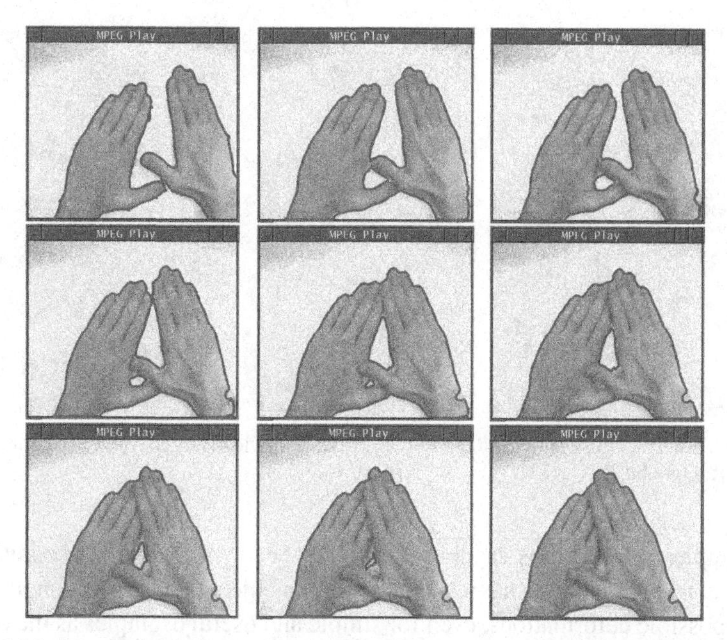

Fig. 3.26. Nine frames of a movie. The first image has been segmented manually. The segmentation over the sequence has been performed with the algorithm described in this section; note the automatic handling of topology changes.

Figure 3.26 shows an example of object tracking. The top left image has, superimposed, the contours of the objects to track. The subsequent images show the contours found by the algorithm. For the sake of space, only one out of every three frames is shown. Note how topological changes are handled automatically. A pioneering topology-independent algorithm for tracking video sequences, based on the general geodesic framework, can be found in Refs. [299 and 301]. In contrast with this approach, that scheme is based on a unique PDE (no morphing flow), deforming the curve toward a (local) geodesic curve, and it is very sensible to spatial and temporal noisy gradients. Because the similarity between frames, the algorithm just described converges very fast. In both Refs. [299 and 301] much more elaborate models are used to track, and, when testing some of the same sequences (e.g., the highway and two-man-walking sequences), a simpler algorithm such as the one here described already achieves satisfactory results. The elaborate models in the work of Paragios and Deriche are needed for more difficult scenes than the ones reported in this section. The CONDENSATION algorithm described in Ref. [37] can also achieve, in theory, topology-free tracking, although to the best of my knowledge real

Fig. 3.27. Tracking example on the *Walking Swedes* movie. (Data courtesy of N. Paragios and R. Deriche.)

examples showing this capability have not been yet reported. In addition, this algorithm requires having a model of the object to track and a model of the possible deformations, even for simple and useful examples as the ones shown below (note that the algorithm here proposed requires no previous or learned information). On the other hand, the outstanding tracking capabilities for cluttered scenes shown with the CONDENSATION scheme cannot be obtained with the simple selections for $\mathcal{F}_i$ and β used for the examples below, and more advanced selections must be investigated.

Additional tracking examples are given in the Figs. 27–29.

Fig. 3.28. Tracking example on the *Highway* movie. (Data courtesy of N. Paragios and R. Deriche.)

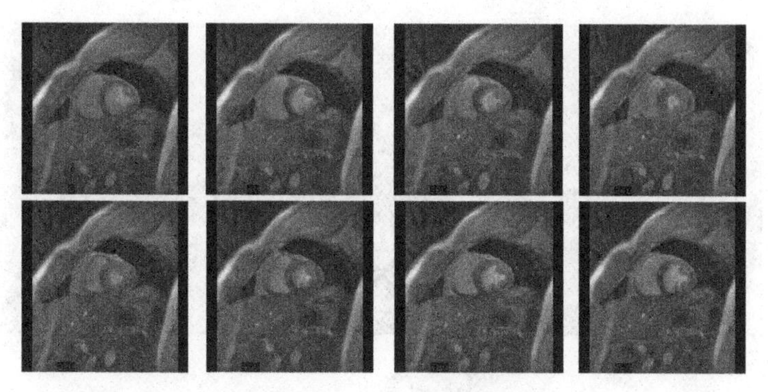

Fig. 3.29. Tracking example on the *Heart* movie. (Data courtesy of A. Tannenbaum.)

3.8. Stereo

We now follow Ref. [137] and show how the stereo problem can also be solved with the geodesic and minimal-area technique (see also [109]).

The problem of stereo is as follows: Recover the geometry of the scene when two or more images from the world are taken simultaneously. Because the internal parameters of the camera are unknown, the problem is essentially one of establishing correspondence between the views. The correspondence problem is usually addressed by setting up a functional and looking for its extrema. Once the correspondence has been achieved, the 3D point is reconstructed by intersecting the corresponding optical rays. See, for example, Refs. [130, 165, and 184] for further geometric details on the old problem of stereo.

Faugeras and Keriven [137] propose to start from some initial 3D surface $\mathcal{S}_0$ and deform it toward the minimization of a given geometric functional. One of the functionals proposed in Ref. [137] is as follows:

$$E(\mathcal{S}, \vec{\mathcal{N}}) = \iint \Psi(\mathcal{S}, \vec{\mathcal{N}})\mathrm{d}a = -\sum_{i,j=1,i\neq j}^{n} \frac{\langle I_i, I_j\rangle}{\langle I_i, I_i\rangle\langle I_j, I_j\rangle}, \qquad (3.73)$$

where i, j run over all the n available images I_i, and $\langle\cdot,\cdot\rangle$ is the cross correlation between the images, which includes the geometry of the perspective projection and the assumption of Lambertian surfaces [137]. Note that this approach consists basically of replacing the edge-dependent metric g that we used for the minimal-surface segmentation approach with a new metric Ψ that favors correlation between the collection of images. Faugeras and Keriven work out the Euler–Lagrange equations for this formulation, and, after embedding in the level-set framework, they obtain the following flow

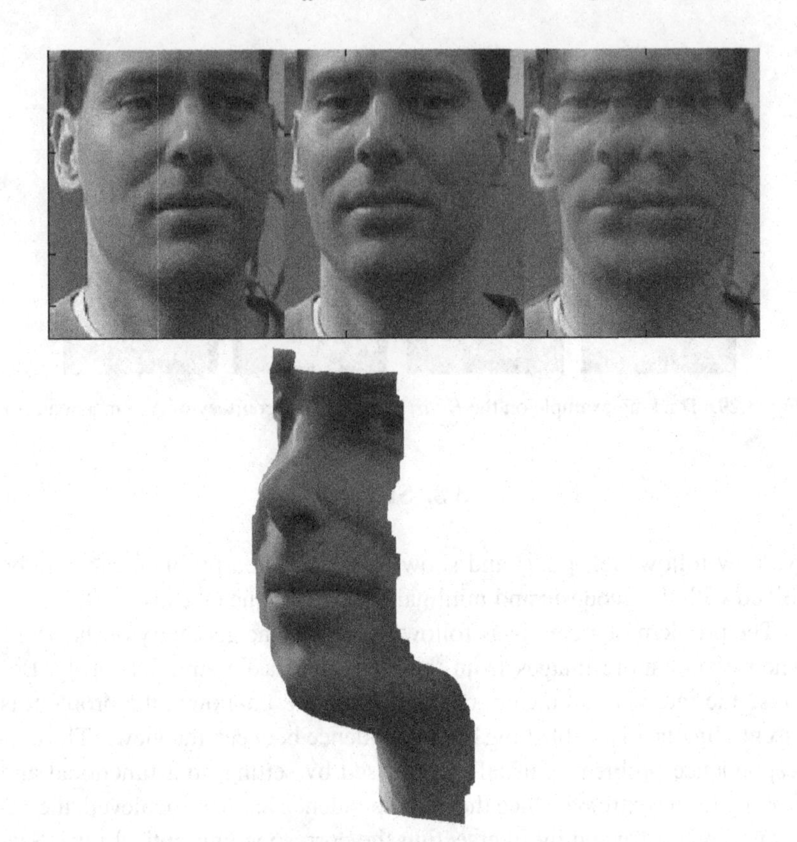

Fig. 3.30. 3D reconstruction from a stereo pair using the geodesic stereo approach. First, the stereo pair is shown, followed by the reconstructed 3D shape.

for $u : \mathbb{R}^3 \rightarrow \mathbb{R}$:

$$u_t = \mathbf{H}\,\|\nabla u\| - \frac{\partial u}{\partial \vec{\mathcal{N}}}[\mathrm{D}\vec{\mathcal{N}} + \mathrm{trace}(\mathrm{D}\vec{\mathcal{N}})\mathbf{I}_3]\,\|\nabla u\| \\ - \mathrm{trace}\left(\frac{\partial^2 u}{\partial \vec{\mathcal{N}} \partial X} + \mathrm{D}\vec{\mathcal{N}} \odot \frac{\partial^2 u}{\partial \vec{\mathcal{N}}^2}\right)\|\nabla u\|,$$

where derivatives are restricted to the tangential plane, $\mathrm{D}\vec{\mathcal{N}}$ is the 3×3 matrix of derivatives of the normal, and $\mathbf{I}_3$ is the 3×3 identity matrix. Examples of this are given in Fig. 3.30. This example was provided by Ronny Kimmel, and it corresponds to a simplification of the model by Faugeras and Keriven that he has recently proposed (additional examples can be found in the home pages of Faugeras and Keriven).

Appendix A

The analog to expression (3.7) when E_0 is a general value is now presented. Note that E_0 gives the difference between E_{int} and E_{ext} in Eq. (3.3). If $E_0 \neq 0$, then, instead of expression (3.7), the following minimization is obtained:

$$\min \int_0^1 \sqrt{2m}\sqrt{E_0 + \lambda g(I)^2}|C'|\mathrm{d}p. \tag{3.74}$$

In order for all the computations after expression (3.7) to hold, the expression above is equivalent to expression (3.7) if

$$g \leftarrow \sqrt{2m}\sqrt{E_0 + \lambda g(I)^2}.$$

As pointed out in Section 3.1, E_0 represents the trade-off between α and λ in Eq. (3.3) (as well as the parameterization), as is clear from the expressions above. Let us further develop this point here for completeness.

Rewriting $E_0 + \lambda g^2(I)$ as a quadratic form $(\sqrt{E_0} + Q)^2$, we can easily to show that $Q = -\sqrt{E_0} + \sqrt{E_0 + \lambda g(I)^2}$, and expression (3.74) becomes

$$\min \left(\int_0^1 Q\mathrm{d}s + \sqrt{E_0}L \right),$$

where L is the Euclidean length of the curve. Because Q is an edge detector like g, we see that basically the minimization problem has an extra term related to the length of the curve. The importance of this length in the minimization is given by the exact value of E_0, manifesting the relation between E_0 and the trade-off parameters α and λ in energy expression (3.3). Note that, as explained before, the Euler–Lagrange of L is κ, and this will appear as an extra term in the corresponding flow if $E_0 \neq 0$. Then, the new geodesic flow will be (compare with Eq. (3.13))

$$\frac{\partial C(t)}{\partial t} = Q(I)\kappa\vec{\mathcal{N}} - (\nabla Q \cdot \vec{\mathcal{N}})\vec{\mathcal{N}} + \sqrt{E_0}\kappa\vec{\mathcal{N}}. \tag{3.75}$$

The extra term appears unrelated to Q, which is the edge-detector part of the algorithm. Therefore, selecting E_0 too big will give too much importance to the minimization of L and may cause the flow to miss the edges. This is clear also from Eq. (3.3), which Eq. (3.75) is trying to minimize. Having $E_0 = 0$ is the only option that makes all the components of the geometric flow that minimizes Eq. (3.3) g dependent, giving a further justification for this selection.

Appendix B

We now compute the Euler–Lagrange of expression (3.8) to obtain the geodesic flow of Eq. (3.14). For the simplification of the notation, we sometimes write $\mathcal{C}(t)$ for the curve $\mathcal{C}(t, p)$, omitting the space parameter q, as well as writing $g(\mathcal{C})$ instead of $g[|\nabla I(\mathcal{C})|]$.

Consider the functional

$$L_R(\mathcal{C}) = \int_0^1 g[\mathcal{C}(t, p)]|\mathcal{C}_q(t, p)|\mathrm{d}p,$$

where $\mathcal{C} : [0, 1] \to \mathbb{R}^2$ is a closed (C^1) curve. Let us compute the first variation of L_R at some closed curve $\mathcal{C}_0$, assumed to be of class C^2. Consider a variation $\mathcal{C}$ of $\mathcal{C}_0$, that is,

$$\begin{aligned} \mathcal{C} : (-\epsilon, \epsilon) \times [0, 1] &\to \mathbb{R}^2 \\ (t, p) &\to \mathcal{C}(t, p) \end{aligned}$$

is a C^2 function of (t, p) such that $\mathcal{C}(0, p) \equiv \mathcal{C}_0$ and $\mathcal{C}(t, 0) = \mathcal{C}(t, 1)$, $t \in (-\epsilon, \epsilon)$ $(\epsilon > 0)$. Assuming a given orientation of $\mathcal{C}$, we compute the derivative of $L_R(\mathcal{C})$ with respect of t, obtaining

$$\begin{aligned} \frac{\mathrm{d}}{\mathrm{d}t} L_R[\mathcal{C}(t)] &= \int_0^1 \frac{\mathrm{d}}{\mathrm{d}t} g[\mathcal{C}(t, p)]|\mathcal{C}_q(t, p)|\mathrm{d}p + \int_0^1 g[\mathcal{C}(t, p)]\frac{\mathrm{d}}{\mathrm{d}t}|\mathcal{C}_q(t, p)|\mathrm{d}p \\ &= \int_0^1 \{\nabla g[\mathcal{C}(t, p)] \cdot \mathcal{C}_t(t, p)\}|\mathcal{C}_q(t, p)|\mathrm{d}p \\ &\quad + \int_0^1 g[\mathcal{C}(t, p)][\vec{\mathcal{T}}(t, p) \cdot \mathcal{C}_{tq}(t, p)]\mathrm{d}p, \end{aligned}$$

where $\vec{\mathcal{T}}(t, p)$ denotes the unit tangent to the curve $\mathcal{C}(t, p)$. Integrating by parts in the second term, we have that the above expression is equal to

$$\begin{aligned} &\int_0^1 \{\nabla g[\mathcal{C}(t, p)] \cdot \mathcal{C}_t(t, p)\}|\mathcal{C}_q(t, p)|\mathrm{d}p \\ &\quad - \int_0^1 \{g[\mathcal{C}(t, p)]\vec{\mathcal{T}}(t, p)\}_q \cdot \mathcal{C}_t(t, p)\mathrm{d}p \\ &= \int_0^1 (\{\nabla g[\mathcal{C}(t, p)] \cdot \mathcal{C}_t(t, p)\}|\mathcal{C}_q(t, p)| \\ &\quad - \{\nabla g[\mathcal{C}(t, p)] \cdot \mathcal{C}_q(t, p)\}[\vec{\mathcal{T}}(t, p) \cdot \mathcal{C}_t(t, p)] \\ &\quad - g[\mathcal{C}(t, p)]\vec{\mathcal{T}}_q(t, p) \cdot \mathcal{C}_t(t, p))\mathrm{d}p \end{aligned}$$

$$= \int_0^1 [(\nabla g[\mathcal{C}(t,p)] \cdot \mathcal{C}_t(t,p) - \{\nabla g[\mathcal{C}(t,p)] \cdot \mathcal{C}_s(t,p)\}[\vec{\mathcal{T}}(t,p) \cdot \mathcal{C}_t(t,p)])|\mathcal{C}_q(t,p)| - g[\mathcal{C}(t,p)]\vec{\mathcal{T}}_q(t,p) \cdot \mathcal{C}_t(t,p)]\mathrm{d}p.$$

Let s denote the arc length of $\mathcal{C}(t)$. Because $\vec{\mathcal{T}}_q = \vec{\mathcal{T}}_s|\mathcal{C}_q|$, parameterizing the curves by arc length, we write the above integral as

$$\int_0^{L[\mathcal{C}(t)]} (\{\nabla g[\mathcal{C}(t,s)] \cdot \mathcal{C}_t(t,s)\} - \{\nabla g[\mathcal{C}(t,s)] \cdot \vec{\mathcal{T}}(t,s)\}[\vec{\mathcal{T}}(t,s) \cdot \mathcal{C}_t(t,s)] - g[\mathcal{C}(t,s)]\vec{\mathcal{T}}_s(t,s) \cdot \mathcal{C}_t(t,s))\mathrm{d}s.$$

To simplify the notation, let us remove the arguments in the expression above, obtaining

$$\frac{\mathrm{d}}{\mathrm{d}t} L_R[\mathcal{C}(t)] = \int_0^{L[\mathcal{C}(t)]} \{\nabla g(\mathcal{C}) - [\nabla g(\mathcal{C}) \cdot \vec{\mathcal{T}}]\vec{\mathcal{T}} - g(\mathcal{C})\vec{\mathcal{T}}_s\} \cdot \mathcal{C}_t \mathrm{d}s.$$

At $t = 0$,

$$\frac{\mathrm{d}}{\mathrm{d}t} L_R[\mathcal{C}(t)]|_{t=0} = \int_0^{L(\mathcal{C}_0)} \{\nabla g(\mathcal{C}_0) - [\nabla g(\mathcal{C}_0) \cdot \vec{\mathcal{T}}]\vec{\mathcal{T}} - g(\mathcal{C}_0)\vec{\mathcal{T}}_s\} \cdot \mathcal{C}_t(0)\mathrm{d}s.$$

Because $\vec{\mathcal{T}}_s = \kappa\vec{\mathcal{N}}$, we have

$$\frac{\mathrm{d}}{\mathrm{d}t} L_R[\mathcal{C}(t)]|_{t=0} = \int_0^{L(\mathcal{C}_0)} \{\nabla g(\mathcal{C}_0) - [\nabla g(\mathcal{C}_0) \cdot \vec{\mathcal{T}}]\vec{\mathcal{T}} - g(\mathcal{C}_0)\kappa\vec{\mathcal{N}}\} \cdot \mathcal{C}_t(0)\mathrm{d}s$$

$$= \int_0^{L(\mathcal{C}_0)} \{[\nabla g(\mathcal{C}_0) \cdot \vec{\mathcal{N}}]\vec{\mathcal{N}} - g(\mathcal{C}_0)\kappa\vec{\mathcal{N}}\} \cdot \mathcal{C}_t(0)\mathrm{d}s.$$

This expression gives the Gateaux derivative (first variation) of L_R at $\mathcal{C} = \mathcal{C}_0$. Then, according to the steepest-descent method, to connect an initial curve $\mathcal{C}_0$ with a local minimum of $L_R(\mathcal{C})$ we should solve the evolution equation

$$\mathcal{C}_t = g(\mathcal{C})\kappa\vec{\mathcal{N}} - [\nabla g(\mathcal{C}) \cdot \vec{\mathcal{N}}]\vec{\mathcal{N}}.$$

This gives Eq. (3.13), that is, the motion of the level sets of Eq. (3.14), minimizing expression (3.8). To compute the motion of the embedding function u, the results in Chapter 2 are used. Following the same steps as before, we can also show that Eq. (3.14) is the flow corresponding to the steepest descent of

$$E(u) = \int_{\mathbb{R}^2} g(\mathcal{X})|\nabla u|\mathrm{d}\mathcal{X}.$$

Exercises

1. Compute $g(I)$, the weight used in the geodesic active contours, for a number of gray-scale images. Repeat the computation for different levels of smoothing applied to the image I and compare them.
2. Implement the geodesic active contours and investigate the influence of the constant balloon forge $g\vec{\mathcal{N}}$ and the doublet force $\nabla g \cdot \nabla u$. (When the Matlab, Maple, or Mathematica programming language and their level-set-finding capabilities are used, a simple implementation can be quickly obtained).
3. Following the computations in Appendix B, compute the corresponding gradient descent flow for the surface area $\iint \mathrm{d}a$ and the weighted surface area $\iint g(I)\mathrm{d}a$.
4. Compute the edges of color images by using the Riemannian formulation presented in this chapter and compare them with edges computed independently in each channel.
5. Create simple, piecewise-constant, examples of color images in which
   a. edges do not appear in some of the color channels and do appear in the Riemannian approach to edges of vector-valued images
   b. edges do not appear in the illuminant channel of a color image and do appear in the Riemannian approach to edges of vector-valued images
   c. edges have opposite directions in each one of the channels.
6. Compute the gradient descent flow corresponding to the energy that maps the multivalued level sets of a multivalued image into those of a single-valued image. Show that the Poisson equation is obtained. Repeat the exercise when only the normalized gradient is used.
7. Compute the weighted distance between two points in an image by using the formulations in Section 3.4. Using the curve evolution approach, compute the weighted distance between a point and the rest of the image. What happens if the point is on the boundary of an object?
8. Implement the affine invariant edge detector. Compare its quality with that of the classical Euclidean gradient.
9. Implement the affine invariant active contours.

CHAPTER FOUR

Geometric Diffusion of Scalar Images

In previous chapters PDEs were presented that deformed curves and surfaces, first according to their intrinsic geometry alone and then according to combinations of their intrinsic geometry and external, image-dependent velocities. In this and following chapters we deal with PDEs applied to full images. We first show how linear PDEs are related to linear, Gaussian filtering, and then extend these models by introducing nonlinear PDEs for image enhancement.

4.1. Gaussian Filtering and Linear Scale Spaces

The simplest and possibly one of the most popular ways to smooth an image $I(x, y): \mathbb{R}^2 \rightarrow \mathbb{R}$ is to filter it with a (radial) Gaussian filter, centered at 0, and with isotropic variance t. We then obtain

$$I(x, y, t) = I(x, t) * G_{0,t}.$$

For different variances t we obtain different levels of smoothing; see Fig. 4.1. This defines a scale space for the image, that is, we get copies of the image at different scales. Note of course that any scale t_1 can be obtained from a scale t_0, where $t_0 < t_1$, as well as from the original images. This is what we denoted in Section 2.5.2 as the causality criteria for scale spaces.

Filtering with a Gaussian has many properties, and as in the curves case discussed in Chap. 3, the scale space can be obtained following a series of intuitive physically based axioms. One of the fundamental properties of Gaussian filtering, which we have already seen when dealing with curves, is that $I(x, y, t)$ satisfies the linear heat flow or Laplace equation:

$$\frac{\partial I(x, y, t)}{\partial t} = \Delta I(x, y, t) = \frac{\partial^2 I(x, y, t)}{\partial x^2} + \frac{\partial^2 I(x, y, t)}{\partial y^2}. \tag{4.1}$$

Fig. 4.1. Smoothing an image with a series of Gaussian filters. Note how the information is gradually being lost when the variance of the Gaussian filter increases from left to right and from top to bottom.

This is very easy to show, first showing that the Gaussian function itself satisfies the Laplace equation (it is the kernel), and then adding the fact that derivatives and filtering are linear operations.

The flow of Eq. (4.1) is also denoted as isotropic diffusion, as it is diffusing the information equally in all directions. It is well know that the "heat" (gray values) will spread, and, in the end, a uniform image, equal to the average of the initial heat (gray values), is obtained. This can be observed in Fig. 4.1. Although this is very good for local reducing noise (averaging is optimal for additive noise; see Section 4.2), this filter also destroys the image content, that is, its boundaries. Our goal in the rest of this chapter is to replace the isotropic diffusion flow with anisotropic ones that remove noise while preserving edges.

4.2. Edge-Stopping Diffusion

We have just seen that we can smooth an image by using the Laplace equation or heat flow, with the image as initial condition. This flow, though, not only removes noise, but also blurs the image. Hummel [191] suggested that the heat flow is not the only PDE that can be used to enhance an image and that in order to keep the scale-space property we need only to make sure that the flow we use holds the maximum principle presented in Chap. 1. Many approaches have been taken in the literature to implement this idea. Although not all of the them really hold the maximum principle, they do share the concept of replacing the linear heat flow with a nonlinear PDE that does not diffuse the image in a uniform way. These flows are normally denoted as anisotropic diffusion, to contrast with the heat flow, that diffuses the image isotropically (equal in all directions).

The first elegant formulation of anisotropic diffusion was introduced by Perona and Malik [310] (see Ref. [146] for very early work in this topic and also Ref. [330] for additional pioneer work), and since then a considerable amount of research has been devoted to the theoretical and practical understanding of this and related methods for image enhancement. Research in this area has been oriented toward the understanding of the mathematical properties of anisotropic diffusion and related variational formulations [17, 75, 200, 310, 421], developing related well-posed and stable equations [5, 6, 75, 167, 271, 330, 421], extending and modifying anisotropic diffusion for fast and accurate implementations, modifying the diffusion equations for specific applications [157], and studying the relations between anisotropic diffusion and other image processing operations [334, 363, 429]. In this section we develop a statistical interpretation of anisotropic diffusion, specifically from the point of view of robust statistics. The Perona–Malik diffusion equation is presented and is shown to be equivalent to a robust procedure that estimates a piecewise-constant image from a noisy input image.

The robust statistical interpretation also provides a means for detecting the boundaries (edges) between the piecewise-constant regions in an image that has been smoothed with anisotropic diffusion. The boundaries between the piecewise-constant regions are considered to be outliers in the robust estimation framework. Edges in a smoothed image are therefore very simply detected as those points that are treated as outliers.

We will also show (following Ref. [34]) that, for a particular class of robust error norms, anisotropic diffusion is equivalent to regularization with an explicit line process. The advantage of the line-process formulation is that we can add constraints on the spatial organization of the edges. We demonstrate that adding such constraints to the Perona–Malik diffusion equation results in a qualitative improvement in the continuity of edges.

4.2.1. Perona–Malik Formulation

As we saw in Section 4.1, diffusion algorithms remove noise from an image by modifying the image by means of a PDE. For example, consider applying the isotropic diffusion equation (the heat equation) discussed in Section 4.1, given by

$$\frac{\partial I(x, y, t)}{\partial t} = \operatorname{div}(\nabla I),$$

with, of course, the original (degraded/noisy) image $I(x, y, 0)$ as the initial condition, where, once again, $I(x, y, 0) : \mathbb{R}^2 \to \mathbb{R}^+$ is an image in the continuous domain, (x, y) specifies spatial position, t is an artificial time parameter, and ∇I is the image gradient.

Perona and Malik [310] replaced the classical isotropic diffusion equation with

$$\frac{\partial I(x, y, t)}{\partial t} = \operatorname{div}[g(\|\nabla I\|)\nabla I], \tag{4.2}$$

where $\|\nabla I\|$ is the gradient magnitude and $g(\|\nabla I\|)$ is an edge-stopping function. This function is chosen to satisfy $g(x) \to 0$ when $x \to \infty$ so that the diffusion is stopped across edges.

As mentioned above, Eq. (4.2) motivated a large number of researchers to study the mathematical properties of this type of equation, as well as its numerical implementation and adaptation to specific applications. The stability of the equation was the particular concern of extensive research, e.g., Refs. [6, 75, 200, 310, and 421]. We should note that the mathematical study of that equation is not straightforward, although it can be shown that if $g(\cdot)$ is computed over a smoothed version of I, the flow is well posed and has a unique solution [75]. Note of course that a reasonable numerical implementation intrinsically smoothes the gradient, and then it is expected to be stable.

In what follows equations are presented that are modifications of Eq. (4.2). We do not discuss the stability of these modified equations because the stability results can be obtained from the mentioned references. Note again that possible stability problems will typically be solved, or at least moderated, by the spatial regularization and temporal delays introduced by the numerical methods for computing the gradient in $g(\|\nabla I\|)$ [75, 200, 293].

Perona–Malik Discrete Formulation. Perona and Malik discretized their anisotropic diffusion equation as follows:

$$I_s^{t+1} = I_s^t + \frac{\lambda}{|\eta_s|} \sum_{p \in \eta_s} g(\nabla I_{s,p}) \nabla I_{s,p}, \tag{4.3}$$

where I_s^t is a discretely sampled image, s denotes the pixel position in a discrete, 2D grid, and t now denotes discrete time steps (iterations). The constant $\lambda \in \mathbb{R}^+$ is a scalar that determines the rate of diffusion, η_s represents the spatial neighborhood of pixel s, and $|\eta_s|$ is the number of neighbors (usually four, except at the image boundaries). Perona and Malik linearly approximated the image gradient (magnitude) in a particular direction as

$$\nabla I_{s,p} = I_p - I_s^t, \quad p \in \eta_s. \tag{4.4}$$

Figure 4.9 in Subsection 4.2.3 shows examples of applying this equation to an image with two different choices for the edge-stopping function $g(\cdot)$. Qualitatively, the effect of anisotropic diffusion is to smooth the original image while preserving brightness discontinuities. As we will see, the choice of $g(\cdot)$ can greatly affect the extent to which discontinuities are preserved. Understanding this is one of our main goals in this chapter.

A Statistical View. Our goal is to develop a statistical interpretation of the Perona–Malik anisotropic diffusion equation. Toward that end, we adopt an oversimplified statistical model of an image. In particular, we assume that a given input image is a piecewise constant function that has been corrupted by zero-mean Gaussian noise with small variance. In Ref. [421], You et al. presented an interesting theoretical (and practical) analysis of the behavior of anisotropic diffusion for piecewise constant images. We will return later to comment on their results.

Consider the image intensity differences $I_p - I_s$ between pixel s and its neighboring pixels p. Within one of the piecewise constant-image regions, these neighbor differences will be small, zero mean, and normally distributed. Hence an optimal estimator for the true value of the image intensity I_s at pixel s minimizes the square of the neighbor differences. This is equivalent to choosing I_s to be the mean of the neighboring intensity values.

The neighbor differences will not be normally distributed, however, for an image region that includes a boundary (intensity discontinuity). Consider,

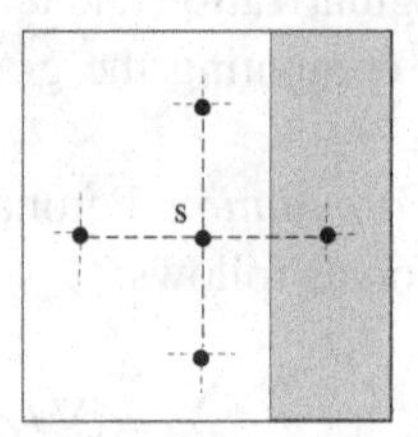

Fig. 4.2. Local neighborhood of pixels at a boundary (intensity discontinuity).

for example, the image region illustrated in Fig. 4.2. The intensity values of the neighbors of pixel s are drawn from two different populations, and in estimating the true intensity value at s we want to include only those neighbors that belong to the same population. In particular, the pixel labeled p is on the wrong side of the boundary so I_p will skew the estimate of I_s significantly. With respect to our assumption of Gaussian noise within each constant region, the neighbor difference $I_p - I_s$ can be viewed as an outlier because it does not conform to the statistical assumptions.

Robust Estimation. The field of robust statistics [172, 186] is concerned with estimation problems in which the data contain gross errors, or outliers.

Many robust statistical techniques have been applied to standard problems in computer vision [259, 320, 353]. There are robust approaches for performing local image smoothing [31], image reconstruction [153, 155], surface reconstruction [368], segmentation [258], pose estimation [225], edge detection [245], structure from motion or stereo [388, 408], optical flow estimation [32, 33, 352], and regularization with line processes [34]. For further details, see Ref. [172] or, for a review of the applications of robust statistics in computer vision, see Ref. [259].

The problem of estimating a piecewise constant (or smooth) image from noisy data can also be posed by use of the tools of robust statistics. We wish to find an image I that satisfies the following optimization criterion:

$$\min_I \sum_{s\in I} \sum_{p\in\eta_s} \rho(I_p - I_s, \sigma) \tag{4.5}$$

where $\rho(\cdot)$ is a robust error norm and σ is a scale parameter that will be discussed further below in Section 4.2.3. To minimize expression (4.5), the intensity at each pixel must be close to those of its neighbors. As we shall see, an appropriate choice of the ρ function allows us to minimize the effect of the outliers $(I_p - I_s)$ at the boundaries between piecewise constant-image regions.

Expression (4.5) can be solved by gradient descent:

$$I_s^{t+1} = I_s^t + \frac{\lambda}{|\eta_s|} \sum_{p \in \eta_s} \psi\left(I_p - I_s^t, \sigma\right), \tag{4.6}$$

where $\psi(\cdot) = \rho'(\cdot)$ and t again denotes the iteration. The update is carried out simultaneously at every pixel s.

The specific choice of the robust error norm or ρ function in expression (4.5) is critical. To analyze the behavior of a given ρ function, we consider its derivative (denoted ψ), which is proportional to the influence function [172]. This function characterizes the bias that a particular measurement has on the solution. For example, the quadratic ρ function has a linear ψ function.

If the distribution of values $(I_p - I_s^t)$ in every neighborhood is a zero-mean Gaussian, then $\rho(x, \sigma) = x^2/\sigma^2$ provides an optimal local estimate of I_s^t. This least-squares estimate of I_s^t is, however, very sensitive to outliers because the influence function increases linearly and without bound (see Fig. 4.3). For a quadratic ρ, I_s^{t+1} is assigned to be the mean of the neighboring intensity values I_p. This selection leads to the isotropic diffusion flow. When these values come from different populations (across a boundary) the mean is not representative of either population and the image is blurred too much. Hence the quadratic gives outliers (large values of $|\nabla I_{s,p}|$) too much influence.

To increase robustness and reject outliers, the ρ function must be more forgiving about outliers; that is, it should increase less rapidly than x^2. For example, consider the following Lorentzian error norm plotted in Fig. 4.4:

$$\rho(x, \sigma) = \log\left[1 + \frac{1}{2}\left(\frac{x}{\sigma}\right)^2\right], \qquad \psi(x, \sigma) = \frac{2x}{2\sigma^2 + x^2}. \tag{4.7}$$

Examination of the ψ function reveals that, when the absolute value of the gradient magnitude increases beyond a fixed point determined by the scale

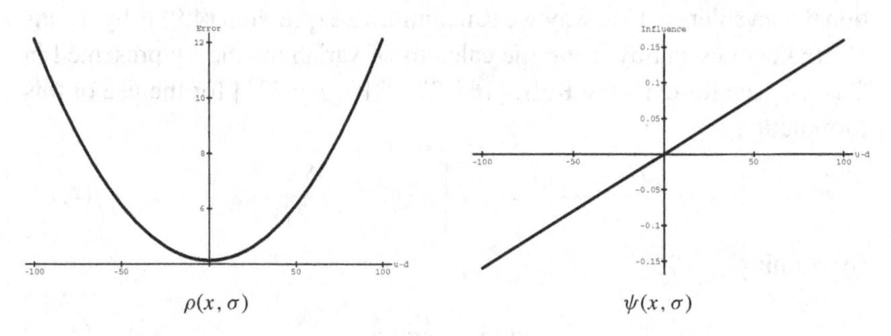

Fig. 4.3. Least-squares (quadratic) error norm.

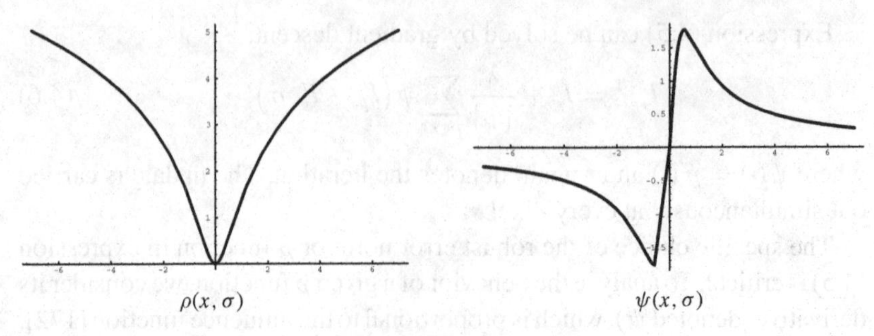

Fig. 4.4. Lorentzian error norm.

parameter σ, its influence is reduced. We refer to this as a redescending influence function [172]. If a particular local difference $\nabla I_{s,p} = I_p - I_s^t$, has a large magnitude then the value of $\psi(\nabla I_{s,p})$ will be small and therefore that measurement will have little effect on the update of I_s^{t+1} in Eq. (4.6).

4.2.2. Robust Statistics and Anisotropic Diffusion

We now explore the relationship between robust statistics and anisotropic diffusion by showing how to convert back and forth between the formulations. Recall the continuous anisotropic diffusion equation:

$$\frac{\partial I(x, y, t)}{\partial t} = \text{div}[g(\|\nabla I\|)\, \nabla I]. \tag{4.8}$$

The continuous form of the robust estimation problem in expression (4.5) can be posed as

$$\min_I \int_\Omega \rho(\|\nabla I\|) \mathrm{d}\Omega, \tag{4.9}$$

where Ω is the domain of the image and where we have omitted σ for notational convenience. One way we can minimize expression (4.9) is by means of gradient descent by using the calculus of variations theory presented in Chap. 1 (see for example Refs. [167, 273, 310, and 421] for the use of this formulation):

$$\frac{\partial I(x, y, t)}{\partial t} = \text{div}\left[\rho'(\|\nabla I\|) \frac{\nabla I}{\|\nabla I\|}\right]. \tag{4.10}$$

By defining

$$g(x) := \frac{\rho'(x)}{x}, \tag{4.11}$$

we obtain the straightforward relation between image reconstruction by means of robust estimation (4.9) and image reconstruction by means of anisotropic diffusion (4.8). You et al. [421] show and make extensive use of this important relation in their analysis.

The same relationship holds for the discrete formulation; compare Eqs. (4.3) and (4.6) with $\psi(x) = \rho'(x) = g(x)\,x$. Note that additional terms will appear in the gradient descent equation if the magnitude of the image gradient is discretized in a nonlinear fashion. We proceed with the discrete formulation as given in Section 4.1. The basic results presented hold for the continuous domain as well.

Perona and Malik suggested two different edge-stopping $g(\cdot)$ functions in their anisotropic diffusion equation. Each of these can be viewed in the robust statistical framework by converting the $g(\cdot)$ functions into the related ρ functions.

Perona and Malik first suggested

$$g(x) = \frac{1}{1 + \frac{x^2}{K^2}} \tag{4.12}$$

for a positive constant K. We want to find a ρ function such that the iterative solution of the diffusion equation and the robust statistical equation are equivalent. Letting $K^2 = 2\sigma^2$, we have

$$g(x)x = \frac{2x}{2 + \frac{x^2}{\sigma^2}} = \psi(x, \sigma), \tag{4.13}$$

where $\psi(x, \sigma) = \rho'(x, \sigma)$. Integrating $g(x)x$ with respect to x gives

$$\int g(x)x \, \mathrm{d}x = \sigma^2 \log\left[1 + \frac{1}{2}\left(\frac{x^2}{\sigma^2}\right)\right] = \rho(x). \tag{4.14}$$

This function $\rho(x)$ is proportional to the Lorentzian error norm introduced in Section 4.1, and $g(x)x = \rho'(x) = \psi(x)$ is proportional to the influence function of the error norm; see Fig. 4.5. Iteratively solving (4.6) with a

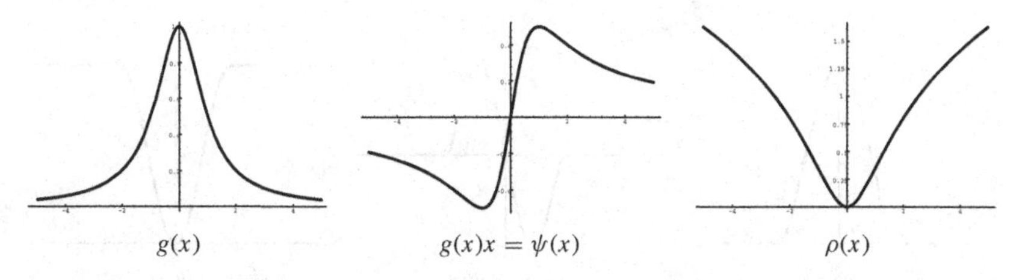

Fig. 4.5. Lorentzian error norm and the Perona–Malik g stopping function.

Lorentzian for ρ is therefore equivalent to the discrete Perona–Malik formulation of anisotropic diffusion. This relation was previously pointed out in Ref. [421] (see also Refs. [34 and 273].

The same treatment can be used to recover a ρ function for the other g function proposed by Perona and Malik

$$g(x) = e^{-\frac{x^2}{k^2}}. \tag{4.15}$$

The resulting ρ function is related to the robust error norm proposed by Leclerc [230]. The derivation is straightforward and is omitted here.

4.2.3. Exploiting the Relationship

The above derivations demonstrate that anisotropic diffusion is the gradient descent of an estimation problem with a familiar robust error norm. What is the advantage of knowing this connection? We argue that the robust statistical interpretation gives us a broader context within which to evaluate, compare, and choose between alternative diffusion equations. It also provides tools for automatically determining what should be considered an outlier (an edge). In this section these connections are illustrated with an example.

Although the Lorentzian is more robust than the L_2 (quadratic) norm, its influence does not descend all the way to zero. We can choose a more robust norm from the robust statistics literature that does descend to zero. The Tukey's biweight, for example, is plotted along with its influence function in Fig. 4.6:

$$\rho(x, \sigma) = \begin{cases} \frac{x^2}{\sigma^2} - \frac{x^4}{\sigma^4} + \frac{x^6}{3\sigma^6}, & |x| \leq \sigma \\ \frac{1}{3}, & \text{otherwise} \end{cases}, \tag{4.16}$$

$$\psi(x, \sigma) = \begin{cases} x[1 - (x/\sigma)^2]^2, & |x| \leq \sigma \\ 0, & \text{otherwise} \end{cases}, \tag{4.17}$$

$$g(x, \sigma) = \begin{cases} \frac{1}{2}[1 - (x/\sigma)^2]^2, & |x| \leq \sigma \\ 0, & \text{otherwise} \end{cases}, \tag{4.18}$$

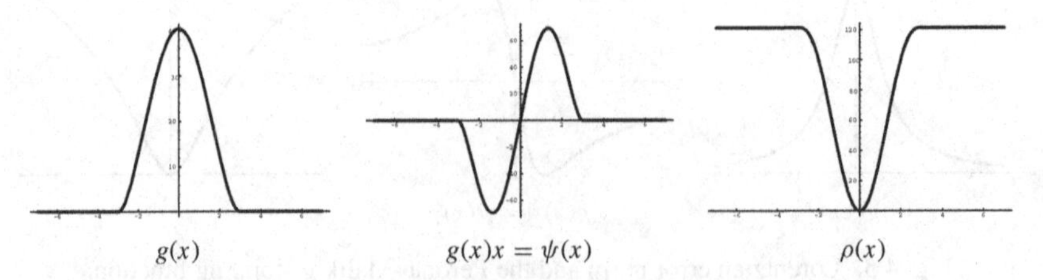

Fig. 4.6. Tukey's biweight.

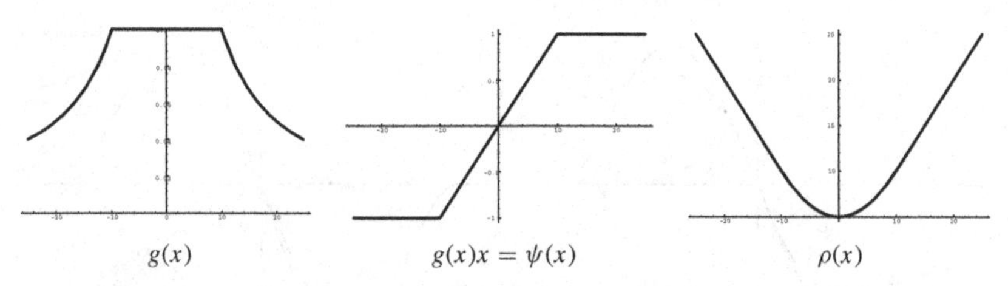

Fig. 4.7. Huber's minmax estimator (modification of the L_1 norm).

Another error norm from the robust statistics literature, Huber's minimax norm [186] (see also Refs. [330] and [421]), is plotted along with its influence function in Fig. 4.7. Huber's minmax norm is equivalent to the L_1 norm for large values. However, for normally distributed data, the L_1 norm produces estimates with higher variance than the optimal L_2 (quadratic) norm, so Huber's minmax norm is designed to be quadratic for small values:

$$\rho(x,\sigma) = \begin{cases} x^2/2\sigma + \sigma/2, & |x| \le \sigma \\ |x|, & |x| > \sigma \end{cases}, \tag{4.19}$$

$$\psi(x,\sigma) = \begin{cases} x/\sigma, & |x| \le \sigma \\ \text{sign}(x), & |x| > \sigma \end{cases}, \tag{4.20}$$

$$g(x,\sigma) = \begin{cases} 1/\sigma, & |x| \le \sigma \\ \text{sign}(x)/x, & |x| > \sigma \end{cases}, \tag{4.21}$$

We would like to compare the influence (ψ function) of these three norms, but a direct comparison requires that we dilate and scale the functions to make them as similar as possible.

First, we need to determine how large the image gradient can be before we consider it to be an outlier. We appeal to tools from robust statistics to automatically estimate the robust scale σ_e of the image as [326]

$$\begin{aligned} \sigma_e &= 1.4826\ \text{MAD}(\nabla I) \\ &= 1.4826\ \text{median}_I[\|\nabla I - \text{median}_I(\|\nabla I\|)\|] \end{aligned} \tag{4.22}$$

where MAD denotes the median absolute deviation and the constant is derived from the fact that the MAD of a zero-mean normal distribution with unit variance is $0.6745 = 1/1.4826$. These parameters can also be computed locally [35].

Second, we choose values for the scale parameters σ to dilate each of the three influence functions so that they begin rejecting outliers at the same value: σ_e. The point where the influence of outliers first begins to decrease occurs when the derivative of the ψ function is zero. For the modified L_1

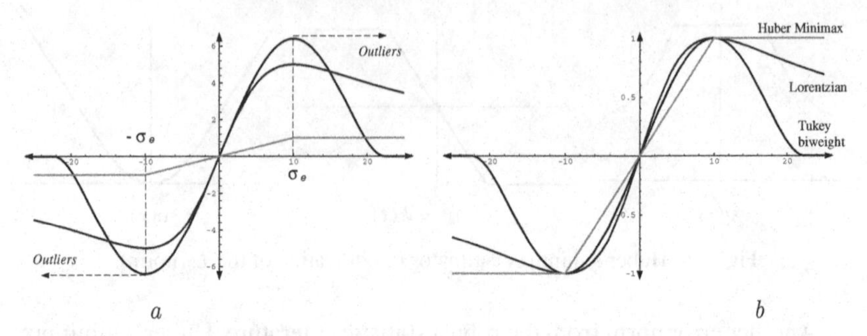

Fig. 4.8. Lorentzian, Tukey, and Huber ψ functions. *a*, values of σ chosen as a function of σ_e so that outlier rejection begins at the same value for each function; *b*, the functions aligned and scaled.

norm this occurs at $\sigma_e = \sigma$. For the Lorentzian norm it occurs at $\sigma_e = \sqrt{2}\,\sigma$ and for the Tukey norm it occurs at $\sigma_e = \sigma/\sqrt{5}$. Defining σ with respect to σ_e in this way we plot the influence functions for a range of values of x in Fig. 4.8a. Note how each function now begins reducing the influence of measurements at the same point.

Third, we scale the three influence functions so that they return values in the same range. To do this we take λ in Eq. (4.3) to be one over the value of $\psi(\sigma_e, \sigma)$. The scaled ψ functions are plotted in Fig. 4.8b.

Now we can compare the three error norms directly. The modified L_1 norm gives all outliers a constant weight of one whereas the Tukey norm gives zero weight to outliers whose magnitude is above a certain value. The Lorentzian (or Perona–Malik) norm is in between the other two. From the shape of $\psi(\cdot)$, we would correctly predict that diffusing with the Tukey norm produces sharper boundaries than diffusing with the Lorentzian (standard Perona–Malik) norm and that both produce sharper boundaries than the modified L_1 norm. We can also see how the choice of function affects the stopping behavior of the diffusion; given a piecewise constant image where all discontinuities are above a threshold, the Tukey function will leave the image unchanged whereas the other two functions will not.

These predictions are borne out experimentally, as can be seen in Fig. 4.9. The figure compares the results of diffusing with the Lorentzian $g(\cdot)$ function and the Tukey g function. The value of $\sigma_e = 10.278$ was estimated automatically with Eq. (4.22) and the values of σ and λ for each function were defined with respect to σ_e, as described above. The figure shows the diffused image after 100 iterations of each method. Observe how the Tukey function results in sharper discontinuities.

Fig. 4.9. Comparison of the Perona–Malik (Lorentzian) function (left) and the Tukey function (right) after 100 iterations. Top, original image; middle, diffused images; bottom, magnified regions of diffused images.

We can detect edges in the smoothed images very simply by detecting those points that are treated as outliers by the given ρ function. Figure 4.10 shows the outliers (edge points) in each of the images, where $|\nabla I_{s,p}| > \sigma_e$.

Finally, Fig. 4.11 illustrates the behavior of the two functions in the limit (shown for 500 iterations). The Perona–Malik formulation continues to smooth the image whereas the Tukey version has effectively stopped.

These examples illustrate how ideas from robust statistics can be used to evaluate and compare different g functions and how new functions can be

Fig. 4.10. Comparison of edges (outliers) for the Perona–Malik (Lorentzian) function (left) and the Tukey function (right) after 100 iterations. Bottom row shows a magnified region.

chosen in a principled way. See Ref. [34] for other robust ρ functions that could be used for anisotropic diffusion. See also Ref. [152] for related work connecting anisotropic diffusion, the mean-field ρ function, and binary line processes.

It is interesting to note that common robust error norms have frequently been proposed in the literature without a mention of the motivation from robust statistics. For example, Rudin et al. [330] proposed a formulation that is equivalent to using the L_1 norm (TV or total variation). As expected from the robust formulation described here and further analyzed in detail by Caselles and colleagues (personal communication) this norm will have a flat image as a unique steady-state solution (see also [76]). You et al. [421] explored a variety of anisotropic diffusion equations and reported better results for some than for others. In addition to their own explanation for this, their results are predicted, following the development presented here, by the robustness of the various error norms they use. Moreover, some of their theoretical results, e.g., Theorem 1 and Theorem 3, are easily interpreted based on the concept of influence functions. Finally, Harris et al. [177] and Mead [257] have used analog VLSI (aVLSI) technology to build hardware

Fig. 4.11. Comparison of the Perona–Malik (Lorentzian) function (left) and the Tukey function (right) after 500 iterations.

devices that perform regularization. The aVLSI circuits behave much like a resistive grid, except that the resistors are replaced with "robust" resistors made up of several transistors. Each such resistive grid circuit is equivalent to using a different robust error norm.

4.2.4. Robust Estimation and Line Processes

In this subsection we derive the relationship between anisotropic diffusion and regularization with line processes. The connection between robust statistics and line processes has been explored elsewhere; see Ref. [34] for details and examples as well as Refs. [38, 85, 152, 154, and 155] for recent related results. Although we work with the discrete formulation here, it is easy to verify that the connections hold for the continuous formulation as well.

Recall that the robust formulation of the smoothing problem was posed as the minimization of

$$E(I) = \sum_{s} E(I_s), \tag{4.23}$$

where

$$E(I_s) = \sum_{p \in \eta_s} \rho(I_p - I_s, \sigma). \tag{4.24}$$

There is an alternative, equivalent formulation of this problem that makes use of an explicit line process in the minimization:

$$E(I, \mathbf{l}) = \sum_s E(I_s, \mathbf{l}), \tag{4.25}$$

where

$$E(I_s, \mathbf{l}) = \sum_{p \in \eta_s} \left[\frac{1}{2\sigma^2}(I_p - I_s)^2 l_{s,p} + P(l_{s,p}) \right] \tag{4.26}$$

and where $l_{s,p} \in \mathbf{l}$ are analog line processes ($0 \leq l_{s,p} \leq 1$) [152, 153]. The line process indicates the presence (l close to 0) or absence (l close to 1) of discontinuities or outliers. The last term, $P(l_{s,p})$, penalizes the introduction of line processes between pixels s and p. This penalty term goes to zero when $l_{s,p} \to 1$ and is large (usually approaching 1) when $l_{s,p} \to 0$.

One benefit of the line-process approach is that the outliers are made explicit and therefore can be manipulated. For example, as we will see in Subsection 4.2.4 we can add constraints on these variables that encourage specific types of spatial organizations of the line processes.

Numerous authors have shown how to convert a line-process formulation into the robust formulation with a ρ function by minimizing over-the-line variables [38, 152, 154], that is

$$\rho(x) = \min_{0 \leq l \leq 1} E(x, l),$$

where

$$E(x, l) = [x^2 l + P(l)].$$

For our purposes here it is more interesting to consider the other direction: Can we convert a robust estimation problem into an equivalent line-process problem? We have already shown how to convert a diffusion problem with a $g(\cdot)$ function into a robust estimation problem. If we can make the connection between robust ρ functions and line processes then we will be able to take a diffusion formulation like the Perona–Malik equation and construct an equivalent line-process formulation.

Then our goal is to take a function $\rho(x)$ and construct a new function $E(x, l) = [x^2 l + P(l)]$ such that the solution at the minimum is unchanged. Clearly the penalty term $P(\cdot)$ will have to depend in some way on $\rho(\cdot)$. By taking derivatives with respect to x and l, it can be shown that the condition on $P(l)$ for the two minimization problems to be equivalent is given by

$$-x^2 = P'\left[\frac{\psi(x)}{2x}\right].$$

By integrating this equation we obtain the desired line process penalty function $P(l)$. See Ref. [34] for details on the explicit computation of this integral. There are a number of conditions on the form of ρ that must be satisfied in order to recover the line process, but, as described in Ref. [34], these conditions do in fact hold for many of the redescending ρ functions of interest.

In the case of the Lorentzian norm, it can be shown that $P(l) = l - 1 - \log l$; see Fig. 4.12. Hence, the equivalent line-process formulation of the Perona–Malik equation is:

$$E(I_s, \mathbf{l}) = \sum_{p\in\eta_s}\left[\frac{1}{2\sigma^2}(I_p - I_s)^2 l_{s,p} + l_{s,p} - 1 - \log l_{s,p}\right]. \tag{4.27}$$

Differentiating with respect to I_s and l gives the following iterative equations for minimizing $E(I_s, \mathbf{l})$:

$$I_s^{t+1} = I_s^t + \frac{\lambda}{|\eta_s|}\sum_{p\in\eta_s} l_{s,p}\nabla I_{s,p}, \tag{4.28}$$

$$l_{s,p} = \frac{2\sigma^2}{2\sigma^2 + \nabla I_{s,p}^2}. \tag{4.29}$$

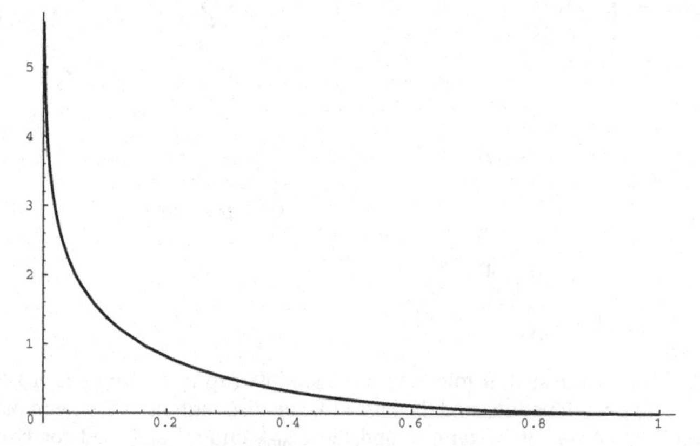

Fig. 4.12. Lorentzian (Perona–Malik) penalty function, $P(l)$, $0 \le l \le 1$.

Note that these equations are equivalent to the discrete Perona–Malik diffusion equations. In particular, $l_{s,p}$ is precisely equal to $g(\|\nabla I_{s,p}\|)$.

Spatial Organization of Outliers. One advantage of the connection between anisotropic diffusion and line processes, obtained through the connection of both techniques to robust statistics, is the possibility of improving anisotropic flows by the explicit design of line processes with spatial coherence. In the classical Perona–Malik flow, which relies on the Lorentzian error norm, there is no spatial coherence imposed on the detected outliers; see Fig. 4.10. Because the outlier process is explicit in the formulation of the line processing (Eq. (4.26)), we can add additional constraints on its spatial organization. Although numerous authors have proposed spatial coherence constraints for discrete line processes [92, 153, 154, 266], we need to generalize these results to the case of analog line processes [34].

We consider two kinds of interaction terms, hysteresis [59] and nonmaximum suppression [268]. Other common types of interactions (for example corners) can be modeled in a similar way. The hysteresis term assists in the formation of unbroken contours, and the nonmaximum suppression term inhibits multiple responses to a single edge present in the data. Hysteresis lowers the penalty for creating extended edges, and nonmaximum suppression increases the penalty for creating edges that are more than one pixel wide.

We consider a very simple neighborhood system as illustrated in Fig. 4.13. We define a new term that penalizes the configurations on the right of the figure and rewards those on the left. This term, E_I, encodes our prior assumptions about the organization of spatial discontinuities,

$$E_I(l_{s,p}) = \alpha \left[-\epsilon_1 \sum_{C_{\text{hyst}}} (1 - l_{s,p})(1 - l_{u,v}) + \epsilon_2 \sum_{C_{\text{supp}}} (1 - l_{s,p})(1 - l_{p,u}) \right],$$

$s\circ \mid p\circ$

$u\circ \mid v\circ$

C_{hyst}

$s\circ \mid p\circ \mid u\circ$

C_{supp}

Fig. 4.13. Cliques for spatial interaction constraints (up to rotation) at a site s. The circles indicate pixel locations and the bars indicate discontinuities between pixels. The C_{hyst} cliques are used for hysteresis, and the C_{supp} cliques are used for nonmaxima suppression.

where the parameters ϵ_1 and ϵ_2 assume values in the interval [0, 1] and α controls the importance of the spatial interaction term. These parameters were chosen by hand to compute the results here reported.

Starting with the Lorentzian norm, the new error term with constraints on the line processes bcomes

$$E(I, \mathbf{l}) = \sum_s E(\nabla I_{s,p}, l_{s,p}), \tag{4.30}$$

where

$$E(\nabla I_{s,p}, l_{s,p}) = \frac{1}{2\sigma^2}\nabla I_{s,p}^2 l_{s,p} + l_{s,p} - 1 - \log l_{s,p} + E_I(l_{s,p}).$$

Differentiating this equation with respect to I and l gives the following update equations:

$$I_s^{t+1} = I_s^t + \frac{\lambda}{|\eta_s|}\sum_{p\in\eta_s} l_{s,p}\nabla I_{s,p}, \tag{4.31}$$

$$l_{s,p}^{t+1} = \frac{2\sigma^2}{2\sigma^2\left[1+\epsilon_1 \sum_{C_{\text{hyst}}} \left(1-l_{u,v}^t\right) - \epsilon_2 \sum_{C_{\text{supp}}} \left(1-l_{p,u}^t\right)\right] + \nabla I_{s,p}^2}. \tag{4.32}$$

Without the additional spatial constraints, the line-process formulation was identical to the original Perona–Malik formulation. In contrast, note here that the value of the line process is dependent on neighboring values.

To see the effect of spatial constraints on the interpretation of discontinuities consider the simple example in Fig. 4.14. The original image is shown on the left. The next column shows values of the line process for the standard Perona–Malik diffusion equation, i.e., without the additional spatial

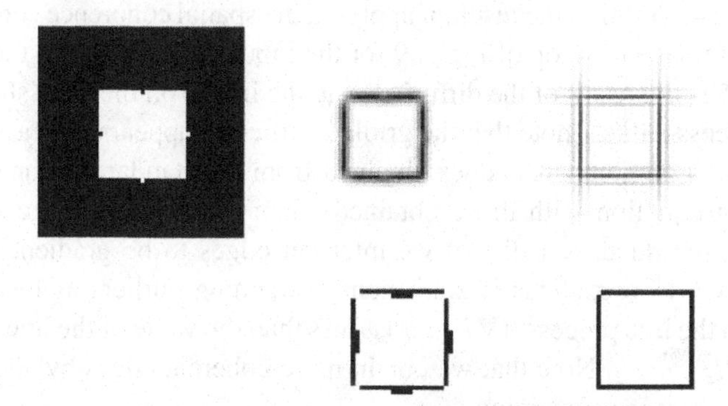

Fig. 4.14. Anisotropic diffusion with spatial organization of outliers. Left, input image; middle, line process for Perona–Malik (bottom – thresholded); right, Perona–Malik line process with spatial coherence (bottom – thresholded).

Fig. 4.15. Anisotropic diffusion with spatially coherent outliers. Left, smoothed image; right: the value of the line process at each point taken to be the product $l_{s,h}l_{s,v}$ of the horizontal and the vertical line processes at s.

coherence constraint. The value of the line process at each point is taken to be the product $l_{s,h}l_{s,v}$ of the horizontal and the vertical line processes. Dark values correspond to likely discontinuities. The bottom image in the column shows a thresholded version of the top image. Note how the line process is diffuse and how the anomalous pixels on the edges of the square produce distortions in the line process.

The column on the right of Fig. 4.14 shows the results when spatial coherence constraints are added. Note how the line process is no longer affected by the anomalous pixels; these are ignored in favor of a straight edge. Note also the hashed, or gridlike, pattern present in the image at the top right. This pattern reflects the simple notion of spatial coherence embodied in E_I that encourages horizontal and vertical edges and discourages edges that are more than one pixel wide.

Figure 4.15 shows the result of applying the spatial coherence constraints on a real image (see top of Fig. 4.9 for the input). The image on the left of Fig. 4.15 is the result of the diffusion, and the image on the right shows the line-process values (note that the gridlike structure appears here as well).

Figure 4.16 compares edges obtained from the standard Perona–Malik diffusion equation with those obtained when the spatial coherence constraints are added. Recall that we interpret edges to be gradient outliers where $|\nabla I_{s,p}| > \sigma_e$. This is equivalent to defining outliers as locations s at which the line process $l(\nabla I_{s,p}, \sigma)$ is less than the value of the line process when $\nabla I_{s,p} = \sigma_e$. Note that we obtain more coherent edges by adding the spatial coherence constraints.

The line-process formulation is more general than the standard anisotropic diffusion equation. Here only some simple examples have been shown of how spatial coherence can be added to diffusion. More sophisticated

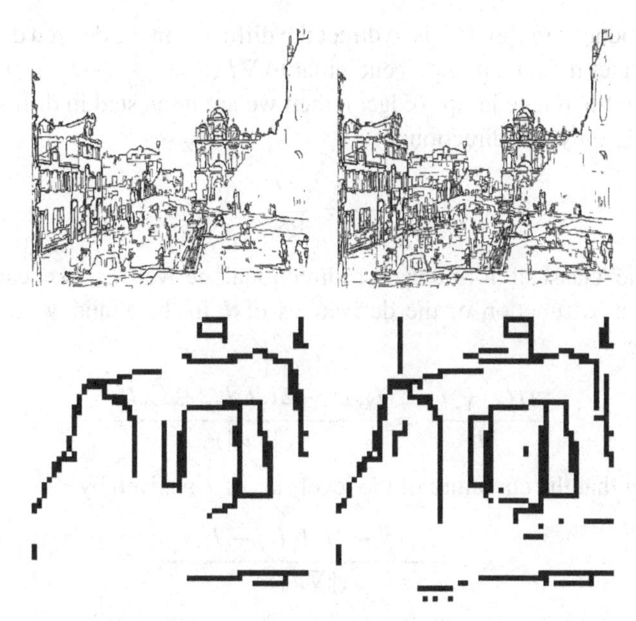

Fig. 4.16. Edges obtained with Perona–Malik (left), and Perona–Malik with additional spatial coherence in the line processes (right). Lower images show details on the gondola.

spatial coherence constraints could be designed to act over larger neighborhoods and/or to encourage edges at a greater variety of orientations.

4.2.5. Remarks on Numerics and Boundary Conditions

Extensive research has been done in the numerical analysis of the different forms of nonlinear diffusion [6, 79, 235, 330, 403, 406]. Both explicit and implicit schemes have been proposed. Because the goal of nonlinear diffusion is to preserve edges, the numerical implementation has to accomplish this goal as well.

With respect to boundary conditions, Newman boundary conditions are normally imposed, as also dictated by the state-of-the-art theoretical results in Ref. [10].

4.3. Directional Diffusion

We have seen that the heat flow diffuses the image in all directions in an equal amount. One possible way to correct this is to stop the diffusion across edges, and this was the technique presented above. An alternative way, introduced in Ref. [6], is to direct the diffusion in the desired direction. If ξ indicates the direction perpendicular to ∇I ($\xi = \frac{1}{\|\nabla I\|}(-I_y, I_x)$), that is, parallel to the image jumps (edges), then we are interested in diffusing the image I in only this direction:

$$\frac{\partial I}{\partial t} = \frac{\partial^2 I}{\partial \xi^2}.$$

Using the classical formulas for directional derivatives, we can write $\partial^2 I / \partial \xi^2$ as a function of the derivatives of I in the x and y directions, obtaining

$$\frac{\partial I(x, y, t)}{\partial t} = \frac{I_{xx} I_y^2 - 2 I_x I_y I_{xy} + I_{yy} I_x^2}{(\|\nabla u\|)^2}.$$

Recalling that the curvature of the level sets of I is given by

$$\kappa = \frac{I_{xx} I_y^2 - 2 I_x I_y I_{xy} + I_{yy} I_x^2}{(\|\nabla u\|)^3},$$

we find that the directional diffusion flow is equivalent to

$$\frac{\partial I(x, y, t)}{\partial t} = \kappa \; \|\nabla u\|, \tag{4.33}$$

which means that the level sets $\mathcal{C}$ of I are moving according to the Euclidean geometric heat flow

$$\frac{\partial \mathcal{C}}{\partial t} = \kappa \vec{\mathcal{N}}.$$

Directional diffusion is then equivalent to smoothing each one of the level sets according to the geometric heat flow.

To further stop diffusion across edges, it is common to modify the flow of Eq. (4.33) to obtain

$$\frac{\partial I(x, y, t)}{\partial t} = g(\|\nabla u\|) \kappa \; \|\nabla u\|, \tag{4.34}$$

where $g(r)$ is such that it goes to zero when $r \to \infty$. Figure 4.17 shows an example of this flow.

We can ask ourselves in which direction is the edge-stopping flows presented before diffusing the image. It is very easy to compute that if η is the direction perpendicular to ξ (parallel to the gradient), then the flow

$$\frac{\partial I}{\partial t} = \operatorname{div}\left[g(\|\nabla I\|) \frac{\nabla I}{\|\nabla I\|} \right]$$

Fig. 4.17. Example of anisotropic diffusion (original on the left and processed on the right).

can be decomposed with the relationship

$$\mathrm{div}[g(\|\nabla I\|)\nabla I] = g(\|\nabla I\|)I_{\xi\xi} + [\|\nabla I\|\ g(\|\nabla I\|)]' I_{\eta\eta},$$

which means that the flow is not only diffusing parallel to the edges, but it might also be performing an inverse diffusion in the direction of the gradient, perpendicular to the edges. This explains the enhancement characteristic of this flow.

4.3.1. Affine Invariant Anisotropic Diffusion

Instead of using the Euclidean geometric heat flow as the basis of the directional diffusion, we can use the affine one. In this case, we obtain

$$\frac{\partial I(x, y, t)}{\partial t} = g\kappa^{1/3}\ \|\nabla I\|, \tag{4.35}$$

which, if we set aside the edge-stopping function g, means that each one of the level sets $\mathcal{C}$ of I is moving according to the affine geometric heat flow

$$\frac{\partial \mathcal{C}}{\partial t} = \kappa^{1/3}\vec{\mathcal{N}}.$$

After the expression is written explicitly for the curvature of the level sets, the affine anisotropic flow becomes

$$\frac{\partial I(x, y, t)}{\partial t} = g\left(I_{xx}I_y^2 - 2I_xI_yI_{xy} + I_{yy}I_x^2\right)^{1/3}. \tag{4.36}$$

Note that, because of the $1/3$ power, the gradient in the denominator is eliminated (this is the only power that achieves this), making the numerical implementation of this flow more stable than its Euclidean counterpart.

If in addition to the stability advantage, we are interested in making the anisotropic diffusion flow completely affine invariant, we must make g affine invariant as well. In Chap. 3 we showed how to compute gradients in an affine invariant form, and Fig. 4.18 shows an example of the affine

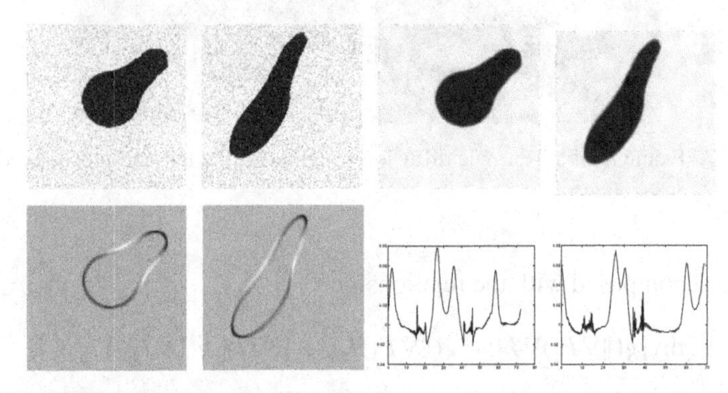

Fig. 4.18. Affine denoising, edge maps, and curvature computation for two affine-related images. The first row shows the two original noisy images (two left figures). Gaussian random noise was added independently to each one of the images after the affine transformation was performed. The subsequent images on the same row show the results of the affine denoising algorithm. The second row shows, on the left, affine invariant edges for the images after affine denoising. The last two figures on the second row show plots of the affine curvature vs. affine arc length of the midrange level set for these images. The affine curvature was computed with implicit functions. Although this is not the best possible way to compute the affine curvature in discrete curves, it is sufficient to show the qualitative behavior of the algorithm. The affine arc length was computed with the relation between affine and Euclidean arc lengths described in the text. The curve was smoothed with the affine geometric heat flow for a small number of steps to avoid large noise in the discrete computations. Note that different starting points were used for both images, and therefore the corresponding plots are shifted.

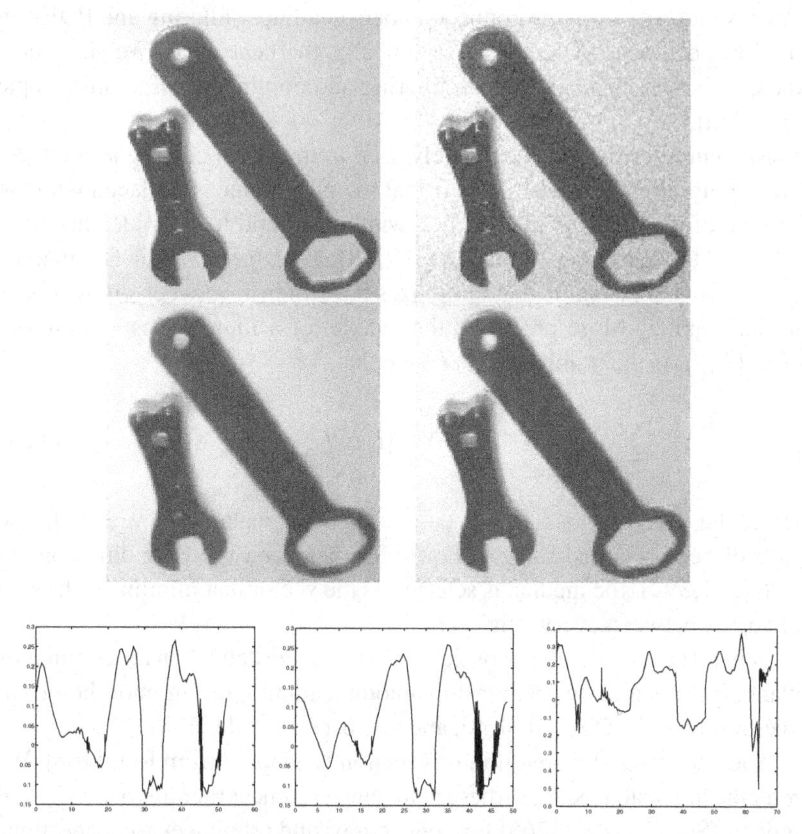

Fig. 4.19. Test of the importance of affine invariant denoising. The first row shows the original image followed by the noisy one. The second row shows the denoising results for the affine and Euclidean flows. The third row shows plots of the affine curvature vs. affine arc length of the midrange level set for the original image and those obtained from the affine and Euclidean denoising algorithms (second row images). The affine curvature was computed with implicit functions. The affine arc length was computed with the relation between affine and Euclidean arc lengths described in the text. In both cases, the curve was smoothed with the corresponding affine and Euclidean geometric heat flows for a small number of steps to avoid large noise in the discrete computations.

anisotropic diffusion equation with this invariant edge detector. Figure 4.19 shows the importance of performing completely affine invariant smoothing of images before the computation of its affine characteristics. Similar techniques were developed by Kimmel (personal communication), who also provided software for generating these figures.

4.3.2. Connections with Other Filters

We have already seen the connection between linear filtering and PDEs (in the form of linear PDEs, more specifically, the heat flow). We show now the connection between median filtering and nonlinear PDEs (anisotropic diffusion).

Median filtering has been widely used in image processing as an edge-preserving filter. The basic idea is that the pixel value is replaced with the median of the pixels contained in a window around it. This idea has been extended to vector-valued images [197, 391, 392], based on the fact that the median is also the value that minimizes the L_1 norm between all the pixels in the window. More precisely, the median of a finite series of numbers $\mathcal{U} = \{I_i\}_{i=1}^{M}$ is the number $I_j \in \mathcal{U}$ such that

$$\sum_{i=1}^{M} \|I_i - I_j\|_1 \leq \sum_{i=1}^{M} \|I_i - I_z\|_1, \quad \forall I_z \in \mathcal{U}. \tag{4.37}$$

Here, the pixels values $\{I_i\}_{i=1}^{N}$ can be either a scalar or a vector. In the case of vector-valued images, Eq. (4.37) has been used for directions as well [391, 392] (the median is selected as the vector that minimizes the sum of angles between every pair in $\mathcal{U}$).

When $I(x) : \mathbb{R}^2 \to \mathbb{R}$ is considered as an image defined in the continuous plane, there is a close relationship among median filtering, morphological operations, and PDEs [71, 168], and this is presented below.

The background material in this section is adapted from Ref. [168]. We refer the interested reader to these notes and references therein for details and proofs. (See also Ref. [265] for connections and references to connections between stack (median) filters and morphology.)

Definition 4.1. *Let $I(x) : \mathbb{R}^2 \to \mathbb{R}$ be the image, a map from the continuous plane to the continuous line. The median filter of u, with support given by a set B of bounded measure, is defined as*

$$\text{med}_B(I) := \inf\left\{\lambda \in \mathbb{R} : \text{measure}[I(x) \leq \lambda] \geq \frac{\text{measure}(B)}{2}\right\}. \tag{4.38}$$

Having this definition in mind, it is easy to prove the following result.

Proposition 4.1.

$$\text{med}_B(I) = \inf_{B' \in \mathcal{B}} \sup_{x \in B'} I(x), \tag{4.39}$$

where

$$\mathcal{B} := \left\{ B' : \text{measure}(B') \geq \frac{\text{measure}(B)}{2},\ B' \subset B \right\}.$$

This relation shows that the median is a morphological filter. This is a relation we will extend to vector-valued images in Section 5.2.

We now report on the relation between median filtering and mean curvature motion, which constitutes the second result we will extend to vector-valued images in Section 5.2.

Theorem 4.1. *Let I be three times differentiable, let $\kappa(I)$ be the curvature of the level set of u, and let $D(x_0, t)$ be the disk centered at x_0 and with radius t. Then,*

$$\text{med}_{D(x_0,t)}(I) = u(x_0) + \frac{1}{6}\kappa(I)\ \|\nabla I\|\ (x_0)t^2 + O(t^{2+1/3}), \quad (4.40)$$

if $\|\nabla I(x_0)\| \neq 0$, and

$$|\text{med}_{D(x_0,t)} - I(x_0)| \leq |\Delta I(x_0)| t^2 + O(t^3) \quad (4.41)$$

otherwise.

This result also leads to the following very interesting relation: If $t \to 0$, that is, the region of support shrinks to a point, iterating the median filter is equivalent to solving the geometric PDE

$$\frac{\partial I}{\partial t} = \kappa\ \|\nabla I\|, \quad (4.42)$$

which is exactly the directional diffusion flow presented above (without the stopping term). Therefore iterated median filtering is basically anisotropic diffusion. From this point of view, median filtering, which is classically used in image processing in its discrete form (and with finite support B), is a nonconsistent numerical implementation of anisotropic diffusion.

This connections follows from an additional relation between curvature motion and Gaussian filtering. Assume that a simple closed curve $\mathcal{C}$ is given (planar curve). Assign the value of one to all the points inside the curve and zero to the points outside of it, obtaining a binary image $I : \mathbb{R}^2 \to [0, 1]$. Filter I with a Gaussian filter of zero mean and variance σ, obtaining I_σ. Because of the maximum principle, $0 \leq I_\sigma \leq 1$. Now set all the points of I_σ greater or equal than $1/2$ back to one and those less than $1/2$ back to zero. It turns out that if $\sigma \to 0$, the iteration of this process (that is, filtering followed by thresholding) converges to the curvature motion

$$\frac{\partial \mathcal{C}}{\partial t} = \kappa \vec{\mathcal{N}}.$$

This filtering process was first proposed by Koenderink [219] as a replacement to the classical Gaussian filtering (note that a threshold nonlinearity is all that is added). Merriman et al. [260, 261] discovered the fascinating relation between this filtering process and curvature motion, a fact that was then proved and extended in Refs. [21, 124, 193, and 194]. Therefore curvature motion is basically the result of a liner filter followed by a static nonlinearity. We should note that this kind of concatenation is very popular to model different biological visual processes and can be used to generate very interesting patterns [332].

4.4. Introducing Prior Knowledge

Anisotropic diffusion applied to the raw image data is well motivated only when the noise is additive and signal independent. For example, if two objects in the scene have the same mean and differ only in variance, anisotropic diffusion of the data is not effective. In addition to this problem, anisotropic diffusion does not take into consideration the special content of the image, that is, in a large number of applications, such as MRI of the cortex and Synthetic Aperture Radar (SAR) data, it is known in advance the number of different types of objects in the scene, and directly applying anisotropic diffusion to the raw data does not take into consideration this important information given a priori.

A possible solution to the problems described above is presented now. The proposed scheme constitutes one of the steps of a complete system for the segmentation of MRI volumes of the human cortex. The technique comprises three steps; see Fig. 4.20. First, the posterior probability of each pixel is computed from its likelihood and a homogeneous prior, i.e., a prior that reflects the relative frequency of each class (white matter, gray matter, and nonbrain in the case of MRI of the cortex), but is the same across all pixels. Next, the posterior probabilities for each class are anisotropically smoothed. Finally, each pixel is classified independently by use of the maximum aposterior probability (MAP) rule. In Fig. 4.20b we use a toy example to motivate the use of this technique. We compare the results of isotropic diffusion on the raw data with those obtained with anisotropic diffusion on the posterior, both followed by a MAP classification. Figure 4.21 compares the classification of cortical white matter with and without the anisotropic smoothing step. Anisotropic smoothing produces classifications

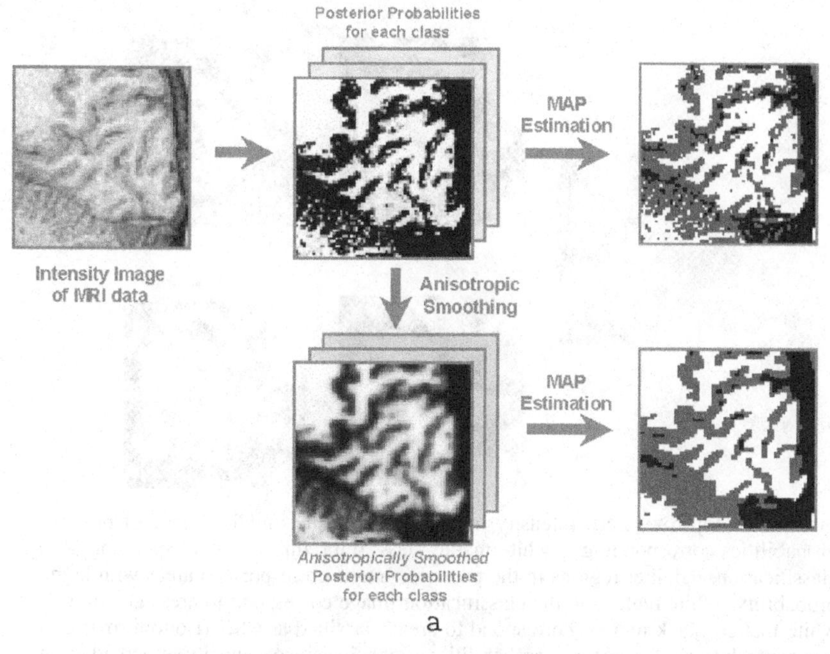

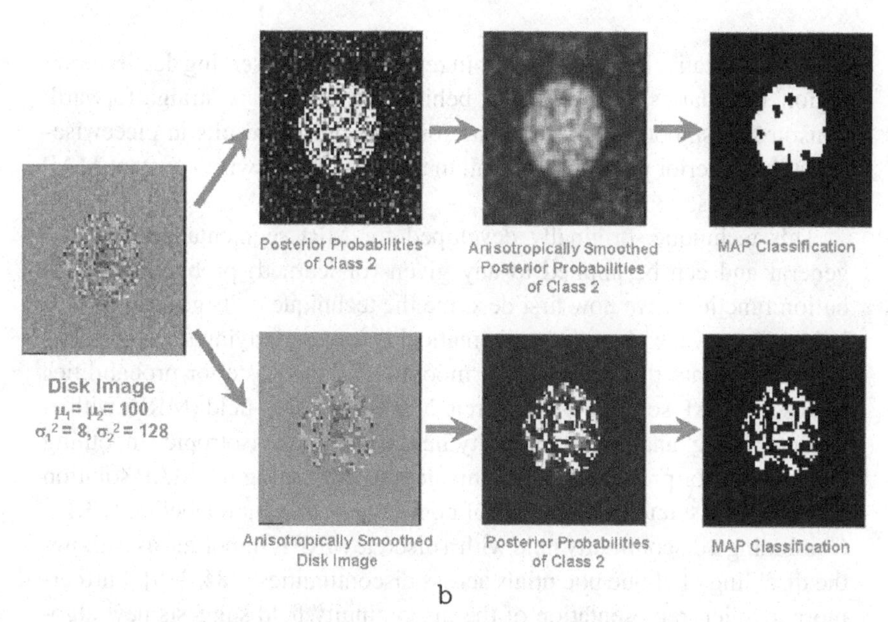

Fig. 4.20. a, Schematic description of the posterior diffusion algorithm; b, Toy example of the posterior diffusion algorithm. Two classes of the same average and different standard deviation are present in the image. The first row shows the result of the proposed algorithm (posterior, diffusion, MAP), and the second row shows the result of classical techniques (diffusion, posterior, MAP).

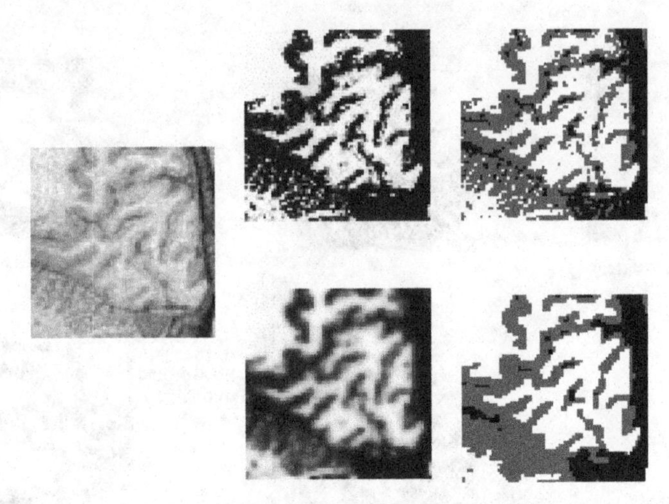

Fig. 4.21. (Top row) Left, Intensity image of MRI data; middle: image of posterior probabilities corresponding to white-matter class; right, image of corresponding MAP classification. Brighter regions in the posterior image correspond to areas with higher probability. White regions in the classification image correspond to areas classified as white matter; black regions correspond to areas classified as CSF. (Bottom row) Left, image of white-matter posterior probabilities after being anisotropically smoothed; right: image of MAP classification computed with smoothed posteriors.

that are qualitatively smoother within regions while preserving details along region boundaries. The intuition behind the method is straightforward. Anisotropic smoothing of the posterior probabilities results in piecewise-constant posterior probabilities that, in turn, yield piecewise-constant MAP classifications.

This technique, originally developed for MRI segmentation, is quite general and can be applied to any given (or learned) probability distribution functions. We now first describe the technique in its general formulation. Then we explore the mathematical theory underlying the technique. We demonstrate that anisotropic smoothing of the posterior probabilities yields the MAP solution of a discrete Markov random field (MRF) with a noninteracting, analog discontinuity field. In contrast, isotropic smoothing of the posterior probabilities is equivalent to computing the MAP solution of a single, discrete MRF by use of continuous-relaxation labeling (CRL). Combining a discontinuity field with a discrete MRF is important as it allows the disabling of clique potentials across discontinuities [384, 385]. Furthermore, explicit representation of the discontinuity field suggests new algorithms that incorporate hysteresis and nonmaximal suppression as shown when the robust diffusion filter was presented before, Section 4.2.4.

We now deal with the scalar case, while later in this book we will show how to diffuse the whole probability vector with coupled PDEs.

4.4.1. The General Technique

Let us assume that it is given to us, in the form of a priori information, the number k of classes (different objects) in the image. In MRI of the cortex for example, these classes would be white matter, gray matter, and cerebral spinal fluid nonbrain. In the case of SAR images, the classes can be object and background. Similar classifications can be obtained for a large number of additional applications and image modalities.

In the first stage, the pixel or voxel intensities within each class are modeled as independent random variables with given or learned distributions. Thus the likelihood $\Pr(V_i = v \mid C_i = c)$ of a particular pixel (or voxel in the case of 3D MRI) V_i belonging to a certain class C_i is given. For example, in the case of normally distributed likelihood, we have

$$\Pr(V_i = v \mid C_i = c) = \frac{1}{\sqrt{2\pi}\sigma_c} \exp\left[-\frac{1}{2}\frac{(v-\mu_c)^2}{\sigma_c^2}\right]. \tag{4.43}$$

Here, i is a spatial index ranging over all pixels or voxels in the image, and the index c stands for one of the k classes. V_i and C_i correspond to the intensity and classification of voxel i, respectively. The mean and the variance (μ_c and σ_c) are given, learned from examples or adjusted by the user.

The posterior probabilities of each voxel's belonging to each class are computed with Bayes' rule:

$$\Pr(C_i = c \mid V_i = v) = \frac{1}{K} \Pr(V_i = v \mid C_i = c) \Pr(C_i = c), \tag{4.44}$$

where K is a normalizing constant independent of c. As in the case of the likelihood, the prior distribution $\Pr(C_i = c)$ is not restricted and can be arbitrarily complex. In a large number of applications, we can adopt a homogeneous prior, which implies that $\Pr(C_i = c)$ is the same over all spatial indices i. The prior probability typically reflects the relative frequency of each class.

After the posterior probabilities are computed (note that we will have now k images), the posterior images are smoothed anisotropically (in two or three dimensions), but preserving discontinuities. The anisotropic smoothing technique applied can be based on the original version proposed by Perona and Malik [310] or any other of the extensions later proposed and already discussed in this chapter. As we have seen, this step involves simulating a discretization of a PDE:

$$\frac{\partial P_c}{\partial t} = \text{div}[g(\|\nabla P_c\|)\nabla P_c], \tag{4.45}$$

where $P_c = \Pr(C = c \mid V)$ stand for the posterior probabilities for class c, the stopping term $g(\|\nabla P_c\|) = \exp[-(\|\nabla P_c\| \,/\eta_c)^2]$, and η_c represents the rate of diffusion for class c. The function $g(\cdot)$ controls the local amount of diffusion such that diffusion across discontinuities in the volume is suppressed. Because we are now smoothing probabilities, to be completely formal, these evolving probabilities should be normalized each step of the iteration to add to one. This problem will be formally addressed and solved later in this book; see Chap. 6

Finally, the classifications are obtained with the MAP estimate after anisotropic diffusion, that is,

$$C_i^* = \arg\max_k \Pr{}^*(C_i = c \mid V_i = v), \tag{4.46}$$

where $\Pr^*(C_i = c \mid V_i = v)$ corresponds to the posterior following anisotropic diffusion.

Recapping, the proposed algorithm has the following steps:

1. Compute the priors and likelihood functions for each one of the classes in the images.
2. Using Bayes' rule, compute the posterior for each class.
3. Apply anisotropic diffusion (combined with normalization) to the posterior images.
4. Use MAP to obtain the classification.

This techniques solves both problems mentioned in the introduction, that is, it can handle nonadditive noise, and, more importantly, it introduces prior information about the type of images being processed. We now describe the relations of this technique with other algorithms proposed in the literature.

4.4.2. Isotropic Smoothing

In this section, the relationship between the MAP estimation of discrete MRFs and CRL is described [325]. This connection was originally made by Li et al. [241]. We review this relationship to introduce the notation that will be used and to point out the similarities between this technique and isotropic smoothing of posterior probabilities.

We specialize our notation to MRFs defined on image grids. Let $\mathcal{S} = \{1, \ldots, n\}$ be a set of sites where each $s \in \mathcal{S}$ corresponds to a single pixel in the image. For simplicity, we assume that each site can take on labels from a common set $\mathcal{L} = \{1, \ldots, k\}$. Adjacency relationships between sites are encoded by $\mathcal{N} = \{\mathcal{N}_i \mid i \in \mathcal{S}\}$, where $\mathcal{N}_i$ is the set of sites neighboring site i. Cliques are then defined as subsets of sites so that any pair of sites in a clique are neighbors. For simplicity, we consider four-neighbor adjacency for images (and eight-neighbor adjacency for volumes) and cliques of sizes no greater than two. By considering each site as a discrete random variable f_i with a probability mass function over $\mathcal{L}$, a discrete MRF $\mathbf{f}$ can be defined over the sites with a Gibbs probability distribution.

If data $d_i \in \mathbf{d}$ is observed at each site i and is dependent on only its label f_i, then the posterior probability is itself a Gibbs distribution and, by the Hammersley–Clifford theorem, also a MRF, albeit a different one [153]: $P(\mathbf{f} \mid \mathbf{d}) = Z^{-1} \times \exp\{-E(\mathbf{f} \mid \mathbf{d})\}$, where

$$E(\mathbf{f} \mid \mathbf{d}) = \sum_{i \in \mathcal{C}_1} V_1(f_i \mid d_i) + \sum_{(i,j) \in \mathcal{C}_2} V_2(f_i, f_j), \tag{4.47}$$

where $V_1(f_i \mid d_i)$ is a combination of the single-site clique potential and the independent likelihood and $V_2(f_i, f_j)$ is the pairwise-site clique potential. The notation (i, j) refers to a pair of sites; thus the sum is actually a double sum. Maximizing the posterior probability $P(\mathbf{f} \mid \mathbf{d})$ is equivalent to minimizing the energy $E(\mathbf{f} \mid \mathbf{d})$.

As a result, the MAP classification problem is one of finding the set of classes $\mathbf{f}^*$ that minimizes

$$\mathbf{f}^* = \arg \min_{\mathbf{f} \in \mathcal{L}^n} E(\mathbf{f} \mid \mathbf{d}). \tag{4.48}$$

This is a combinatorial problem because $\mathcal{L}^n$ is discrete. There are a variety of solution techniques to this problem, some of which are stochastic, like simulated annealing, whereas others are deterministic [30].

Continuous Relaxation Labeling: The CRL approach to solving this problem was introduced by Li et. al. [241]. In CRL, the class (label) of each site i is represented by a vector $p_i = [p_i(f_i) \mid f_i \in \mathcal{L}]$ subject to the constraints: (1) $p_i(f_i) \geq 0$ for all $f_i \in \mathcal{L}$, and (2) $\sum_{f_i \in \mathcal{L}} p_i(f_i) = 1$. Within this framework, the energy $E(\mathbf{f} \mid \mathbf{d})$ to be minimized is rewritten as

$$E(\mathbf{p} \mid \mathbf{d}) = \sum_{i \in \mathcal{C}_1} \sum_{f_i \in \mathcal{L}} V_1(f_i \mid d_i) p_i(f_i) + \sum_{(i,j) \in \mathcal{C}_2} \sum_{(f_i, f_j) \in \mathcal{L}^2} V_2(f_i, f_j) p_i(f_i) p_j(f_j).$$

Note that when $p_i(f_i)$ is restricted to $\{0, 1\}$, $E(\mathbf{p} \mid \mathbf{d})$ reverts to its original counterpart $E(\mathbf{f} \mid \mathbf{d})$. Hence CRL embeds the actual combinatorial problem into a larger, continuous, constrained minimization problem.

The constrained minimization problem is typically solved by iterating two steps: (1) gradient computation and (2) normalization and update. The first step decides the direction that decreases the objective function and the second updates the current estimate while ensuring compliance with the constraints. A review of the normalization techniques that have been proposed are summarized in Ref. [241]. Ignoring the need for normalization, CRL is similar to traditional gradient descent: $p_i^{t+1}(f_i) \leftarrow p_i^t(f_i) - \frac{\partial E(\mathbf{p} \mid \mathbf{d})}{\partial p_i^t(f_i)}$ where

$$\frac{\partial E(\mathbf{p} \mid \mathbf{d})}{\partial p_i^t(f_i)} \doteq V_1(f_i \mid d_i) + 2 \sum_{j:(i,j) \in \mathcal{C}_2} \sum_{f_j \in \mathcal{L}} V_2(f_i, f_j) p_j^t(f_j) \tag{4.49}$$

and the superscripts t, $t + 1$ denote iteration numbers. The notation $j : (i, j)$ refers to a single sum over j such that (i, j) are pairs of sites belonging to a clique. Barring the different normalization techniques that could be used, Eq. (4.49) is found in the update equations of various CRL algorithms [133, 192, 325]. There are, however, two differences. First, in most CRL problems, the first term of Eq. (4.49) is absent and thus proper initialization of $\mathbf{p}$ is important. We also omit this term in the rest of the section to emphasize the similarity with CRL. Second, CRL problems typically involve maximization; thus, $V_2(f_i, f_j)$ would represent consistency as opposed to potential, and the update equation would add instead of subtract the gradient.

Isotropic Smoothing: A convenient way of visualizing the above operation is as isotropic smoothing. Because the sites represent pixels in an image, for each class f_i, $p_i(f_i)$ can be represented by an image (of posterior probabilities) such that k classes imply k such image planes. Together, these k planes form a volume of posterior probabilities. Each step of Eq. (4.49) then essentially replaces the current estimate $p_i^t(f_i)$ with a weighted average of the neighboring assignment probabilities $p_j^t(f_j)$. In other words, the volume of posterior probabilities is linearly filtered. If the potential functions $V_2(f_i, f_j)$ favor similar labels, then the weighted average is essentially low pass among sites with common labels and high pass among sites with differing labels.

This notion is best illustrated with a simple example. Consider a classification problem with three classes; i.e., $f_i \in \{1, 2, 3\}$ and the volume is made up of three planes. Define $V_2(f_i, f_j) = -1/4$ when $f_i = f_j$ and sites i and j are four-neighbors; for example, when $f_i = 2$ and i is the middle site, $V_2(2, f_j)$ has the following values:

$$\begin{bmatrix} 0 & 0 & 0 \\ 0 & 0 & 0 \\ 0 & 0 & 0 \end{bmatrix} \begin{bmatrix} 0 & -1/4 & 0 \\ -1/4 & 0 & -1/4 \\ 0 & -1/4 & 0 \end{bmatrix} \begin{bmatrix} 0 & 0 & 0 \\ 0 & 0 & 0 \\ 0 & 0 & 0 \end{bmatrix}.$$
$$f_j = 1 \qquad\qquad f_j = 2 \qquad\qquad f_j = 3$$

Thus the penalty is smaller when adjacent pixels have the same class than when their classes differ. When these penalties are replaced with the update equation, $p_i^t(f_i)$ (with $f_i = 2$) is replaced with a linear combination of $p_j^t(f_j)$ with weights equal to

$$\begin{bmatrix} 0 & 0 & 0 \\ 0 & 0 & 0 \\ 0 & 0 & 0 \end{bmatrix} \begin{bmatrix} 0 & 1/4 & 0 \\ 1/4 & 1 & 1/4 \\ 0 & 1/4 & 0 \end{bmatrix} \begin{bmatrix} 0 & 0 & 0 \\ 0 & 0 & 0 \\ 0 & 0 & 0 \end{bmatrix}.$$
$$f_j = 1 \qquad\qquad f_j = 2 \qquad\qquad f_j = 3$$

As a result, the posterior probabilities for each class are smoothed during each step of the iteration. In this example, each of the three planes of posterior probabilities is low-pass filtered separately.

4.4.3. Anisotropic Smoothing

As we have seen, isotropic smoothing causes significant blurring, especially across region boundaries. A solution to this problem is to smooth adaptively such that smoothing is suspended across region boundaries and takes place only within region interiors. In this section, we show that although isotropic smoothing of posterior probabilities is the same as CRL of a MRF, anisotropic smoothing of posterior probabilities is equivalent to CRL of a MRF supplemented with a (hidden) analog discontinuity field. We also demonstrate that this method could also be understood as incorporating a robust consensus-taking scheme within the framework of CRL.

We extend the original MRF problem to include a noninteracting, analog discontinuity field on a displaced lattice. Thus the new energy to be minimized is

$$E(\mathbf{f}, \mathbf{l}) = \sum_{(i,j)\in C_2} \left[\frac{1}{2\sigma^2} V_2(f_i, f_j) \cdot l_{i,j} + (l_{i,j} - 1 - \log l_{i,j}) \right], \quad (4.50)$$

where $V_1(f_i)$ has been dropped for simplicity, as the discontinuity field does not interact with it. The individual sites in the discontinuity field $\mathbf{l}$ are denoted by $l_{i,j}$, which represent either the horizontal or the vertical separation between sites i and j in $\mathcal{S}$. When $l_{i,j}$ is small, indicating the presence of a discontinuity, the effect of the potential $V_2(f_i, f_j)$ is suspended; meanwhile, the energy is penalized by the second term in Eq. (4.50). There are a variety of penalty functions that could be derived from the robust estimation framework (see Ref. [34]). The penalty function in Eq. (4.50) was derived from the Lorentzian robust estimator.

The minimization of $E(\mathbf{f}, \mathbf{l})$ is now over both $\mathbf{f}$ and $\mathbf{l}$. Because the discontinuity field is noninteracting, we can minimize $\mathbf{l}$ analytically by computing the partial derivatives of $E(\mathbf{f}, \mathbf{l})$ with respect to $l_{i,j}$ and setting that to zero. Doing so and inserting the result back into $E(\mathbf{f}, \mathbf{l})$ gives us

$$E(\mathbf{f}) = \sum_{(i,j)\in\mathcal{C}_2} \log\left[1 + \frac{1}{2\sigma^2} V_2(f_i, f_j)\right]. \tag{4.51}$$

Rewriting this equation in a form suitable for CRL, we get

$$E(\mathbf{p}) = \sum_{(i,j)\in\mathcal{C}_2} \log\left[1 + \frac{1}{2\sigma^2} \sum_{(f_i, f_j)\in\mathcal{L}^2} V_2(f_i, f_j) p_i(f_i) p_j(f_j)\right]. \tag{4.52}$$

Anisotropic Smoothing: To compute the update equation for CRL, we take the derivative of $E(\mathbf{p})$ with respect to $p_i(f_i)$:

$$\frac{\partial E(\mathbf{p})}{\partial p_i(f_i)} \doteq \sum_{j:(i,j)\in\mathcal{C}_2} w_{i,j} \left[\sum_{f_j\in\mathcal{L}} V_2(f_i, f_j) p_j(f_j)\right] \tag{4.53}$$

where

$$w_{i,j} = 2\sigma^2 \Bigg/ \left[2\sigma^2 + \sum_{(f_i, f_j)\in\mathcal{L}^2} V_2(f_i, f_j) p_i(f_i) p_j(f_j)\right]. \tag{4.54}$$

The term $w_{i,j}$ encodes the presence of a discontinuity. If $w_{i,j}$ is constant, then Eq. (4.53) reverts to the isotropic case. Otherwise, $w_{i,j}$ either enables or disables the penalty function $V_2(f_i, f_j)$. This equation is similar to the anisotropic diffusion equation proposed by Perona and Malik [310]. However, the image difference between sites i and j in the equation of Perona and Malik is, in our case, replaced with a discrete version: $\sum_{f_j\in\mathcal{L}} V_2(f_i, f_j) p_j(f_j)$. The stopping term $w_{i,j}$ is also the same except that the magnitude of the image gradient is again replaced with a discrete counterpart. Figure 4.22 compares isotropic and anisotropic diffusion of the posterior and shows $w_{i,j}$.

Robust Continuous Relaxation Labeling: Each iteration of CRL can be viewed as a consensus-taking process [407]. Neighboring pixels vote

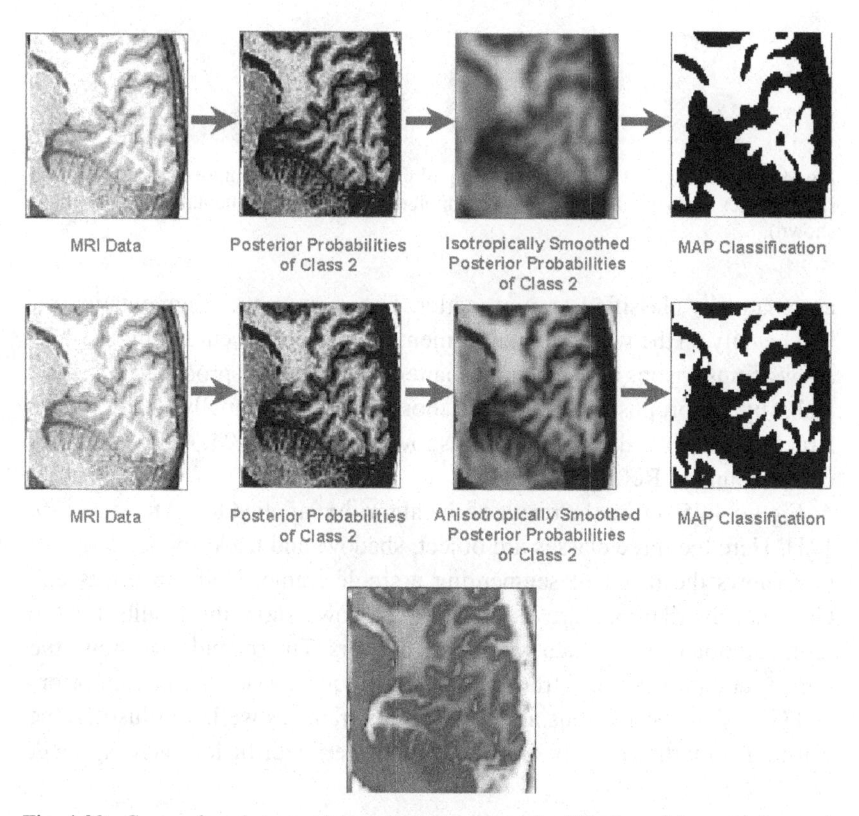

Fig. 4.22. Comparison between isotropic and anisotropic diffusion of the posterior, and the variable weight $w_{i,j}$ (last row) is shown for one of the classes in the anisotropic case.

on the classification of a central pixel based on their current assignment probabilities $p_j(f_j)$, and their votes are tallied by a weighted sum. The weights used are the same throughout the image; thus pixels on one side of a region boundary may erroneously vote for pixels on the other side. Following the robust formulation of anisotropic diffusion, anisotropic smoothing of the posterior probabilities can be regarded as implementing a robust voting scheme because votes are tempered by $w_{i,j}$, which estimates the presence of a discontinuity.

4.4.4. Discussion

The anisotropic smoothing scheme was first proposed and used to segment white matter from MRI data of a human cortex. Pixels at a given distance from the boundaries of the white-matter classification were then

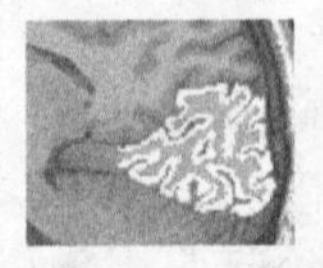
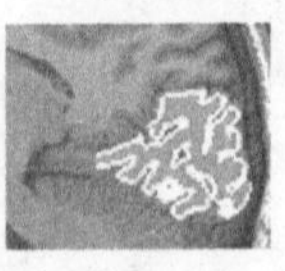

Fig. 4.23. The two left images show manual gray-matter segmentation results; the two right images show the automatically computed gray-matter segmentation (same slices shown).

automatically classified as gray-matter. Thus gray-matter segmentation relied heavily on the white-matter segmentation's being accurate. Figure 4.23 shows comparisons between gray-matter segmentations produced automatically by the proposed method and those obtained manually. A 3D reconstruction based on this technique is shown in the Fig. 4.24. More examples can be found in Ref. [384].

Figure 4.25 shows the result of the algorithm applied to SAR data [170, 171]. Here the three classes are object, shadow, and background. The first row shows the result of segmenting a single frame. Uniform priors and Gaussian distributions are used. The next rows show the results for the segmentation of video data with learned priors. The second row shows the segmentation of the first frame of a video sequence (left). Uniform priors and Gaussian distributions are used for this frame as well. To illustrate the learning importance, only two steps of posterior diffusion were applied.

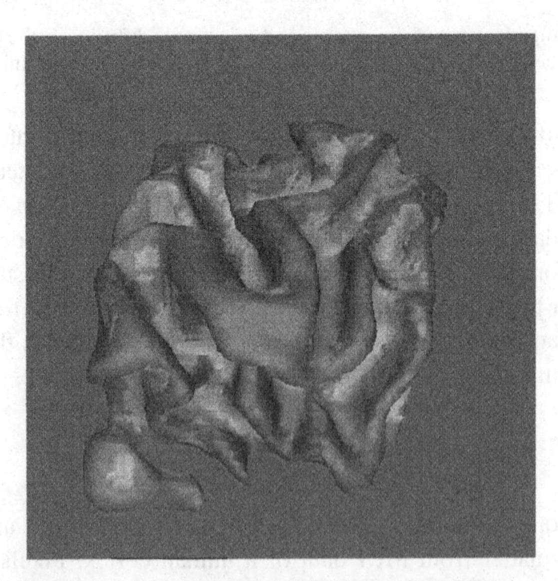

Fig. 4.24. 3D reconstruction of the human white matter from MRI, based on posterior diffusion techniques.

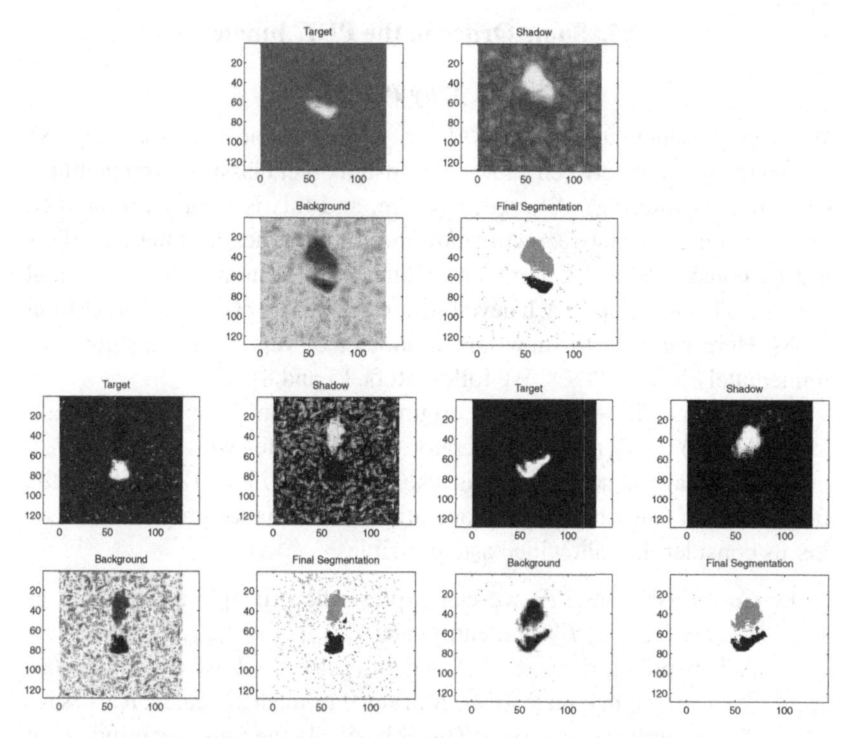

Fig. 4.25. Segmentation of SAR data (scalar and video).

After this frame, smooth posteriors of frame i are used as priors of frame $i + 1$. The figure on the right shows the result, once again with only two posterior smoothing steps, for the eight frames in the sequence.

When applied to MRI, the technique being proposed bears some superficial resemblance to schemes that anisotropically smooth the raw image before classification [157]. Besides the connection between our technique and MAP estimation of MRF's, which is absent in schemes that smooth the image directly, there are other important differences, as Ref. [157] suffers from the common problems of diffusion raw data detailed in the introduction. We should note again that applying anisotropic smoothing on the posterior probabilities is feasible even when the class likelihoods are described by general probability mass functions (and even multivariate distributions!).

The equivalence between anisotropic smoothing of posterior probabilities and MRFs with discontinuity fields also offers a solution to the problems of edge handling and missing data. These two issues can be treated in the same manner as in traditional regularization. Solving of the latter implies that MAP classification can be obtained even at locations where the pixel values are not provided.

4.5. Some Order in the PDE Jungle

4.5.1. Why PDEs?

We have presented a number of PDEs for anisotropic diffusion and have shown the relation between a few of them and other classical discrete filters (like median filtering). As argued before, and this is clearly exemplified by the connection between curvature motion and median filtering, PDEs can be considered as the iteration of filtering operations with very small support. This is the approach developed in the book by Guichard and Morel [168]. Here we want to show how a number of very simple assumptions immediately lead to PDEs. We follow Refs. [5 and 8].

Let $I_0(x, y) : \mathbb{R}^2 \to \mathbb{R}$ be the original image that we plan to precess. Let $I_t(x, y, t) := T_t[I_0]$ be an operator that associates with the image I_0 a processed image at scale t. Let us also define $T_{t,h}[I] := T_t[I_h]$. T_t can be for example a Gaussian filtering operation with variance proportional to t. Let us consider the following basic principles:

1. *Causality:* Using T_t, we can express the principle of causality as $T_{t+h} = T_{t+h,t}T_t$, $T_{t,t} =$ identity, and $T_{t+h,t}[\hat{I}] > T_{t+h,t}[I]$ for all $\hat{I} > I$. This principle means that to get to scale $t + h$ we can either start from the original scale (0) or from any intermediate scale t. It also says that the order is preserved. This is basically the causality principle of not adding information, which we have already discussed a number of times in this book.
2. *Regularity:* Let $I(y) = \frac{1}{2}\{[A(x - y), x - y] + (p, x - y) + c\}$ be a quadratic form of $\mathbb{R}^N$. There exists a function $F(A, p, x, c, t)$, continuous with respect to A, such that
$$\frac{T_{t+h,t}I - I}{h} \to_{h\to 0} F(A, p, x, c, t).$$
3. *Morphological invariance:* $gT_{t+h,t}[I] = T_{t+h,t}[gI]$ for any change of contrast given by g, a nondecreasing continuous function.
4. *Euclidean invariance:* The operator is invariant to Euclidean transformations (rotations and translations of the image).
5. *Affine invariance:* The operator is invariant to affine transformations.
6. *Linearity:* $T_{t+h,t}[aI + b\hat{I}] = aT_{t+h,t}[I] + bT_{t+h,t}[\hat{I}]$.

From these properties, we can derive a number of interesting results:

Theorem 4.2. *If an operator holds the principles of causality, Euclidean invariance, and linearity, then as the operator scale t goes to zero, I satisfies*

$$\frac{\partial I}{\partial t} = \Delta I.$$

In other words, $I(t) := T_t[I_0]$ obeys the heat equation. This result shows that, as mentioned before, Gaussian filtering, which is equivalent to the heat flow, is a direct result from simple assumptions (and the unique one holding those). Of course, we are more interested in what happens if we do not require linearity. Let us consider the flow

$$\frac{\partial I}{\partial t} = F(\nabla^2 I, \nabla I, I, x, y, t). \tag{4.55}$$

Then we have the following theorems.

Theorem 4.3. *If the operator T_t is causal and regular, then $I(x, y, t) = T_t[I(x, y)]$ is a viscosity solution of Eq. (4.55), where F, given by the regularity axiom, is nondecreasing with respect to the first argument $\nabla^2 I$.*

Theorem 4.4. *If the operator is causal, regular, Euclidean invariant, and morphological, then I obeys the equation*

$$\frac{\partial I}{\partial t} = \|\nabla I\| \, F\left[\operatorname{div}\left(\frac{\nabla I}{\|\nabla I\|}\right), t\right].$$

Important examples of this theorem are $F =$ constant, which gives the constant Euclidean motion (dilation and erosion), and $F(a, b) = a$, which gives curvature motion (directional diffusion).

Finally, we have the following theorem.

Theorem 4.5. *If the operator is causal, regular, affine invariant, and morphological, then I obeys the equation*

$$\frac{\partial I}{\partial t} = \|\nabla I\| \left[t \operatorname{div}\left(\frac{\nabla I}{\|\nabla I\|}\right)\right]^{1/3},$$

which is the level-set form of the affine invariant heat flow.

Note that we have obtained that there is only one affine invariant flow that follows the basic principles of multiscale analysis. This result can also be proved from the invariant geometric flows classification given in Chap. 2 (derivatives higher than two can already violate the causality principle, for example).

Let us conclude by mentioning that although the above results were presented for images, similar results hold if we consider shapes, defined by their boundary curves, and look at flows such as those in Chap. 2.

4.5.2. Are All the PDEs the Same?

Now we first extend the formulation of the geodesic active contours given in Chap. 3 to the full segmentation/simplification of images. This will lead us to show a number of connections between the PDEs described so far in this book.

Observe once again the level-set flow corresponding to the single-valued geodesic snakes introduced in Chap. 3, given by

$$\frac{\partial u}{\partial t} = |\nabla u| \operatorname{div}\left[g(I)\frac{\nabla u}{|\nabla u|}\right],$$

where g is the edge-detection map. Two functions (maps from $\mathbb{R}^2$ to $\mathbb{R}$) are involved in this flow, the image I and the auxiliary level-set one u. Assume now that $u \equiv I$, that is, the auxiliary level-set function is the image itself. The equation becomes

$$\frac{\partial I}{\partial t} = |\nabla I| \operatorname{div}\left[g(I)\frac{\nabla I}{|\nabla I|}\right]. \tag{4.56}$$

A number of interpretations can be given to Eq. (4.56). First, each level set of the image I moves according to the geodesic active-contour flow, being smoothly attracted by the term ∇g_{gray} to areas of high gradient. This gives the name of self-snakes to the flow. An example is provide in Fig. 4.26.

Furthermore, Eq. (4.56) can be rewritten as

$$\frac{\partial I}{\partial t} = \mathcal{F}_{\text{diffusion}} + \mathcal{F}_{\text{shock}},$$

where

$$\mathcal{F}_{\text{diffusion}} := g(I)\,|\nabla I| \operatorname{div}\left(\frac{\nabla I}{|\nabla I|}\right),$$

$$\mathcal{F}_{\text{shock}} := \nabla g(I) \cdot \nabla I.$$

Fig. 4.26. Gray-level self-snakes. Note how the image is simplified while preserving the main structures.

The term $\mathcal{F}_{\text{diffusion}}$ is the anisotropic diffusion flow proposed in Ref. [6] and described in Section 4.3 as

$$I_t = g(I)|\nabla I|\text{div}\left(\frac{\nabla I}{|\nabla I|}\right) = g I_{\xi\xi},$$

where ξ is perpendicular to ∇I and $g_{\text{gray}}(I)$ is, for example, selected as in the snakes. Let us explain the second term in Eq. (4.56). This is straightforward from the geometric interpretation of $\nabla g \cdot \nabla u$. The term $\nabla g \cdot \nabla I$ pushes toward valleys of high gradients, acting as the shock filter introduced in Ref. [293] for image deblurring. Therefore the flow $I_t = \mathcal{F}_{\text{shock}}$ is a shock filter. We then obtain that the self-snake given by Eq. (4.56) corresponds to a combination of anisotropic diffusion and shock filtering, further supporting its possible segmentation properties (for another combination of anisotropic diffusion and shock filtering see Ref. [7]).

Let us now make a direct connection between the self snakes and Perona–Malik flow. As explained before, Perona and Malik [310] proposed the flow

$$\frac{\partial I}{\partial t} = \text{div}\,[\hat{g}(I)\nabla I], \tag{4.57}$$

where $\hat{g}$ is an edge-stopping function, as before. We have already seen the connection of this flow with energy-minimization approaches in Chap. 3. Now, comparing Eq. (4.57) with Eq. (4.56), we note that the difference between the two flows is given by the gradient term $|\nabla I|$. Although this term can be compensated for by means of the selection of $\hat{g}$ (or g), it is important to note it when comparing the self-snakes with Perona–Malik flow. The term $|\nabla I|$ affects both the diffusion part and the shock one. This connection shows the somehow amazing relation between Terzopoulos snakes and Perona–Malik anisotropic diffusion. These two algorithms, one developed for object detection and the other for image enhancement, not only have been two of the most influential works in the area, but actually they are closely related mathematically (from Terzopoulos snakes to the geodesic active contours to the self-snakes into Perona–Malik). This is depicted in Fig. 4.27.

Recapping, this connection actually shows that Perona–Malik flow is like moving the image level sets according to Terzopoulos snakes.

4.5.3. Is There Any Connection with Wavelets?

Readers interested in mathematical approaches to image processing might be wondering whether the PDE approach is related to other mathematical

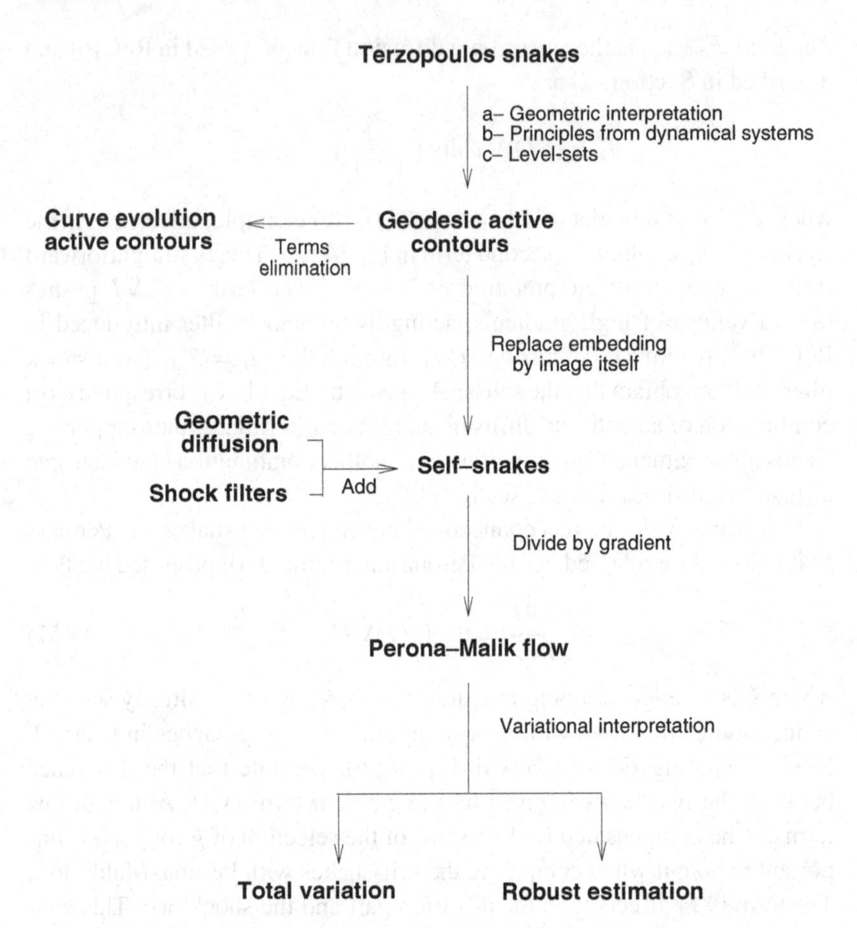

Fig. 4.27. Connections among some of the basic PDEs used in image processing.

techniques that are popular in image processing. We have seen the connection of anisotropic diffusion with robust estimation and the connection between PDEs and energy formulations (variational problems) by means of gradient descent flows. We have also seen that anisotropic diffusion applied to posterior probabilities is closely related to MRFs. The connection between anisotropic diffusion and wavelet theory is now described [253]. This is based on the results in Ref. [77].

As we have seen before in Section 4.2 anisotropic diffusion, or more specifically edge-stopping diffusion, can be obtained as the gradient descent of a robust minimization problem. Therefore anisotropic diffusion is closely related to variational problems. Indeed, (robust) anisotropic diffusion is searching for a (local) minima of a variational problem. Let us generalize this variational problem. We are basically looking for an image $\tilde{I}$ that minimizes

$$\|I - I_0\|_2 + \lambda \, \|I\|_Y, \tag{4.58}$$

where $\|\cdot\|_2$ stands for the L_2 norm between the reconstructed image I and the noisy one I_0 and $\|\cdot\|_Y$ stands for the norm of the reconstructed signal in a smoothness space. The first term looks for the similarity of the reconstructed signal to the original one (this term was ignored when we were dealing with robust edge-stopping diffusion) and the second term deals with the smoothness of the solution. Examples of Y used in the literature were given before in Section 4.2 when we were dealing with robust anisotropic diffusion, and they include L_2 and L_1 (total variation or bounded variation [330]). (Not all the norms presented before define smoothness spaces.)

One way to solve Eq. (4.58) is to compute its gradient descent flow and numerically search for a local minima. This is what we have discussed so far in this book. However, for a number of function spaces, the norm can be expressed in terms of the wavelets coefficients of I. In other words, I can be expanded in terms of its wavelet coefficients, and then the norm $\|I\|_Y$ is equivalent to a norm of these wavelet coefficients.

Let us consider Besov spaces $B_q^{\alpha}(L_p)$ (roughly speaking, this means that α derivatives of the function are in L_p). For $Y = B_1^1(L_1)$, which is very close to the bounded variation space used in Ref. [330], we find that minimizing Eq. (4.58) is obtained by shrinking the wavelet coefficients by an amount $\lambda/2$, reducing to the famous wavelet shrinkage algorithm of Donoho and Johnstone [112]. Thus wavelet shrinkage is like diffusion in the $B_1^1(L_1)$ space. Other norms give other types of filtering operations to the wavelet coefficients.

Exercises

1. For a set of images, compute the edge-stopping functions given by the Lorentzian, Huber, L_1, and Tukey norms and compare them.
2. Repeat the exercise above, this time performing edge-stopping diffusion on the images. Test with both small and large numbers of iterations. What happens to the image when the number of iterations is very large?

3. Implement the directional diffusion flow, without an edge-stopping function, and the median filtering technique, and compare the results.
4. Implement the curvature flow of a planar curve following the level-set technique and following the filtering plus thresholding approach and compare the results.
5. Recreate the toy example used to exemplify the posterior diffusion description.

CHAPTER FIVE

Geometric Diffusion of Vector-Valued Images

In this chapter the results on geometric diffusion are extended to vector-valued images. A number of different formulations are presented. The first one is based on directional diffusion, extending the flow described in Section 4.3. Then the relation between median filtering and anisotropic diffusion is extended to the vectorial case. The chapter concludes with an extension of the self-snake contours that leads to an additional formulation of vector-valued PDEs. Additional vector-valued flows will be introduced in Chap. 6. Additional applications of coupled PDEs can be found in [307].

5.1. Directional Diffusion of Multivalued Images

5.1.1. Anisotropic Diffusion of Multivalued Images

The anisotropic diffusion flow in Section 4.3 is based on smoothing the image in the direction ξ of minimal change in the image, that is, parallel to the edges (level sets) or perpendicular to the gradient of the image. In Subsection 3.3.1 we introduced the concept of direction of maximal and minimal change for vector-valued images $I : \mathbb{R}^2 \to \mathbb{R}^m$, and this is enough to formulate the corresponding anisotropic diffusion flow. Diffusion of the image occurs normal to the direction of maximal change θ_+ (i.e., in the direction perpendicular to the multivalued edge), which, in our case, is given by θ_-. In this way, salient edges are preserved, avoiding as much as possible diffusion across them. Thus we obtain

$$\frac{\partial I}{\partial t} = \frac{\partial^2 I}{\partial \theta_-^2}, \tag{5.1}$$

which is a system of coupled PDEs. The coupling results from the fact that θ_- depends on all the components of the image I.

To have control over the local diffusion coefficient we add a factor g similar to the one used for single-valued images. As we want to reduce the amount of diffusion near the edges, we require that g be close to zero when $\lambda_+ \gg \lambda_-$. The general evolution equation is

$$\frac{\partial I}{\partial t} = g(\lambda_+, \lambda_-)\frac{\partial^2 I}{\partial \theta_-^2}. \tag{5.2}$$

A suitable choice for g is any decreasing function of the difference $(\lambda_+ - \lambda_-)$.

Because a function of the form $f = f(\lambda_+ - \lambda_-)$ is the analog multivalued extension of $f = f(\|\nabla I\|^2)$ for single-valued images ($m = 1$), image processing algorithms for single-valued images based on $\|\nabla I\|^2$ can be extended to multivalued images when the square magnitude of the gradient is replaced with $\lambda_+ - \lambda_-$. For example, the work on robust diffusion can be extended to vector-valued images. The analog functional for vector-valued images would be $\int_\Omega \rho(\lambda_+, \lambda_-)\, dx dy$. Functionals of this type were further studied by Blomgren and Chan [40].

In Ref. [344] there is a complete discussion of the connections between this work and those in Refs. [76 and 409]. The use of systems of coupled PDEs for different computer vision applications such as stereo and motion can be found in the work of VanGool and colleagues in Ref. [324].

5.1.2. Experimental Results

An example of vector-valued anisotropic diffusion for color data is given in Fig. 5.1. The image is represented in the CIE 1976 $L^*a^*b^*$ space [415], which is an approximately uniform color space. On the top left we see the original image, a noisy version is shown on the top right, and the filtered image is depicted on the bottom-left corner. The bottom-right figure shows the result of applying the algorithm of Ref. [76]. Another example of the algorithm is shown Fig. 5.2. Here the noise is introduced by JPEG compression 32:1. The original is shown on the left, an expanded region of the JPEG-compressed image next, followed by the filtered image. Additional examples can be found in Ref. [344].

Figure 5.3 is an example in which the image was first locally represented in four frequency bands (low–low, low–high, high–low, high–high). This yields a vector-valued representation of the image. The multivalued image was processed in this space, and subsequently the associated real-valued image was reconstructed. This is the same kind of procedure applied to segment textured images in Chap. 3.

Fig. 5.1. Anisotropic diffusion of a noisy color image. See text.

5.2. Vectorial Median Filter

An additional alternative to define anisotropic diffusion for vector-valued images is to exploit the connection between median filtering and curvature motion in the scalar case. If we can extend the definition of median filtering to the vectorial case, then we might obtain PDEs from the asymptotic behavior of this filter. We proceed to investigate this now [73].

In order to use the definition analogous to the one given by Eq. (4.38), we need to impose an order in the vector space. This order is necessary to

Fig. 5.2. Anisotropic diffusion of a JPEG-compressed color image. See text.

Fig. 5.3. Vector image processing in a wavelet like decomposition.

obtain the vectorial extensions of the results presented in Section 5.1. In the comments subsection and Chap. 6 we discuss the use of other definitions to obtain median-type operations for vector-valued images.

To be able to compare two vectors, we assume a lexicographic order. Lexicographic order has recently been used in vector-valued morphology as well; see Ref. [179] for the most recently published results. Given two N-dimensional vectors I and $\hat{I}$, we say that $I \geq \hat{I}$ if and only if $I = \hat{I}$, or $I_1 > \hat{I}_1$, or $I_i = \hat{I}_i$, for all $1 \leq i < j \leq N$ and $I_j > \hat{I}_j$. This order means that we compare the coordinates in order until we find the first component that is different between the vectors. It is well known that the lexicographic order is a total order and any two vectors in $\mathbb{R}^N$ can be compared. Given a set of vectors $\Lambda \in \mathbb{R}^N$ we define its supremum (infimum) as the least of the upper bounds (respectively, the greatest of the lower bounds). That is, $I^* = \sup \Lambda$ means that $I \leq I^*$ for all $I \in \Lambda$ and if $I \leq \hat{I}$ for all $I \in \Lambda$ then also $I^* \leq \hat{I}$. Analogous relations hold for $\inf \Lambda$ with $\geq$ instead of $\leq$. With this order in mind, we see that the definition given by Eq. (4.38) is consistent for the vectorial case as well, with minor modifications.

Definition 5.1. *Let $I(x) : \mathbb{R}^2 \to \mathbb{R}^N$ be the image, a map from the continuous plane to the continuous space $\mathbb{R}^N$. The vector median filter of u, with* *support given by a set B of bounded measure, is defined as (over lines stands for set closure)*

$$\mathrm{med}_B(I) := \inf \overline{\left\{ \lambda \in \mathbb{R}^N : \mathrm{measure}[x \in B : I(x) \leq \lambda] \geq \frac{\mathrm{measure}(B)}{2} \right\}}. \tag{5.3}$$

Remark: It is necessary to define the vector median filter over the closure of a set to guarantee that if a series of vectors I_ϵ is greater or equal than a fix vector $\hat{I}$, the limit of the series is also greater or equal than $\hat{I}$. This property does not hold in the general case if the closure is omitted (e.g., $\hat{I} = [0, 1000]$ and $I = [\epsilon, 0], \epsilon \to 0$).

We proceed below to develop the main results of this section concerning the relations among vector median filtering (with a lexicographic order), morphological operators, and geometric PDEs. We should note that for the developments below, it is enough to consider $N = 2$, that is, a 2D vector. The operations we show for $N = 2$ will hold as well for $N > 2$, where we relate the component $i + 1$ to the component i in the same way we will relate the second component ($i = 2$) to the first one ($i = 1$) in the developments below.

We should also note that in many cases there is no natural lexicographic order among all the vector components, but there is just a relation of importance between one vector and the rest. For example, in color representations such as Lab and Yuv, the first component is usually more important than the other two, but there is no natural order between the last two by themselves. In this case, the system is treated as a collection of 2D vectors of the form $(I_1, I_i), 1 < i \leq N$.

Vector Median Filtering as a Morphological Operator. The connection between scalar median filtering and curvature motion described in Chap. 4 contains implicitly the fact that if I is a scalar function, we may restrict the sets in $\mathcal{B}$ to be level sets of I. This is an obvious fact: if $I : \mathbb{R}^2 \to \mathbb{R}$ is a measurable function and B is a subset of $\mathbb{R}^2$ of finite measure we always have

$$\mathrm{med}_B(I) = \inf_{B' \in \mathcal{S}} \sup_{x \in B'} I(x), \tag{5.4}$$

where

$$\mathcal{S} := \{[I \leq b] \cap B : b \in \mathbb{R},\ \mathrm{measure}([I \leq b] \cap B) \geq \tfrac{\mathrm{measure}(B)}{2}\}.$$

A formula similar to Eq. (5.4) also holds in the vectorial case.

Proposition 5.1. *Let* $I = (I_1, I_2) : \mathbb{R}^2 \to \mathbb{R}^2$ *be a bounded measurable function such that* $\text{measure}([I_1 = \alpha]) = 0$ *for all* $\alpha \in \mathbb{R}$ *and* B *a subset of* $\mathbb{R}^2$ *of finite measure. Then*

$$\text{med}_B(I) = \inf_{B' \in \mathcal{G}} \sup_{x \in B'} I(x), \tag{5.5}$$

where

$$\mathcal{G} := \{[I \leq \lambda] \cap B : \lambda \in \mathbb{R}^2,\ \text{measure}([I \leq \lambda] \cap B) \geq \tfrac{\text{measure}(B)}{2}\}.$$

Proof: For simplicity, for any function v from $\mathbb{R}^2$ to $\mathbb{R}^N$, $N = 1, 2$, we denote by $[v \leq \lambda]$ the set $\{x \in B : v(x) \leq \lambda\}$, $\lambda \in \mathbb{R}^N$. To prove this proposition, we first observe that the infimum on the right-hand side of Eq (5.5) is indeed attained. We have that

$$\varphi := \inf \overline{\{\sup_{x \in B'} I(x) : B' \in \mathcal{G}\}} = \inf\{\sup_{x \in B'} I(x) : B' \in \mathcal{G}\}. \tag{5.6}$$

Because the infimum in the lexicographic order of a closed set is attained, then $\varphi = (\varphi_1, \varphi_2) \in \overline{\{\sup_{x \in B'} I(x) : B' \in \mathcal{G}\}}$. Let $B'_n := [I \leq \lambda'_n] \in \mathcal{G}$ be such that $\lambda_n := \sup_{B'_n} I \to \varphi$. Observe that $B'_n = [I \leq \lambda_n]$. Let $\lambda_n = (\lambda_{n,1}, \lambda_{n,2})$. Let $\lambda^*_{n,1} = \sup\{\lambda_{n,1}, \lambda_{n+1,1}, \ldots, \}$, $\lambda^*_{n,2} = \sup\{\lambda_{n,2}, \lambda_{n+1,2}, \ldots, \}$. Then $\lambda^*_{n,1} \downarrow \varphi_1$, $\lambda^*_{n,2} \downarrow \varphi_2$. Thus we have $\lambda^*_n = (\lambda^*_{n,1}, \lambda^*_{n,2}) \geq \lambda_n$, λ^*_n is decreasing, and $\lambda^*_n \to \varphi$. It follows that $[I_1 \leq \varphi_1] = \cap_n [I_1 \leq \lambda^*_{n,1}]$. Because $\text{measure}([I_1 = \alpha]) = 0$ for all $\alpha \in \mathbb{R}$ and $[I \leq \lambda] = [I_1 < \lambda_1] \cup [I_1 = \lambda_1, I_2 \leq \lambda_2]$ for all $\lambda = (\lambda_1, \lambda_2) \in \mathbb{R}^2$, then $\text{measure}([I \leq \lambda]) = \text{measure}([I_1 \leq \lambda_1])$ for all $\lambda \in \mathbb{R}^2$. As a consequence, we have that

$$\begin{aligned}\text{measure}([I \leq \varphi]) &= \text{measure}([I_1 \leq \varphi_1]) = \lim_n \text{measure}([I_1 \leq \lambda^*_{n,1}]) \\ &= \lim_n \text{measure}([I \leq \lambda^*_n]) \geq \limsup_n \text{measure}([I \leq \lambda_n]) \\ &\geq \frac{\text{measure}(B)}{2}.\end{aligned}$$

Now, by the definition of φ, we have

$$\varphi \leq \sup_{[I \leq \varphi]} I \leq \varphi.$$

This justifies Eq. (5.6). In a similar way, we can show that

$$\text{med}_B(I) \in \mathcal{F}_B := \left\{\lambda : \text{measure}[x \in B : I(x) \leq \lambda] \geq \frac{\text{measure}(B)}{2}\right\} \tag{5.7}$$

and

$$\mathrm{med}_B(I) = \inf\{\lambda : \lambda \in \mathcal{F}_B\}.$$

Now, because $\mathrm{measure}([I \leq \varphi]) \geq \frac{\mathrm{measure}(B)}{2}$, we have $\mathrm{med}_B(I) \leq \varphi$. Because also $\mathrm{measure}([I \leq \mathrm{med}_B(I)]) \geq \frac{\mathrm{measure}(B)}{2}$ then $\varphi \leq \sup_{[I \leq \mathrm{med}_B(I)]} I \leq \mathrm{med}_B(I)$. Both inequalities prove Proposition 5.1. □

For simplicity, we shall assume in what follows that $I = (I_1, I_2) : \mathbb{R}^2 \to \mathbb{R}^2$ is a continuous function and $\mathrm{measure}([I_1 = \alpha]) = 0$ for all $\alpha \in \mathbb{R}$. As above we shall write $[v \leq \lambda]$ to mean $[v \leq \lambda] \cap B$.

Proposition 5.2. *Assume that* $I = (I_1, I_2) : \mathbb{R}^2 \to \mathbb{R}^2$ *is a continuous function,* $\mathrm{measure}([I_1 = \alpha]) = 0$ *for all* $\alpha \in \mathbb{R}$ *and* B *is a compact subset of* $\mathbb{R}^2$. *Then*

$$\mathrm{med}_B(I) = \begin{pmatrix} \mathrm{med}_B I_1 \\ \inf_{[x \in B : I_1(x) = \mathrm{med}_B I_1]} I_2(x) \end{pmatrix} \tag{5.8}$$

Proof: Let

$$\mu := \begin{pmatrix} \mathrm{med}_B I_1 \\ \inf_{[x \in B : I_1(x) = \mathrm{med}_B I_1]} I_2(x) \end{pmatrix}. \tag{5.9}$$

Because we assumed that $\mathrm{measure}([I_1 = \alpha]) = 0$ for all $\alpha \in \mathbb{R}$ we have that $\mathrm{measure}([I \leq \mu]) \geq \frac{\mathrm{measure}(B)}{2}$. Then, by definition of $\mathrm{med}_B(I)$ we have that $\mathrm{med}_B(I) \leq \mu$. Now, let $\lambda = (\lambda_1, \lambda_2) \in \mathcal{F}_B$. Because $\mathrm{measure}([I_1 \leq \lambda_1]) = \mathrm{measure}([I \leq \lambda]) \geq \frac{\mathrm{measure}(B)}{2}$, then $\mathrm{med}_B(I_1) \leq \lambda_1$. If $\mathrm{med}_B(I_1) < \lambda_1$, then $\mu \leq \lambda$. Thus we may assume that $\mathrm{med}_B(I_1) = \lambda_1$. Because $[I \leq \lambda] = [I_1 < \lambda_1] \cup [I_1 = \lambda_1, I_2 \leq \lambda_2]$, if $\inf_{[x \in B : I_1(x) = \mathrm{med}_B I_1]} I_2(x) > \lambda_2$, then $[I \leq \lambda] = [I_1 < \lambda_1]$. Because I_1 is continuous then $\mathrm{med}_B(I_1) \leq \sup_{[I \leq \lambda]} I_1 < \lambda_1$. This contradiction proves that $\inf_{[x \in B : I_1(x) = \mathrm{med}_B I_1]} I_2(x) \leq \lambda_2$. Thus $\mu \leq \lambda$. We conclude that $\mu \leq \mathrm{med}_B(I)$. Therefore $\mu = \mathrm{med}_B(I)$. □

Proposition 5.2 means that for the first component of the vector, the median is as in the scalar case, whereas for the second one, we obtain the result by looking for the infimum over all the pixels of I_2 corresponding to the positions on the image plane where the median of the first component is obtained. This result is expected, as the first component already selects the whole possible vectors, and then the positions in the plane where the second component can select from are determined. Note also that new vectors, and then new pixel values, are not created, as expected from a median filter.

Let us now give an alternative definition of the vector median.

Definition 5.2. *Let $I = (I_1, I_2) : \mathbb{R}^2 \to \mathbb{R}^2$ be a measurable function and $B \subseteq \mathbb{R}^2$ be of finite measure. Define*

$$\operatorname{med}^*_B(I) = \inf_{B' \in \mathcal{H}} \sup_{x \in B'} I(x), \tag{5.10}$$

where

$$\mathcal{H} := \{[I_1 \leq b] \cap B : b \in \mathbb{R},\ \text{measure}([I_1 \leq b] \cap B) \geq \tfrac{\text{measure}(B)}{2}\}$$

Proposition 5.3. *Let $I = (I_1, I_2) : \mathbb{R}^2 \to \mathbb{R}^2$ be a continuous function and B be a compact subset of $\mathbb{R}^2$. Then*

$$\operatorname{med}^*_B(I) = \begin{pmatrix} \operatorname{med}_B I_1 \\ \sup_{[x \in B : I_1(x) = \operatorname{med}_B I_1]} I_2(x) \end{pmatrix}. \tag{5.11}$$

To prove this proposition we need the following simple Lemma.

Lemma 5.1. *Let $\pi_i : \mathbb{R}^2 \to \mathbb{R}$ be the projection of $\mathbb{R}^2$ onto the i coordinate. Let $\Lambda \subseteq \mathbb{R}^2$. Then*

1. $\pi_1(\sup \Lambda) = \sup \pi_1(\Lambda)$.
2. *If $\pi_1(\overline{I}) = \pi_1(\sup \Lambda)$ for some $\overline{I} \in \Lambda$ then*

$$\pi_2(\sup \Lambda) = \sup\{\pi_2(I) : I \in \Lambda, \quad \pi_1(I) = \pi_1(\sup \Lambda)\}. \tag{5.12}$$

In particular, if Λ is compact we always have Eq. (5.12). A similar statement holds for the infimum.

Proof of Proposition 5.3: Let

$$q = \begin{pmatrix} \operatorname{med}_B I_1 \\ \sup_{[x \in B : I_1(x) = \operatorname{med}_B I_1]} I_2(x) \end{pmatrix}.$$

Because $\text{measure}([I_1 \leq \operatorname{med}_B(I_1)]) \geq \frac{\text{measure}(B)}{2}$, we have that $\operatorname{med}^*_B(I) \leq \sup_{[x \in B : I_1(x) \leq \operatorname{med}_B I_1]} I$. Now, observe that if $I_1(x) = \operatorname{med}_B I_1$ then $I_2(x) \leq \sup_{[I_1 = \operatorname{med}_B I_1]} I_2$. From this relation it is then simple to show that we have $\sup_{[x \in B : I_1(x) \leq \operatorname{med}_B I_1]} I \leq q$. Hence $\operatorname{med}^*_B(I) \leq q$. To prove the opposite inequality, observe that if $b \in \mathbb{R}$ is such that $\text{measure}([I_1 \leq b]) \geq \frac{\text{measure}(B)}{2}$ then $b \geq \operatorname{med}_B(I_1)$ and, in consequence, $[I_1 \leq \operatorname{med}_B(I_1)] \subseteq [I_1 \leq b]$. Because, by Lemma 5.1, $q = \sup_{[I_1 \leq \operatorname{med}_B(I_1)]} I$, we have $q \leq \sup_{[I_1 \leq b]} I$ for all $b \in \mathbb{R}$ such that $\text{measure}([I_1 \leq b]) \geq \frac{\text{measure}(B)}{2}$. Hence $q \leq \operatorname{med}^*_B(I)$. Observe that the infimum in Eq. (5.11) is attained. □

Asymptotic Behavior and Coupled Geometric PDE's. Because $\pi_1(\operatorname{med}_B I) = \pi_1(\operatorname{med}^*_B I) = \operatorname{med}_B I_1$, then Theorem 4.1 describes the asymptotic behavior of the first coordinate of the vector median $\operatorname{med}_{D(x_0,t)} I$ for a three-time differentiable function $I : \mathbb{R}^2 \to \mathbb{R}^2$, where $D(x_0, t)$ is the disk of radius $t > 0$ centered at $x_0 \in \mathbb{R}^2$. Let us describe the asymptotic behavior of the second coordinate of $\mathrm{med}_{D(x_0,t)} I$. The formula will be written explicitly only when $DI_2(x_0) \cdot DI_1^{\perp}(x_0) \neq 0$. ($DI := \nabla I$ and $\langle DI, DI^{\perp} \rangle = 0$, whereas $\|DI\| = \|DI^{\perp}\|$.) If $DI_2(x_0) \cdot DI_1^{\perp}(x_0) = 0$, the formula can be written with the Taylor expansion of I_2 given in Proposition 5.4.

Proposition 5.4. *Let $I : \mathbb{R}^2 \to \mathbb{R}^2$ be three-time differentiable, $I = (I_1, I_2)$. Assume that $DI_1(x_0) \neq 0$. Then*

$$\begin{aligned} I_2(x) = I_2(x_0) &+ \langle x - x_0, e_1 \rangle DI_2(x_0) \cdot e_1 + \frac{1}{6} t^2 \kappa(I_1)(x_0) DI_2(x_0) \cdot e_2 \\ &- \frac{1}{2} \kappa(I_1)(x_0) \langle x - x_0, e_1 \rangle^2 DI_2(x_0) \cdot e_2 \\ &+ \frac{1}{2} \langle D^2 I_2(x_0) e_1, e_1 \rangle \langle x - x_0, e_1 \rangle^2 + o(t^2), \end{aligned} \tag{5.13}$$

for $x \in [I_1 = \mathrm{med}_{D(x_0,t)} I_1] \cap D(x_0, t)$, where $e_1 = \frac{DI_1^{\perp}(x_0)}{|DI_1(x_0)|}$, $e_2 = \frac{DI_1(x_0)}{|DI_1(x_0)|}$, such that $\{e_1, e_2\}$ is positive oriented. In particular, if $DI_2(x_0) \cdot DI_1^{\perp}(x_0) \neq 0$, then

$$\begin{aligned} \pi_2\big[\mathrm{med}_{D(x_0,t)} I\big] &= \sup_{[I_1 = \mathrm{med}_{D(x_0,t)} I_1] \cap D(x_0,t)} I_2 \\ &= I_2(x_0) - t \left| DI_2(x_0) \cdot \frac{DI_1^{\perp}(x_0)}{|DI_1(x_0)|} \right| + O(t^2). \end{aligned} \tag{5.14}$$

Similarly,

$$\begin{aligned} \pi_2\big[\mathrm{med}^*_{D(x_0,t)} I\big] &= \inf_{[I_1 = \mathrm{med}_{D(x_0,t)} I_1] \cap D(x_0,t)} I_2 \\ &= I_2(x_0) + t \left| DI_2(x_0) \cdot \frac{DI_1^{\perp}(x_0)}{|DI_1(x_0)|} \right| + O(t^2). \end{aligned} \tag{5.15}$$

Before giving the proof of this proposition, let us observe that the above formula for I_2 coincides with the asymptotic expansion of Eq. (4.40) if $I_2 = I_1$. Indeed, if $x \in [I_1 = \mathrm{med}_{D(x_0,t)} I_1] \cap D(x_0, t)$, using that $DI_2(x_0) \cdot e_1 = 0$ and $DI_2(x_0) \cdot e_2 = |DI_1(x_0)|$, we have

$$\begin{aligned} I_2(x) = I_2(x_0) &+ \frac{1}{6} t^2 \kappa(I_1)(x_0) |DI_1(x_0)| \\ &- \frac{1}{2} \kappa(I_1)(x_0) \langle x - x_0, e_1 \rangle^2 |DI_1(x_0)| \\ &+ \frac{1}{2} \langle D^2 I_2(x_0) e_1, e_1 \rangle \langle x - x_0, e_1 \rangle^2 + o(t^2). \end{aligned}$$

Because $I_2 = I_1$ and $|DI_1(x_0)|\kappa(I_1)(x_0) = \langle D^2 I_2(x_0)e_1, e_1\rangle$, the last two terms in the above expression cancel each other and we have

$$I_2(x) = I_1(x) = \mathrm{med}_{D(x_0,t)} I_1 = I_2(x_0) + \frac{1}{6}t^2\kappa(I_1)(x_0)|DI_1(x_0)| + o(t^2),$$

an expression consistent with Eq. (4.40).

More generally, if $DI_2(x_0) \cdot e_1 = 0$, we may write

$$\begin{aligned} I_2(x) = I_2(x_0) &+ \frac{1}{6}t^2\kappa(I_1)(x_0)|DI_1(x_0)| \\ &- \frac{1}{2}\kappa(I_1)(x_0)\langle x - x_0, e_1\rangle^2 DI_2(x_0) \cdot e_2 \\ &+ \frac{1}{2}\langle D^2 I_2(x_0)e_1, e_1\rangle\langle x - x_0, e_1\rangle^2 + o(t^2) \end{aligned}$$

for $x \in [I_1 = \mathrm{med}_{D(x_0,t)} I_1] \cap D(x_0, t)$. Now, observe that $DI_2(x_0) \cdot e_2 = \pm|DI_2(x_0)|$. In particular, if $DI_2(x_0) \cdot e_2 = |DI_2(x_0)| \neq 0$, because $\frac{DI_2(x_0)}{|DI_2(x_0)|}$ is collinear to e_1, the expression for $I_2(x)$ can be reduced to

$$\begin{aligned} I_2(x) = I_2(x_0) + \frac{1}{6}t^2\kappa(I_1)(x_0)|DI_1(x_0)| + \frac{1}{2}|DI_2(x_0)|[\kappa(I_2)(x_0) \\ - \kappa(I_1)(x_0)]\langle x - x_0, e_1\rangle^2 + o(t^2). \end{aligned}$$

The value of the corresponding infimum or supremum depends on the sign of the terms containing $\langle x - x_0, e_1\rangle^2$ and we shall not write them explicitly.

Proof: Let $x \in [I_1 = \mathrm{med}_{D(x_0,t)} I_1] \cap D(x_0, t)$. Obviously

$$x - x_0 = \langle x - x_0, e_1\rangle e_1 + \langle x - x_0, e_2\rangle e_2.$$

To compute $\langle x - x_0, e_2\rangle$ we expand I_1 in Taylor series up to the second order and write the identity $I_1(x) = \mathrm{med}_{D(x_0,t)} I_1$ as

$$\begin{aligned} I_1(x_0) + \langle DI_1(x_0), x - x_0\rangle + \frac{1}{2}\langle D^2 I_1(x_0)(x - x_0), x - x_0\rangle \\ + o(t^2) = \mathrm{med}_{D(x_0,t)} I_1. \end{aligned}$$

Using Eq. (4.40), we have

$$\begin{aligned} \langle DI_1(x_0), x - x_0\rangle = &\frac{1}{6}t^2\kappa(I_1)(x_0)|DI_1(x_0)| \\ &- \frac{1}{2}\langle D^2 I_1(x_0)(x - x_0), x - x_0\rangle + o(t^2). \end{aligned}$$

Thus we may write

$$x - x_0 = \langle x - x_0, e_1\rangle e_1 + \frac{1}{6}t^2\kappa(I_1)(x_0)e_2 - \frac{\langle D^2 I_1(x_0)(x - x_0), x - x_0\rangle}{2|DI_1(x_0)|}e_2 + o(t^2).$$

Introducing this expression for $x - x_0$ into the right-hand side, we obtain

$$x - x_0 = \langle x - x_0, e_1\rangle e_1 + \frac{1}{6}t^2\kappa(I_1)(x_0)e_2 - \frac{\langle D^2 I_1(x_0)e_1, e_1\rangle}{2|DI_1(x_0)|}\langle x - x_0, e_1\rangle^2 e_2 + o(t^2),$$

an expression that can be written as

$$x - x_0 = \langle x - x_0, e_1\rangle e_1 + \frac{1}{6}t^2\kappa(I_1)(x_0)e_2 - \frac{1}{2}\kappa(I_1)(x_0)\langle x - x_0, e_1\rangle^2 e_2 + o(t^2),$$

because $\kappa(I_1)(x_0) = \frac{\langle D^2 I_1(x_0)e_1, e_1\rangle}{|DI_1(x_0)|}$. Introducing this into the Taylor expansion of I_2,

$$I_2(x) = I_2(x_0) + \langle DI_2(x_0), x - x_0\rangle + \frac{1}{2}\langle D^2 I_2(x_0)(x - x_0), x - x_0\rangle + o(t^2),$$

we obtain Eq. (5.13), the first part of the proposition.

Now, from the formula above we have for $x \in [I_1 = \text{med}_{D(x_0,t)} I_1] \cap D(x_0, t)$,

$$x - x_0 = \langle x - x_0, e_1\rangle e_1 + O(t^2).$$

In particular, the curve $[I_1 = \text{med}_{D(x_0,t)} I_1] \cap D(x_0, t)$ intersects the axis e_2 at some point $\overline{x}$ such that $\overline{x} - x_0 = O(t^2)$. We also deduce that

$$\sup\cdots\langle x - x_0, e_1\rangle = \sup\cdots|x - x_0| + O(t^2) = t + O(t^2),$$

where the suprema are taken in $[I_1 = \text{med}_{D(x_0,t)} I_1] \cap D(x_0, t)$. Taking the supremum of the last expression for I_2 on $[I_1 = \text{med}_{D(x_0,t)} I_1] \cap D(x_0, t)$, we obtain Eq. (5.15). In a similar way we deduce Eq. (5.14). □

In analogy to the scalar case, this result can also lead us to deduce the following result: Modulo the different scales of the two coordinates, when the median filter is iterated, and $t \to 0$, the second component of the vector, $I_2(x)$, is moving its level sets to follow those of the first component $I_1(x)$, which are by themselves moving with curvature motion. This is expressed with the equation

$$\frac{\partial I_2(x,t)}{\partial t} = \pm \left| \frac{DI_2(x,t)}{|DI_2(x,t)|} \cdot \frac{DI_1^{\perp}(x,t)}{|DI_1(x,t)|} \right| |DI_2(x,t)|, \tag{5.16}$$

where the sign depends on the exact definition of the median being used. Deriving this equation from the asymptotic result presented above is much more complicated than in the scalar case, and this is beyond the scope of this book. In spite of the very attractive notation, this is not a well-defined PDE because the right-hand side is not defined when $DI_1(x) = 0$ (see remark below), and certainly this happens in images (and in those that are solutions of mean curvature motion). Note that $DI_1(x) = 0$ means that there is no level-set direction at that place, and then the level sets of I_2 have nothing to follow. (It is mentioned in passing that the regularity of DI_1 would be an additional problem to study in the analysis of Eq. (5.16).) Anyway, this equation clarifies the meaning of the vector median and gives it a very intuitive interpretation. The next terms in the Taylor expansion of I_2 depend on curvatures of I_1 and I_2. These terms play a role when the previous one is zero, and, in particular, this will happen when the level sets of both components of the vector are equal. The precise form of $\text{med}_{D(x_0,t)} I$ and $\text{med}^*_{D(x_0,t)} I$ can be deduced from the asymptotic expansion for I_2 previous to the proof of Proposition 5.4 and we shall not write them explicitly. Let us only mention that if $I_2 = I_1$ on a neighborhood of a point, then I_2 moves with curvature motion, as expected.

Remark: To illustrate the case when $DI_1(x_0) = 0$ and x_0 is nondegenerate, we first assume that $x_0 = (0, 0)$ and $I_1(x_1, x_2) = Ax_1^2 + Bx_2^2, x = (x_1, x_2) \in \mathbb{R}^2$, $A, B > 0$. It is immediate to compute

$$\begin{aligned} \text{med}_{D(x_0,t)} I_1 &= \inf \left\{ \alpha \in \mathbb{R} : \text{measure}([I_1 \le \alpha]) \ge \frac{\text{measure}(D(x_0,t))}{2} \right\} \\ &= \frac{t^2\sqrt{AB}}{2}. \end{aligned}$$

The set $X \equiv [I_1 = \frac{t^2\sqrt{AB}}{2}] \cap D(x_0, t) = [(x_1, x_2) \in D(x_0, t) : \sqrt{\frac{A}{B}} x_1^2 + \sqrt{\frac{B}{A}} x_2^2 = \frac{t^2}{2}]$. Again, it is straightforward to obtain

$$\sup_X I_2(x_1, x_2) = I_2(0,0) + \frac{t}{\sqrt{2}} \left(\sqrt{\frac{A}{B}} I_{2x}^2 + \sqrt{\frac{B}{A}} I_{2y}^2 \right)^{1/2} + o(t).$$

Consider now the case in which I_1 is constant in a neighborhood of x_0.

Suppose that $I_1 = \alpha$ in $D(x_0, t)$. Then by using either Eq. (4.38) or Eq. (4.39) we conclude that

$$\mathrm{med}_{D(x_0,t)} I = \begin{pmatrix} \alpha \\ \sup_{D(x_0,t)} I_2(x) \end{pmatrix}.$$

In this case,

$$\sup_{D(x_0,t)} I_2(x) = I_2(x_0) + t|DI_2(x_0)| + o(t).$$

We conclude that there is no common simple expression for all cases. Therefore, in contrast with the scalar case, the asymptotic behavior of the median filter when the gradient of the first component, $I_1(x)$, is zero is not uniquely defined and decisions need to be taken when the equation is implemented (see below).

Projected Mean-Curvature Motion

If we set aside for a moment the requirement for an inf–sup morphological operator and start directly from the definitions in Eqs. (5.8) and (5.11) [instead of Eqs. (5.5) and (5.10)], we obtain an interesting alternative to the median filtering of vector-valued images (we once again consider only 2D vectors).

Definition 5.3. *Let $I = (I_1, I_2) : \mathbb{R}^2 \to \mathbb{R}^2$ be a continuous function and $B \subseteq \mathbb{R}^2$ a compact subset. Define*

$$\mathrm{med}_B^{**}(I) := \begin{pmatrix} \mathrm{med}_B I_1 \\ \mathrm{med}_{[x \in B: I_1(x) = \mathrm{med}_B I_1]} I_2(x) \end{pmatrix} \tag{5.17}$$

In contrast with the previous definitions, we here consider also the median of the second component, restricted to the positions where the first component achieved its own median value.

The asymptotic expansion of Eq. (5.13) in Proposition 5.4 is of course general. By Replacing sup with med at the end of the proof we obtain that the expression analogous to Eqs. (5.14) and (5.15) for med** is

$$\pi_2[\mathrm{med}^{**}_{D(x_0,t)} I] = I_2(x_0) + t^2 \left[DI_2(x,t) \cdot \kappa_{I_1}(x,t) \frac{DI_1(x,t)}{|DI_1(x,t)|} \right] + o(t^2).$$

Note that the time scale of this expression is t^2, as in the scalar case (Theorem 4.1), and then as in the asymptotic expansion of the first component of the vector. This is in contrast with a time scale of t for the expressions having inf or sup in the second component (Eqs. (5.14) and (5.15)).

The PDE corresponding to the expression above, and therefore to the second component of the vector, is

$$\frac{\partial I_2(x,t)}{\partial t} = \left[\kappa_{I_1}(x,t)\,\frac{DI_1(x,t)}{|DI_1(x,t)|}\cdot\frac{DI_2(x,t)}{|DI_2(x,t)|}\right]|DI_2(x,t)|. \quad (5.18)$$

This equation shows that the level set of the second component I_2 are moving with the same geometric velocity as those of the first one I_1, meaning mean-curvature motion (the projection reflects the well-known fact that tangential velocities do not affect the geometry of the motion). Short-term existence of this flow can be derived from the results in Ref. [127].

Using Lemma 5.1 we can show that med**, as defined in Eq. (5.17), is also a morphological inf–sup operation of the type of, Eqs. (5.5) and (5.10). This time, the set over which the inf–sup operations are taken is given by

$$\begin{aligned}\mathcal{R} := \Bigg\{ &\lambda = (\lambda_1, \lambda_2) : \text{measure}([I_1 \le \lambda_1]) \ge \frac{\text{measure}(B)}{2},\\ &[I_1 \le \lambda_1] \ne \emptyset,\ \text{measure}([I_1 = \lambda_1, I_2 \le \lambda_2])\\ &\ge \frac{\text{measure}([I_1 = \lambda_1] \cap B)}{2}\Bigg\}.\end{aligned}$$

Recapping, the second component of the vector can be obtained by means of inf, sup, or med operations over a restricted set. In all the cases, the filter is a morphological inf–sup operation, computed over different structuring elements sets, and in all the cases a corresponding asymptotic behavior and PDE interpretation can be given. In the case of the med operation, the asymptotic expansion of the vector components have all the same scale, and the second-component level set are just moving with the geometric velocity of the first-component ones. In the other cases, the level sets of the second component move toward those of the first component. In all the cases then, the level sets of the second component follow those of the first one, as expected from a lexicographic order.

Figure 5.4 shows an example of the theoretical results presented in this section.

Comments. To obtain the results here reported, we have assumed a lexicographic order that permits us to compare vectors. If we do not want to use this assumption, we will not have an order, and then an inf–sup type of operation. Therefore both the positive and the negative results reported are a direct consequence of imposing and order in $\mathbb{R}^N$. To avoid this, we need to follow a different approach to compute the median filter, for example, Eq. (4.37). We should note that for continuous signals, minimizing the L_1

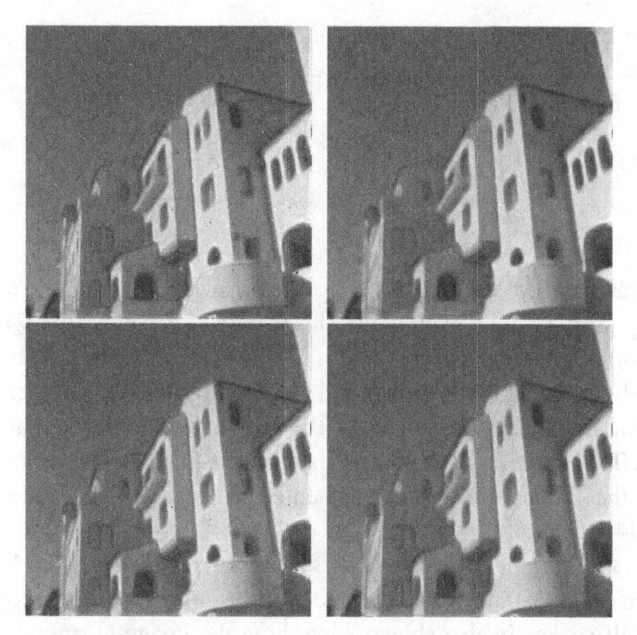

Fig. 5.4. Simulated results of the theoretical connections derived in this section. The top right shows the result of alternating Eqs. (5.8) and (5.11) for one step with a 3×3 discrete support (as these equations correspond to erosion and dilation, respectively, alternating them constitutes an opening filter). The bottom figures show results of the vectorial PDE derived from the mean-curvature motion for the first component and projected mean-curvature motion for the rest (after 2 and 20 iterations, respectively). All computations were performed on the Lab color space.

norm of a vector is equivalent to the independent minimization of each one of its components, reducing then the problem to the scalar case, in which, for example, each plane is independently enhanced by means of mean-curvature motion. Note that in the classical discrete case, because the median belongs to the finite set of vectors in the window, the vectorial case is not reduced to a collection of scalar cases. Therefore, in order to have continuous equations that are coupled, we need to look for a different approach. This will be presented in Chap. 6, when we deal with the diffusion of directions.

5.3. Color Self-Snakes

In the same way as we extended the geodesic active contours to the self-snakes, we now present this extension for vector-valued images. As before, $I : \mathbb{R}^2 \rightarrow \mathbb{R}^m$; we now obtain a number of possible flows. We derive the first one by combining the metric in the color-snake level-set model of Eq. (3.54) with the single-valued self-snakes of Eq. (4.56), obtaining a system of coupled PDEs of the form ($i = 1, \ldots, m$)

$$\frac{\partial I_i}{\partial t} = |\nabla I_i| \operatorname{div}\left(g_{\text{color}}(\lambda_+, \lambda_-) \frac{\nabla I_i}{|\nabla I_i|} \right). \tag{5.19}$$

In this case, the color interaction is given by g_{color}, affecting the diffusion-stopping term g_{color} as well as the shock-type one $\nabla g_{\text{color}}(I) \cdot \nabla I_i$. A different formulation can be obtained if ∇I_i is replaced with the direction of maximal change $(\cos\theta_+, \sin\theta_-)$ and $|\nabla I_i|$ with the color gradient $f(\lambda_\pm)$. The color self-snakes are related to the previously commented algorithms as well. The equation is of course also related to the vector-valued extension of the Perona–Malik flow presented in Ref. [409] and the work in Ref. [344].

An example of the color self-snakes is presented in Fig. 5.5 (see Ref. [333] for more examples), in which the effect of the color shock-type component of the self-snakes is also shown. Note how the image is simplified at the steady-state solution of the flow.

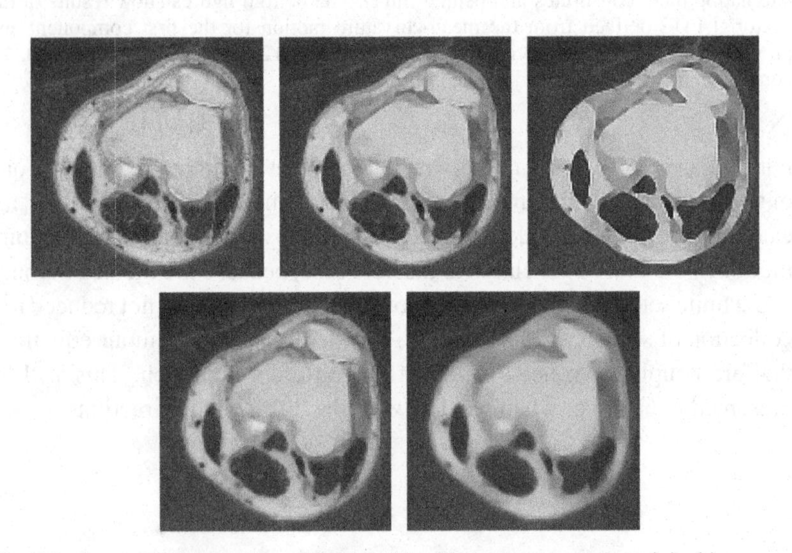

Fig. 5.5. Example of the color self-snakes. The first row presents the original image on the left and two steps of the color self-snakes; the figure on the rights is the steady-state solution of the flow. The last row shows the two same steps again of the color self-snakes without the shock-type part. Note that this flow will continue to smooth the image.

Exercises

1. Show that the diffusion direction in Chambolle's approach [76] to vector-valued diffusion is exactly the one given by the direction of minimal change of the vectorial edges.
2. Propose extensions to Perona–Malik edge-stopping diffusion for vector-valued images.
3. Compare vector-valued diffusion with the results of independent diffusion in each one of the components. Do this both for color images and for vector data obtained from multiscale decompositions.
4. Show that the vector median filtering algorithms are morphological operations like erosion and dilation.

CHAPTER SIX

Diffusion on Nonflat Manifolds

In a number of disciplines, directions provide a fundamental source of information. Examples in the area of computer vision are (2D, 3D, and 4D) gradient directions, optical flow directions, surface normals, principal directions, and color. In the color example, the direction is given by the normalized vector in the color space. Frequently, these data are available in a noisy fashion, and there is a need for noise removal. In addition, it is often desired to obtain a multiscale-type representation of the directional data, similar to those representations obtained for gray-level images [218, 306, 310, 413] and previously described in this book. Addressing these issues is the goal of this chapter (see Ref. [380]).

Image data, as well as directions and other sources of information, are not always defined on the $\mathbb{R}^2$ plane or $\mathbb{R}^3$ space. They can be, for example, defined over a surface embedded in $\mathbb{R}^3$. It is also important then to extend the scale space and diffusion frameworks to general data, defined on general (not necessarily flat) manifolds. In other words, we want to deal with maps between two general manifolds and be able to isotropically and anisotropically diffuse them. This will make it possible for example to denoise data defined on 3D surfaces. Although we are particularly interested in this chapter in directions defined on $\mathbb{R}^2$ (nonflat data defined on a flat manifold), the framework presented here applies to the general case as well.

An $\mathbb{R}^n$ direction defined on an image in $\mathbb{R}^2$ is given by a vector $I(x, y, 0) : \mathbb{R}^2 \to \mathbb{R}^n$ such that the Euclidean norm of $I(x, y, 0)$ is equal to one, that is,

$$\sqrt{\sum_{i=1}^{n} I_i^2(x, y, 0)} = 1,$$

where $I_i(x, y, 0) : \mathbb{R}^2 \to \mathbb{R}$ are the components of the vector. We can simplify the notation by considering $I(x, y, 0) : \mathbb{R}^2 \to S^{n-1}$, where S^{n-1} is the unit ball in $\mathbb{R}^n$. This implicitly includes the unit norm constraint. (Any nonzero vector can be transformed into a direction by normalization. For zero vectors, the unit norm constraint has to be relaxed, and a norm less or equal to one needs to be required.) When smoothing the data or computing a multiscale representation $I(x, y, t)$ of a direction $I(x, y, 0)$ (t stands for the scale), it is crucial to maintain the unit norm constraint, which is an intrinsic characteristic of directional data, that is, the smoothed direction $\hat{I}(x, y, 0) : \mathbb{R}^2 \to \mathbb{R}^n$ must also satisfy

$$\sqrt{\sum_{i=1}^{n} \hat{I}_i^2(x, y, 0)} = 1$$

or $\hat{I}(x, y, 0) : \mathbb{R}^2 \to S^{n-1}$. The same constraint holds for a multiscale representation $I(x, y, t)$ of the original direction $I(x, y, 0)$. This is what makes the smoothing of directions different from the smoothing of ordinary vectorial data as in Refs. [40, 344, 370, 405, 409, and 416]; see Chap. 5. The smoothing is performed in S^{n-1} instead of $\mathbb{R}^n$.

Directions can also be represented by the angle(s) the vector makes with a given coordinate system, denoted in this chapter as orientation(s). In the 2D case for example, the direction of a vector (I_1, I_2) can be given by the angle θ that this vector makes with the x axis [we consider $\theta \in [0, 2\pi)$]: $\theta = \arctan(I_2/I_1)$ (with the corresponding sign considerations to have the map in the $[0, 2\pi)$ interval). There is of course a one-to-one map between a direction vector $I(x, y) : \mathbb{R}^2 \to S^1$ and the angle function $\theta(x, y)$. Using this relation, Perona [309] well motivated the necessity for orientation and direction diffusion and transformed the problem of 2D direction diffusion into a 1D problem of angle or orientation diffusion (see additional comments in Section 6.2 below). Perona then proposed PDE-based techniques for the isotropic smoothing of 2D orientations; see also Refs. [161 and 402] and the general discussion of these methods in Ref. [309]. Smoothing orientations instead of directions solve the unit norm constraint, but add a periodicity constraint. Perona showed that a simple heat flow (Laplacian or Gaussian filtering) applied to the $\theta(x, y)$ image, together with special numerical attention, can address this periodicity issue. This pioneering approach theoretically it applies to only smooth data (indeed to vectors with coordinates in the Sobolev space $W^{1,2}$), thereby disqualifying edges. The straightforward extension of this to S^{n-1} would be to consider $n - 1$ angles and smooth each one of these as a scalar image. The natural coupling is then missing, yielding a set of decoupled PDEs.

As Perona pointed out in his work, directional data are just one example of the diffusion of images representing data beyond flat manifolds. Extensions to Perona's work, by use of intrinsic metrics on the manifold, can be found in Refs. [80 and 370]. In Ref. [80] Chan and Shen explicitly deal with orientations (no directions) and present the L_1 norm as well as many additional new features, contributions on discrete formulations, and connections with the now described approach. In Ref. [370] Sochen et. al. do not deal with orientations or directions, although their framework is valid for approaching this problem as well. In Ref. [329] Rudin and Osher also mention the minimization of the L_1 norm of the divergence of the normalized image gradient (curvature of the level sets). They do this in the framework of image denoising, without addressing the regularization and analysis of directional data or presenting examples. Note that not only the data (e.g., directions) can go beyond flat manifolds, but their domain can also be nonflat (e.g., the data can be defined on a surface). The harmonic framework here described addresses the nonflatness of both manifolds, that is, with the general framework introduced here we can obtain isotropic and anisotropic diffusion and scale spaces for any function mapping two manifolds (see also Ref. [370], a framework also permitting this). We can for example diffuse (and denoise) data on a 3D surface or diffuse posterior probability vectors [305]. In this chapter the general framework is presented and details are given in the case of directions defined on the plane; other cases are described elsewhere, [24, 237].

From the original unit norm vectorial image $I(x, y, 0) : \mathbb{R}^2 \to S^{n-1}$ we construct a family of unit norm vectorial images $I(x, y, t) : \mathbb{R}^2 \times [0, \tau) \to S^{n-1}$ that provides a multiscale representation of directions. The method intrinsically takes care of the normalization constraint, eliminating the need to consider orientations and develop special periodicity preserving numerical approximations. Discontinuities in the directions are also allowed by the algorithm. The approach follows results from the literature on harmonic maps in liquid crystals, and $I(x, y, t)$ is obtained from a system of coupled PDEs that reduces a given (harmonic) energy. Energies giving both isotropic and anisotropic flows will be described. Because of the large amount of literature on the subject of harmonic maps applied to liquid crystals, a number of relevant theoretical results can immediately be obtained.

Before the details of the direction diffusion framework are given its main unique characteristics are described:

1. It includes both isotropic and anisotropic diffusion.
2. It works for directions in any dimension.

3. It supports nonsmooth data.
4. The general framework also includes directions and general image data defined on general manifolds, e.g., surface normals, principal directions, and images on 3D surfaces.
5. It is based on a substantial amount of existing theoretical results that help to answer a number of relevant computer vision questions.

6.1. The General Problem

Let $I(x, y, 0) : \mathbb{R}^2 \to S^{n-1}$ be the original image of directions, that is, this is a collection of vectors from $\mathbb{R}^2$ to $\mathbb{R}^n$ such that their unit norm is equal to one, i.e., $\|I(x, y, 0)\| = 1$, where $\|\cdot\|$ indicates Euclidean length. $I_i(x, y, 0) : \mathbb{R}^2 \to \mathbb{R}$ stands for each one of the n components of $I(x, y, 0)$. We search for a family of images, a multiscale representation, of the form $I(x, y, t) : \mathbb{R}^2 \times [0, \tau) \to S^{n-1}$, and once again we use $I_i(x, y, t) : \mathbb{R}^2 \to \mathbb{R}$ to represent each one of the components of this family. Let us define the component gradient ∇I_i as

$$\nabla I_i := \frac{\partial I_i}{\partial x}\vec{x} + \frac{\partial I_i}{\partial y}\vec{y}, \tag{6.1}$$

where $\vec{x}$ and $\vec{y}$ are the unit vectors in the x and the y directions respectively. From this,

$$\|\nabla I_i\| = \left[\left(\frac{\partial I_i}{\partial x}\right)^2 + \left(\frac{\partial I_i}{\partial y}\right)^2\right]^{1/2} \tag{6.2}$$

gives the absolute value of the component gradient.

The component Laplacian is given by

$$\Delta I_i = \frac{\partial^2 I_i}{\partial x^2} + \frac{\partial^2 I_i}{\partial y^2}. \tag{6.3}$$

We are also interested in the absolute value of the image gradient, given by

$$\|\nabla I\| := \left\{\sum_{i=1}^{n}\left[\left(\frac{\partial I_i}{\partial x}\right)^2 + \left(\frac{\partial I_i}{\partial y}\right)^2\right]\right\}^{1/2}. \tag{6.4}$$

Having this notation, we are now ready to formulate the framework. The problem of harmonic maps in liquid crystals is formulated as the search for the solution to

$$\min_{I:\mathbb{R}^2\rightarrow S^{n-1}} \iint_{\Omega} \|\nabla I\|^p \, dxdy, \tag{6.5}$$

where Ω stands for the image domain and $p \geq 1$. This variational formulation can be rewritten as

$$\min_{I:\mathbb{R}^2\rightarrow \mathbb{R}^n} \iint_{\Omega} \|\nabla I\|^p \, dxdy, \tag{6.6}$$

such that

$$\|I\| = 1. \tag{6.7}$$

This is a particular case of the search for maps I between Riemannian manifolds (M, g) and (N, h), which are critical points (that is, minimizers) of the harmonic energy

$$E_p(I) = \int_M \|\nabla_M I\|^p \, \mathrm{dvol} M, \tag{6.8}$$

where $\|\nabla_M I\|$ is the length of the differential in M. In our particular case, M is a domain in $\mathbb{R}^2$ and $N = S^{n-1}$, and $\|\nabla_M I\|$ reduces to Eq. (6.4). The critical points of Eq. (6.8) are called p-harmonic maps (or simply harmonic maps for $p = 2$). This is in analogy to the critical points of the Dirichlet energy $\int_\Omega \|\nabla f\|^2$ for real-valued functions f, which are called harmonic functions.

The general form of the harmonic energy, normally from a 3D surface (M) to the plane (N) with $p = 2$ (the most classical case, e.g., Refs. [118 and 119]), was successfully used for example in computer graphics to find smooth maps between two given (triangulated) surfaces (again, normally a surface and the complex or real plane); see, e.g., Refs. [14, 114, 169, and 425]. In this case, the search is indeed for the critical point, that is, for the harmonic map between the surfaces. This can be done for example by means of finite elements [14, 169]. In our case, the problem is different. We already have a candidate map, the original (normally noisy or with many details at all scales) image of directions $I(x, y, 0)$, and we want to compute a multiscale representation or regularized/denoised version of it, that is, we are not (just) interested in the harmonic map between the domain in $\mathbb{R}^2$ and S^{n-1} (the critical point of the energy), but are interested in the process of computing this map by means of PDEs. More specifically, we are interested in the gradient-descent-type flow of the harmonic energy [Eq. (6.8)]. This is partially motivated by the fact that, as we have already seen, the basic diffusion equations for multiscale representations and denoising of gray-valued images are obtained as well as gradient descent flows acting on real-valued data; see, for example, Refs. [36, 310, 330, and 421]. Isotropic diffusion (linear heat flow) is just the gradient descent of the L_2 norm of the image gradient, whereas anisotropic diffusion can be interpreted as the gradient descent flow of more robust functions acting on the image gradient.

Because we have an energy formulation, it is straightforward to add additional data-dependent constraints on the minimization process, e.g., preservation of the original average; see for example, Ref. [330] for examples for gray-valued images. In this case we might indeed be interested in the critical point of the modified energy, which can be obtained as the steady-state solution of the corresponding gradient descent flow. Because our goal in this chapter is to describe the general framework for direction diffusion, we will not add these type of constraints in the examples in Section 6.4. These constraints are normally closely tied to both the specific problem and the available information about the type of noise present in the image. In Ref. [80], data terms are added.

For the most popular case of $p = 2$, the Euler–Lagrange equation corresponding to Eq. (6.8) is a simple formula based on Δ_M, the Laplace–Beltrami operator of M, and $A_N(I)$, the second fundamental form of N (assumed to be embedded in $\mathbb{R}^k$) evaluated at I; see, e.g., Refs. [118, 119, 169, and 376]:

$$\Delta_M I + A_N(I)\langle \nabla_M I, \nabla_M I \rangle = 0. \tag{6.9}$$

This leads to a gradient-descent-type flow, that is,

$$\frac{\partial I}{\partial t} = \Delta_M I + A_N(I)\langle \nabla_M I, \nabla_M I \rangle. \tag{6.10}$$

In the following sections, the gradient descent flows are presented for our particular energy (6.5), that is, for M being a domain in $\mathbb{R}^2$ and N equal to S^{n-1}. We concentrate on the cases of $p = 2$, isotropic, and $p = 1$, anisotropic (or in general $1 \leq p < 2$). The use of $p = 2$ corresponds to the classical heat flow from the linear scale-space theory [218, 413], whereas the case $p = 1$ corresponds to the total-variation flow studied in Ref. [330]. For data such as surface normals, principal directions, or simple gray values in three dimensions, M is a surface in three dimensions and the general flow of Eq. (6.10) is used. This flow can be implemented by classical numerical techniques to compute ∇_M, Δ_M, $A_N(I)$ on triangulated surfaces; see, e.g., Refs. [14, 169, and 187]. The harmonic-map framework can also be extended to data defined on implicit surfaces by adaptation of the variational level-set approach described in Section 2.3 in this book [24] and to data mapping into arbitrary implicit surfaces [237].

Most of the literature on harmonic maps deals with $p = 2$ in Eq. (6.8) or expression (6.5), the linear case. Some more recent results are available for $1 < p < \infty$, $p \neq 2$, [87, 102]; very few results deal with the case $p = 1$ [158]. A number of theoretical results, both for the variational formulation and its corresponding gradient descent flow, which are relevant to the multiscale representation of directions, will be given in the following sections as well. References [118 and 119] are an excellent source of information for regular harmonic maps, and Ref. [175] contains a comprehensive review of singularities of harmonic maps (check also Ref. [376], a classic on harmonic maps). A classical paper for harmonic maps in liquid crystals, that is, the particular case of expression (6.5) (or, in general, M is a domain in $\mathbb{R}^n$ and $N = S^{n-1}$), is Ref. [48].

6.2. Isotropic Diffusion

It is easy to show (see Appendix) that, for $p = 2$, the gradient descent flow corresponding to Eq. (6.6) with the constraint of Eq. (6.7) is given by the set of coupled PDEs:

$$\frac{\partial I_i}{\partial t} = \Delta I_i + I_i \, \|\nabla I\|^2, \quad 1 \leq i \leq n. \tag{6.11}$$

This system of coupled PDEs defines the isotropic multiscale representation of $I(x, y, 0)$, which is used as initial data to solve Eq. (6.11). (Boundary conditions are also added in the case of finite domains.)

This result can also be obtained directly from general Euler–Lagrange equations (6.9) and (6.10). The Laplace–Beltrami operator Δ_M and manifold gradient ∇_M become the regular Laplace and gradient respectively, as we are working on $\mathbb{R}^2$ (the same for $\mathbb{R}^n$). The second fundamental form $A_N(I)$ of the sphere (in any dimension) is I.

The first part of Eq. (6.11) comes from the variational form, and the second part comes from the constraint (see, for example, Ref. [377]). As expected, the first part is decoupled between components I_i and is linear, whereas the coupling and nonlinearity come from the constraint.

If $n = 2$, that is, we have 2D directions, then it is easy to show that for (smooth data) $I(x, y) = [\cos\theta(x, y), \sin\theta(x, y)]$, the energy in expression (6.5) becomes $E_p(\theta) := \iint_\Omega (\theta_x^2 + \theta_y^2)^{p/2} \mathrm{d}x\mathrm{d}y$. For $p = 2$ we then obtain the linear heat flow on θ ($\theta_t = \Delta\theta$) as the corresponding gradient descent flow, as expected from the results in Ref. [309]. The precise equivalence between the formulation of the energy in terms of directions, $E_p(I)$, and the formulation in terms of orientations, $E_p(\theta)$, is true only for $p \geq 2$ [29, 158]. In spite of this connection, and as was pointed out in Ref. [158], using directions and orientations is not fully equivalent. Observe for example the map $v(X) = \frac{X}{\|X\|}$ defined on the unit ball B^2 of $\mathbb{R}^2$. On the one hand, this map has finite E_p energy if and only if $1 \leq p < 2$. On the other hand, the map cannot be defined with an angle function $\theta(x, y)$, θ being in the Sobolev space $W^{1,p}(B^2)$. As pointed out in Ref. [158], the only obstruction to this representation is the index of the vector of directions. Thus smooth directions can lead to nonsmooth orientations (Perona's goal in the periodic formulations he proposes is in part to address this issue). This problem then gives an additional advantage to selecting a direction-based representation instead of an orientation-based one.

For the isotropic case, $p = 2$, we have the following important results from the literature on harmonic maps:

Existence: Existence results for harmonic mappings were already reported in Ref. [120] for a particular selection of the target manifold N (nonpositive curvature). Struwe [376] showed, in one of the classical papers in the area, that for initial data with finite energy (as measured by Eq. (6.8)), M a 2D dimensional manifold with $\partial M = \emptyset$ (manifold without boundary), and $N = S^{n-1}$, there is a unique solution to the general gradient descent flow. Moreover, this solution is regular with the exception of a finite number of isolated points and the harmonic energy is decreasing in time. If the initial energy is small, the solution is completely regular and converges to a constant value. (These results actually hold for any compact N.) This uniqueness result was later extended to manifolds with smooth $\partial M \neq \emptyset$ and for weak solutions [144]. Recapping, there is a unique weak solution to Eq. (6.11) (weak solutions defined in natural spaces, $H^{1,2}(M, N)$ or $W^{1,2}(M, N)$), and the set of possible singularities is finite. These solutions decrease the harmonic energy. The result is not completely true for M with dimension greater than two, and this was investigated for example in Ref. [86]. Global weak solutions exist for example for $N = S^{n-1}$, although there is no uniqueness for the general initial value problem [101]. Results on the regularity of the solution, for a restricted suitable class of weak solutions, to the harmonic flow for high-dimensional manifolds M into S^{n-1} have been recently reported [88, 139]. In this case, it is assumed that the weak solutions hold a number of given energy constraints.

Singularities in two dimensions: If $N = S^1$ and the initial and the boundary conditions are well behaved (smooth, finite energy), then the solution of the harmonic flow is regular. This is the case for example for smooth $2D$ image gradients and $2D$ optical flow.

Singularities in three dimensions: Unfortunately, for $n = 3$ in Eq. (6.11) (that is $N = S^2$, $3D$ vectors), smooth initial data can lead to singularities in finite time [84]. Chang et al. showed examples in which the flow of Eq. (6.11), with initial data $I(x, y, 0) = I_0(x, y) \in C^1(D^2, S^2)$ (D^2 is the unit disk on the plane) and boundary conditions $I(x, y, t)|_{\partial D^2} = I_0|_{\partial D^2}$, develops singularities in finite time. The idea is to use as original data I_0, a function that covers S^2 more than once in a certain region. From the point of view of the harmonic energy, the solution is giving up on regularity in order to reduce energy.

Singularities topology: Because singularities can occur, it is then interesting to study them [48, 175, 323]. For example, Brezis et al. [48] studied the value of the harmonic energy when the singularities of the critical point are prescribed (the map is from R^3 to S^2 in this case). (Perona suggested looking at this line of work to analyze the singularities of the orientation diffusion flow.) Qing [323] characterized the energy at the singularities. A recent review on the singularities of harmonic maps was prepared by Hardt [175]. (Singularities for more general energies are studied for example in Ref. [315].) The results there reported can be used to characterize the behavior of the multiscale representation of high-dimensional directions, although these results mainly address the shape of the harmonic map, that is, the critical point of the harmonic energy and not the flow. Of course, for the case in which M is of two dimensions, which corresponds to Eq. (6.11), we have Struwe's results, mentioned above.

6.3. Anisotropic Diffusion

The picture becomes even more interesting for the case $1 \leq p < 2$. Now the gradient descent flow corresponding to expression (6.6), in the range $1 < p < 2$ (and formally for $p = 1$), with constraint (6.7) is given by the set of coupled PDEs:

$$\frac{\partial I_i}{\partial t} = \operatorname{div}\left(\|\nabla I\|^{p-2}\, \nabla I_i\right) + I_i\, \|\nabla I\|^p, \quad 1 \leq i \leq n. \qquad (6.12)$$

This system of coupled PDEs defines the anisotropic multiscale representation of $I(x, y, 0)$, which is used as the initial datum to solve Eq. (6.11). In contrast with the isotropic case, now both terms in Eq. (6.12) are nonlinear and include coupled components. Formally, we can also explicitly write the case $p = 1$, giving

$$\frac{\partial I_i}{\partial t} = \operatorname{div}\left(\frac{\nabla I_i}{\|\nabla I\|}\right) + I_i \, \|\nabla I\|, \quad 1 \leq i \leq n, \tag{6.13}$$

although the formal analysis and interpretation of this case is much more delicate than the description presented in the Appendix.

The case of $p \neq 2$ in Eq. (6.8) has been less studied in the literature, e.g. [176]. When M is a domain in $\mathbb{R}^m$ and $N = S^{n-1}$, the function $v(X) := \frac{X}{\|X\|}$, $X \in \mathbb{R}^m$, is a critical point of the energy for $p \in \{2, 3, \ldots, m-1\}$, for $p \in [m-1, m)$ (this interval includes the energy case that leads to Eq. (6.12)), and for $p \in [2, m - 2\sqrt{m-1}]$ [175]. For $n = 2$ and $p = 1$, the variational problem has also been investigated in Ref. [158], in which Giaquinta et al. addressed, among other things, the correct spaces to perform the minimization (in the scalar case, $BV(\Omega, \mathbb{R})$ is used) and the existence of minimizers. Of course, we are more interested in the results for the flow of Eq. (6.12), and not just in its corresponding energy. Some results exist for $1 < p < \infty$, $p \neq 2$, showing in a number of cases the existence of local solutions that are not smooth. To the best of my knowledge, the case of $1 \leq p < 2$, and in particular $p = 1$, has not been fully studied for the evolution equation.

Following the framework for robust anisotropic diffusion we can generalize expression (6.5) and study problems of the form

$$\min_{I:\mathbb{R}^2 \to S^{n-1}} \iint_\Omega \rho(\|\nabla I\|) \mathrm{d}x \mathrm{d}y, \tag{6.14}$$

where ρ is now a robust function like the Tukey biweight.

6.4. Examples

In this section a number of illustrative examples are presented for the harmonic flows for $p = 2$ (isotropic) and $p = 1$ (anisotropic) presented above. These examples will mainly show that the proposed framework produces for directional data the same qualitative behavior that is well known and studied for scalar images. One of the advantages of directional diffusion is that, although advanced specialized numerical techniques to solve expression (6.5) and its corresponding gradient descent flow have been developed (e.g., Ref. [4]), as a first approximation we can basically use the algorithms developed for isotropic and anisotropic diffusion without the unit norm constraint to implement Eqs. (6.11) and (6.12) [99]. Although, as stated before, these PDEs preserve the unit norm (that is, the solutions are vectors in S^{n-1}), numerical errors might violate the constraint (recent developments by Osher and Vese might solve this numerical problem [295]). Therefore, between every two steps of the numerical implementation of these equations we add a renormalization step [99]. Basically, a simple time-step iteration is used for Eq. (6.11), whereas for Eq. (6.12) we incorporate the edge-capturing technique developed in Ref. [330] (we always use the maximal time step that ensures stability).

For the examples shown below, a number of visualization techniques are used (note that the visualization of vectors is an active research area in the graphics community):

1. Arrows. Drawing arrows that indicate the vector direction is very illustrative, but can be done only for sparse images and they are not very informative for dense data such as gradients or optical flow. Therefore we use arrows either for toy examples or to show the behavior of the algorithm in local areas.
2. Hue-Saturation-Valve (HSV) color mapping. We use the HSV color map (applied to orientation) to visualize whole images of directions while being able to also illustrate details such as small noise.
3. Line integral convolution (LIC) [55]. LIC is based on locally integrating the values of a random image at each pixel. The integration is done in the line corresponding to the direction at the pixel. The LIC technique gives the general form of the flow, while the color map is useful to detect small noise in the direction (orientation) image.

All these visualization techniques are used for vectors in S^1. We also show examples for vectors in S^2. In this case, we consider these vectors as red–green–(RGB) vectors, and color is used to visualize the results.

Figure 6.1 shows a number of toy examples to illustrate the general ideas introduced in this chapter. The first row shows, by use of LIC, an image with two regions having two different (2D) orientations on the left (original), followed by the results of isotropic diffusion for 200, 2000, and 8000 iterations (scale space). Three different steps of the scale space are shown by arrows in the next row. Note how the edge in the directional data is being smoothed out. The horizontal and the vertical directions are being smoothed out to converge to the diagonal average. This is in contrast with the edges in the third row. This row shows the result of removing noise in the directional data. The original noisy image is shown first, followed by the results with isotropic and anisotropic smoothing. Note how the anisotropic

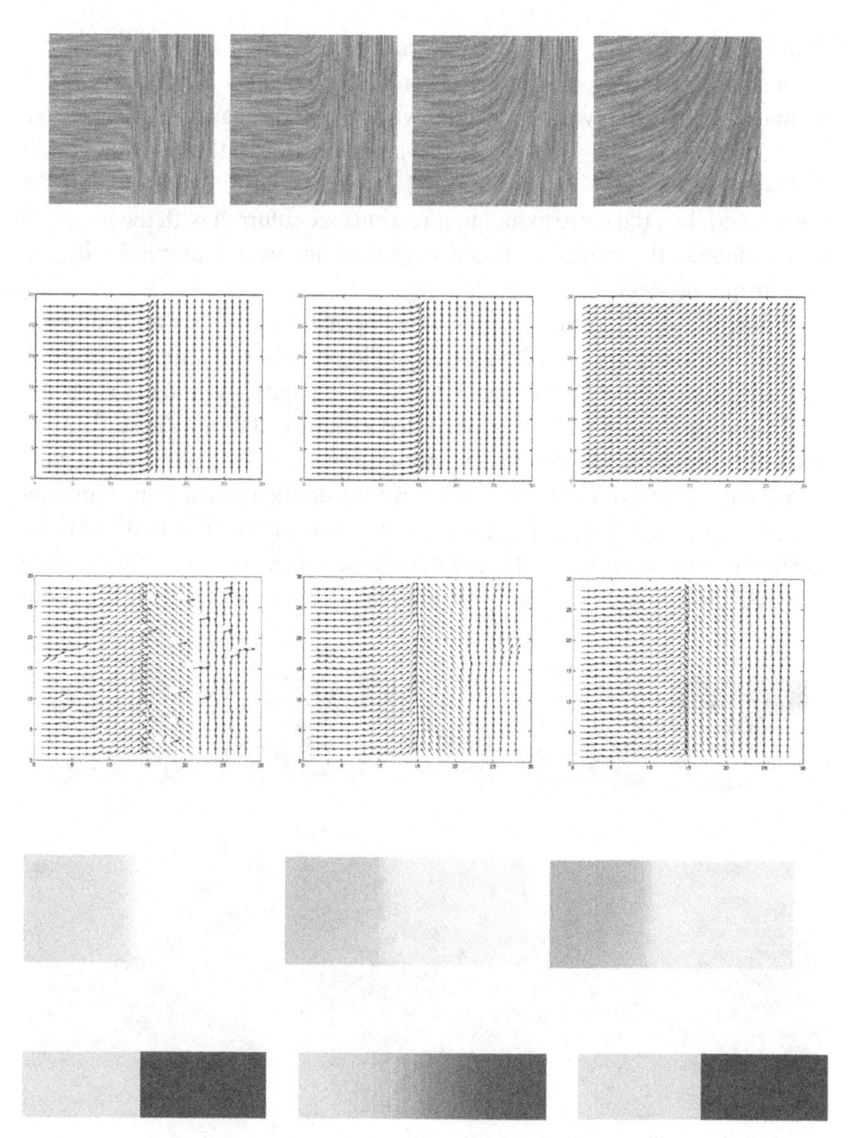

Fig. 6.1. Toy examples illustrating the ideas in this chapter. See text for details.

flow gets rid of the noise (outliers) while preserving the rest of the data, whereas the isotropic flow also affects the data themselves while removing the noise. Note that because the discrete theory developed in Ref. [309] applies to only small changes in orientation, theoretically it cannot be applied to the images we have seen so far; all of them contain sharp discontinuities in the directional data. The last two rows of Fig. 6.1 deal with 3D directions. In this case, we interpret the vector as RGB coordinates and use color to visualize them. First two steps are shown in the scale space for the isotropic flow (original on the left). Then the last row shows the original image followed by the results of isotropic and anisotropic diffusion. Note how the colors, and then the corresponding directions, get blurred with the isotropic flow, whereas the edges in direction (color) are well preserved with the anisotropic process.

Figure 6.2 shows results for simulated optical flow. This simulated optical flow data was computed with the public domain software described in Ref. [33] (downloaded from M. Black's home page), stopping at early annealing stages to have enough noise for experimentation. These data are used here for illustration purposes only. Alternatively, we could include the harmonic energy as a regularization term inside the optical flow computation, that is, combined with the optical flow constraint. The first figure on the first row shows a frame of the famous Yosemite movie. Next, in the same row, from left to right is shown the original optical flow direction, the result

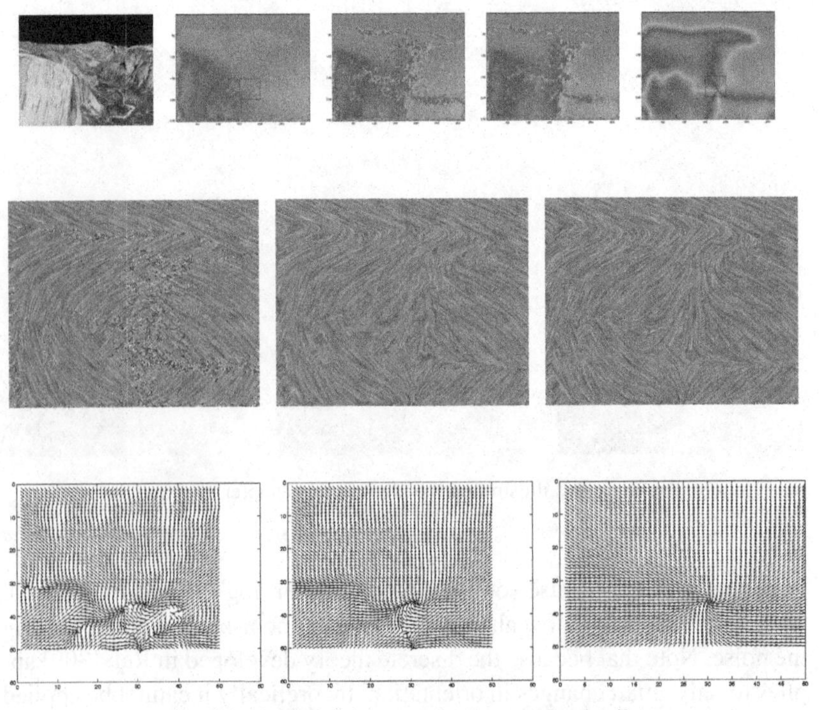

Fig. 6.2. Optical flow example. See text for details.

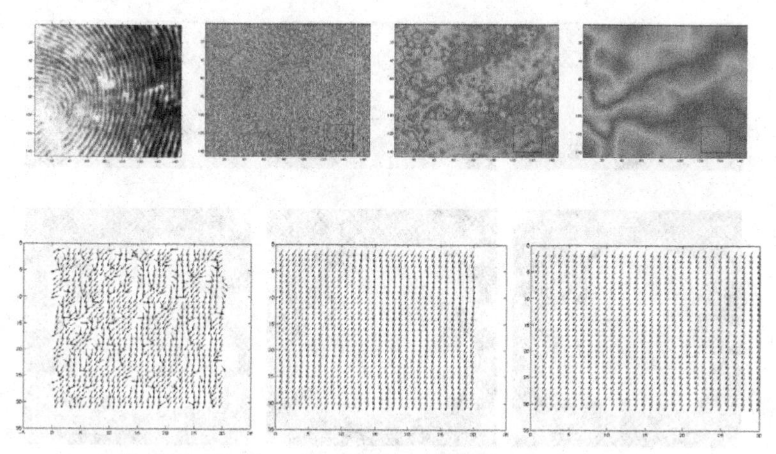

Fig. 6.3. Gradient direction example. See text for details.

of the isotropic flow, the result of the anisotropic flow, and the result of the isotropic flow for a large number of iterations. The HSV color map is used. In the next row, LIC is used to show again the three middle figures of the first row. In the third row, arrows are used to show a blowup of the marked region corresponding to the isotropic flow for 20, 60, and 500 iterations. Note how the noise in the optical flow directions is removed.

Figure 6.3 deals with 90°-rotated gradient directions in a fingerprint image. After the raw data for the image are shown, the color-map visualization technique is used to present a number of steps (scale space) of the isotropic flow. This is followed by arrows for a blowup of the marked region (last row, from left to right, original, 20 steps, and 200 steps, respectively). Note in the arrow images how the noise and details (small scales) are removed and progressively the averaged orientation for the fingerprint is obtained.

Figure 6.4 presents two examples for color images, that is, 3D directions defined on $\mathbb{R}^2$. Both rows show the original image, followed by the noisy image and the enhanced one. In the first row, noise is added to only the 3D RGB directions representing the chroma (the RGB vector normalized to a unit norm vector), while the magnitude of the vector (brightness) was kept untouched. The enhanced image was then obtained when the proposed direction diffusion flow was applied to the chroma, while the original magnitude was kept. This experiment shows that when the magnitude is preserved (or well reconstructed; see below), direction diffusion for chroma denoising produces an image practically indistinguishable from the original one. In the second row, the noise was added to the original image, resulting in both

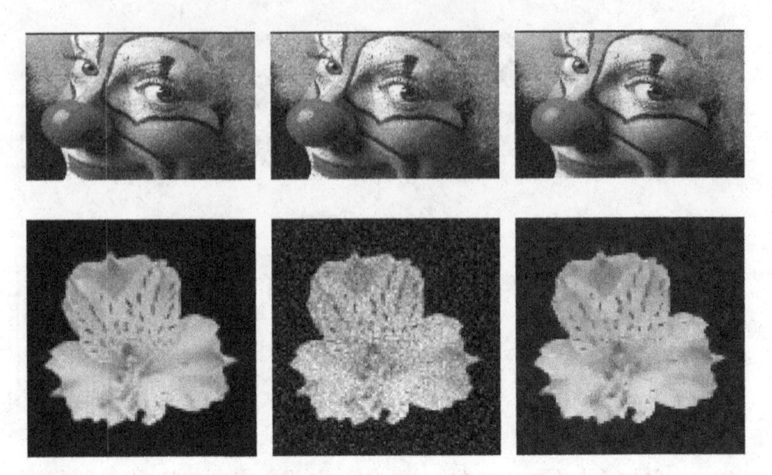

Fig. 6.4. Denoising of a color image by 3D directional diffusion. See text for details.

noisy chroma (direction) and brightness (magnitude) of the RGB vector. The directions are processed with the diffusion flow, and the magnitude is processed with the scalar anisotropic flow of Ref. [36]. Isotropic and anisotropic direction diffusion flows on the chroma produce similar results as long as the magnitude is processed with an edge-preserving denoising algorithm. See Ref. [381] for additional examples, comparisons with the literature, and details on the applications of direction diffusion to color images.

6.5. Vector Probability Diffusion

We have already seen in Section 4.4 that we can smooth posterior probabilities for improving classification results by means of MAP. In Section 4.4 we anisotropically smoothed each posterior and added a normalization step to guarantee that the posteriors add to one. Two main difficulties are encountered in this. First, each posterior probability (one per class) is independently diffused, thereby ignoring the intrinsic correlation between them. Second, because of the independent processing, the posterior probabilities are not guaranteed to add to one even after a very short diffusion time. To overcome this, we need to normalize the posterior probabilities after each discrete iteration, normalization that has a nontrivial effect on the diffusion process itself.

To solve these problems we need then to find a set of coupled PDEs that guarantees that the vector of posterior probabilities remains a probability vector, that is, with all its components positive and adding to one. This is a particular and simple case of harmonic maps, in which instead of having the data defined on a hypersphere as in the direction diffusion case, we have data that are defined on an hyperplane. We develop this now. Moreover, as we shall see, the numerical implementation of this system of coupled PDEs also preserves these properties, thereby removing the necessity to project back into the semihyperplane. This approach then overcomes both difficulties mentioned above and can be directly incorporated into the segmentation technique, replacing the component-by-component diffusion. Although the work is here discussed in the framework of posterior diffusion for image classification, it is clear that the technique can be applied to the diffusion of other probabilities in other applications as well.

6.5.1. Problem Formulation and Basic Equations

Assume that a vector of a posteriori probabilities p, mapping the image domain Ω in $\mathbb{R}^2$ to the manifold $\mathcal{P} = \{p \in \mathbb{R}^m : \|p\|_1 = 1, p_i \geq 0\}$, is given. Each component p_i of p equals the posterior probability of a class $c_i \in \mathcal{C} = \{c_i : i = 1, \ldots, m\}$. These posterior probabilities can be obtained for example by means of Bayes' rule.

If we view the vectors p as a vector field, one possible way to add spatial coherence into the classification process is to diffuse the distance between points in $\mathcal{P}$, propagating the information in the probability space, before a pixelwise MAP decision is made. Inspired by the harmonic-map theory described above, the distance between two differential adjacent points in $\mathcal{P}$ depends on $\|\nabla p\| := \sqrt{\sum_{i=1}^m \|\nabla p_i\|^2}$. This is the gradient of the probability vector. Giving a function $\rho : \mathbb{R} \to \mathbb{R}$ (we will later discuss different selections of ρ), we proceed to solve the following minimization process:

$$\min_{p \in \mathcal{P}} J_\rho, \quad J_\rho := \int_\Omega \rho(\|\nabla p\|) \mathrm{d}\Omega. \tag{6.15}$$

Note that the minimization is restricted to the semihyperplane $\mathcal{P}$.

From this, the system of coupled diffusion equations is obtained by means of the gradient descent flow corresponding to this energy.

Proposition 6.1. *The gradient descent of J_ρ restricted to $\mathcal{P}$ is given by*

$$\frac{\partial p}{\partial t} = \nabla \cdot \left(\frac{\rho'(\|\nabla p\|)}{\|\nabla p\|} \nabla p \right). \tag{6.16}$$

This equation is an abbreviated notation for a set of PDEs of the form

$$\frac{\partial p_i}{\partial t} = \nabla \cdot \left[\frac{\rho'(\|\nabla p\|)}{\|\nabla p\|} \nabla p_i \right], \quad i = 1, \ldots, m.$$

Proof: We easily obtain this by computing the Euler–Lagrange and projecting it into $\mathcal{P}$. □

Note again that the minimization is performed in the probability space $\mathcal{P}$ (a semihyperplane), and the system of equations (6.16) guarantees that $p(t) \in \mathcal{P}$ for all t. We have therefore obtained a system of coupled PDEs that preserves the unit L_1 norm that is characteristic of probability vectors (the components $p_i(t)$ are positive as well).

It can be shown that $\sum_{i=1}^{m} \|\nabla p_i\|^2 = 2\|\nabla p_1\|^2 - 2\sum_{i=2}^{m}\sum_{j=i+1}^{m} \nabla p_i \cdot \nabla p_j$. Therefore, when the number of classes is $m = 2$, $\|\nabla p\|^2 = 2\|\nabla p_i\|^2, i = 1, 2$, and the method is equivalent to separately applying the diffusion to each posterior. For $m \geq 3$, the second term is like the correlation between different components of p. This term is not present if we diffuse each posterior probability on its own. This coupling is important to improve the classification results.

If we select ρ as the L_2 norm, $\rho(x) = x^2$, then Eq. (6.16) becomes the well-known linear heat equation $\frac{\partial p}{\partial t} = \nabla^2 p$, which isotropically diffuses p. It is interesting to note that the heat equation that was previously used to denoise signals as well to generate the so-called scale spaces preserves the diffusion in the probability semihyperplane $\mathcal{P}$. This is expected, as it is well known that this equation holds the maximum principle and preserves linear combinations.

The isotropic flow has no coupling between the probability components and does not respect boundaries. Therefore, as classically done for scalar diffusion [330], a more robust norm is selected. For example, we can select the L_1 norm, $\rho(x) = |x|$, obtaining

$$\frac{\partial p}{\partial t} = \nabla \cdot \left(\frac{\nabla p}{\|\nabla p\|} \right), \tag{6.17}$$

which is clearly anisotropic, with a conduction coefficient controlled by $\|\nabla p\|$. Once again, from the basic invariants of this flow, the preservation of the vector on the $\mathcal{P}$ space was expected.

6.5.2. Numerical Implementation

The vector probability diffusion equations are to be implemented on a square lattice. The vector field p is then $p_{j,k}$, where j, k is the position in the lattice. (For simplicity, we do not write the subscript i indicating the probability components.) First we rewrite Eq. (6.16) in a simpler way, $(\partial p/\partial t) = \nabla \cdot (g\nabla p),\ \ g = \rho'(\|\nabla p\|)/\|\nabla p\|$, and then we apply a standard numerical scheme (assuming that $\Delta x = \Delta y = 1$):

$$\frac{p_{j,k}^{t+\Delta t} - p_{j,k}^{t}}{\Delta t} = g_{j+\frac{1}{2},k}\left(p_{j+1,k}^{t} - p_{j,k}^{t}\right) - g_{j-\frac{1}{2},k}\left(p_{j,k}^{t} - p_{j-1,k}^{t}\right)$$
$$+ g_{j,k+\frac{1}{2}}\left(p_{j,k+1}^{t} - p_{j,k}^{t}\right) - g_{j,k-\frac{1}{2}}\left(p_{j,k}^{t} - p_{j,k-1}^{t}\right). \quad (6.18)$$

The condition on Δt for stability is easily computed to be

$$\Delta t \leq \frac{1}{4 \max_{j,k}\{g_{j+\frac{1}{2},k}, g_{j-\frac{1}{2},k}, g_{j,k+\frac{1}{2}}, g_{j,k-\frac{1}{2}}\}}. \quad (6.19)$$

To allow the existence of discontinuities in the solution we use the approximation of the gradient developed in Ref. [293].

Proposition 6.2. *If Δt fulfills stability condition (6.19) then $p_{j,k}^{t+\Delta t} \geq 0$ and $\|p_{j,k}^{t+\Delta t}\|_1 = 1$, so the evolution given by Eq. (6.18) lives always in the manifold $\mathcal{P}$. Moreover, if $p_m^t := \min\{p_{j+1,k}^t, p_{j-1,k}^t, p_{j,k+1}^t, p_{j,k-1}^t\}$ and $p_M^t := \max\{p_{j+1,k}^t, p_{j-1,k}^t, p_{j,k+1}^t, p_{j,k-1}^t\}$, the solution satisfies a maximum (minimum) principle: $p_m^t \leq p_{j,k}^{t+\Delta t} \leq p_M^t$.*

We then conclude that also discrete equation (6.18) lives in the manifold $\mathcal{P}$, and there is no need for a projection back into the semihyperplane, in contrast with the scalar approach by Teo et al. [384], as discussed before. To complete the implementation details, we need to address possible problems when $\|\nabla p\|$ vanishes or becomes very small. As is standard in the literature, we define $\|\nabla p\|_\beta = \sqrt{\beta^2 + \|\nabla p\|^2}$, and use this instead of the traditional gradient. [Stability condition (6.19) becomes $\Delta t \leq \beta/4$.) To select the value of β we propose to look at β as a lower diffusion scale, because in the discrete case probability differences lower than β will be diffused with a conduction coefficient approximately inverse to β. We set the value of the lower scale β in the range $[0.001, 0.01]$.

6.5.3. Examples

Examples are now presented of the vector probability diffusion approach, here presented as applied to image segmentation. As for the scalar case, first, posterior probabilities are computed for each class by Bayes' rule. These are diffused with the anisotropic vector probability diffusion flow (instead of the scalar one used in Ref. [384] and described in Section 4.4), and then the MAP decision rule is applied. We will compare these results with those obtained with the scalar formulation in Section 4.4 to show the improvements obtained with the vectorial approach.

Fig. 6.5. Left to right and top to bottom: Original image, classification without diffusion, results of vector probability diffusion for 18 and 25 iterations, respectively, and results of the scalar approach for 10 and 15 iterations.

The first example is based on synthetic data. The original image, containing four classes, is shown on the top left of Fig. 6.5. We then add Gaussian noise, and the objective is to segment the image back into four classes. The figure then shows the classification results without posterior diffusion, followed by the classification results corresponding to two time steps for the vector posterior diffusion approach and two time steps for the scalar approach. Using the correlation between the posterior probabilities, as done by the vector approach introduced here and not by the scalar one in Ref. [384], is of particular importance when one of the classes has less weight than the others. As shown by this example, this class (white dots) will be mostly missed by the scalar approach. Comparing the classification errors for class 4 (white dots) in Fig. 6.6, we see that the lowest classification error is with vector probability diffusion, 15.82% against 18.46% for the scalar approach. Furthermore, the classification error is more stable with the vectorial approach, and for the same average error (approximately 10%), the lowest classification error for class 4 is obtained with the vectorial method, 26.95% against 61.04% (see Fig. 6.6).

Figure 6.7 shows the results of the vectorial approach applied to the segmentation of video SAR data. Three classes are considered, {shadow, object, background}, each one modeled as a Gaussian distribution. The first image in the sequence is segmented by hand to obtain an estimation of the parameters of the distributions. For the rest of the images we perform

Iter.	Class 1	Class 2	Class 3	Class 4	Average
0	53.77	26.67	53.68	27.34	46.65
10	47.63	17.47	47.40	22.46	39.70
15	33.26	4.76	33.30	17.77	26.02
18	24.01	2.33	24.83	15.82	18.95
20	19.75	1.50	20.88	17.77	15.78
25	12.16	0.81	14.39	26.95	10.70
5	53.70	26.51	53.61	27.34	46.55
7	50.02	18.74	49.85	24.90	41.85
10	35.06	6.24	35.03	18.46	27.69
15	17.48	1.51	18.99	24.32	14.40
20	18.11	0.76	13.14	61.04	10.34

Fig. 6.6. Classification errors per class and average for the example in the figure above. The first row corresponds to the classical MAP approach, in which no diffusion has been applied to the posterior probabilities. The next five rows correspond to four different time steps of the vector probability diffusion approach presented here, followed by five time steps for the scalar approach of Teo et al. [384].

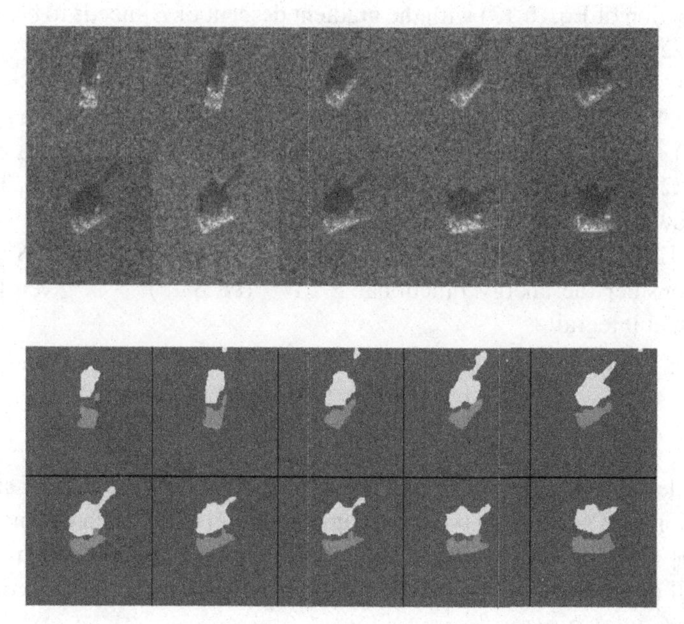

Fig. 6.7. Original images and classification results with MAP vector probability diffusion.

the following process: (1) The diffused posteriors of the previous image are used as priors, and after 30 iterations the new means and variances are estimated; (2) to obtain a better segmentation that does not depend on the previous priors, a new diffusion is applied by use of uniform priors. In this way the small details are captured; (3) steps (1) and (2) are iterated and in the last iteration the idea of (1) is used.

As we can see in Fig. 6.7 the results are very good, and the small details are captured. These results outperform those previously reported, in which the scalar approach was used.

Appendix

We now derive the Euler–Lagrange equation for $p = 2$, Section 6.2. Similar computations can be used to derive the corresponding flows for $1 < p < +\infty$. The case $p = 1$ requires a different treatment [16, 158], and Eq. (6.13) has to be seen as an evolution equation decreasing the energy E_1. The precise connection of Eq. (6.13) with the gradient descent of E_1 needs to be further analyzed.

Let Ω be a domain in $\mathbb{R}^2$ with smooth boundary and $n \geq 2$. As usual, $W^{1,2}(\Omega, \mathbb{R}^n)$ denotes the Sobolev space of functions $I(x_1, x_2) \in L^2(\Omega)$, whose distributional derivatives $\frac{\partial I}{\partial x_j} \in L^2(\Omega)$, $j = 1, 2$. Let $W^{1,2}(\Omega, S^{n-1}) = \{I(x_1, x_2) \in W^{1,2}(\Omega, \mathbb{R}^n) : I(x_1, x_2) \in S^{n-1} a.e.\}$. Then it is known that functions $W^{1,2}(\Omega, S^{n-1})$ can be approximated by smooth maps [29]. If Ω is a square, the same result is true for $W^{1,2}(\Omega, S^1)$ [29]. We consider the energy functional $E : W^{1,2}(\Omega, S^{n-1}) \to \mathbb{R}$ given by the Dirichlet integral

$$E(I) = \int_\Omega \|\nabla I\|^2 \, \mathrm{d}x, \quad I \in W^{1,2}(\Omega, S^{n-1}). \tag{6.20}$$

Now, let $I \in W^{1,2}(\Omega, S^{n-1})$ and let $\varphi \in C_0^\infty(\Omega, \mathbb{R}^n)$, the space of smooth functions in Ω vanishing outside a compact subset of Ω. Then we know that $\|I(x)\| = 1$ almost everywhere in Ω and, thus, for $|t|$ small enough $I + t\varphi$ is well defined and hence $I(t) := \frac{I + t\varphi}{\|I + t\varphi\|} \in W^{1,2}(\Omega, S^{n-1})$. Let us compute $\frac{d}{dt} E[I(t)]|_{t=0}$. For that we observe that

$$\partial_{x_j} \|I + t\varphi\| = \frac{1}{\|I + t\varphi\|} \langle I + t\varphi, \partial_{x_j} I + t\partial_{x_j} \varphi \rangle. \tag{6.21}$$

Thus

$$\partial_{x_j} I(t) = \frac{\partial_{x_j} I + t\partial_{x_j}\varphi}{\|I + t\varphi\|} - \frac{1}{\|I + t\varphi\|^3}\langle I + t\varphi, \partial_{x_j} I + t\partial_{x_j}\varphi\rangle(I + t\varphi).$$

On the other hand, observe that from the constraint $\langle I, I\rangle = 1$ a.e, it follows that $\langle I, \partial_{x_j} I\rangle = 0$. Because

$$\frac{1}{\|I + t\varphi\|} = 1 - t\langle I, \varphi\rangle + O(t^2),$$

where $O(t^2)$ means a quantity that is bounded by Ct^2, for some constant $C > 0$, then we may write

$$\partial_{x_j} I(t) = \partial_{x_j} I + tQ_j + O(t^2),$$

where

$$Q_j = -\langle \varphi, \partial_{x_j} I\rangle I - \langle I, \partial_{x_j}\varphi\rangle I + \partial_{x_i}\varphi - \langle I, \varphi\rangle \partial_{x_j} I.$$

Thus

$$\begin{aligned} \frac{\mathrm{d}}{\mathrm{d}t} E[I(t)]|_{t=0} &= 2\int_\Omega \sum_{i=1}^{2} \langle \partial_{x_j} I, Q_j\rangle \\ &= 2\int_\Omega \sum_{j=1}^{2} \langle \partial_{x_j} I, \partial_{x_j}\varphi\rangle - \langle I, \varphi\rangle \sum_{j=1}^{2} \langle \partial_{x_j} I, \partial_{x_j} I\rangle. \end{aligned}$$

Because $\|\nabla I\|^2 = \sum_{j=1}^{2} \langle \partial_{x_j} I, \partial_{x_j} I\rangle$, we may write the Euler–Lagrange equation corresponding to Eq. (6.20) as

$$\Delta I + I\ \|\nabla I\|^2 = 0 \quad \text{in } \Omega, \tag{6.22}$$

where ΔI denotes the Laplacian on each coordinate of I.

Exercises

1. Implement the isotropic and the anisotropic diffusion flows introduced in this chapter for direction denoising.
2. Experimentally show that the harmonic energy decreases with the direction diffusion flow.
3. Study the effect of the numerical normalization step in the harmonic energy. Compute the harmonic energy before and after the numerical normalization.

4. Compare the results of the isotropic direction diffusion flow with the standard isotropic diffusion flow, independently applied to each vector component, followed by a normalization step to keep the unit norm.

5. Compare the component-by-component anisotropic diffusion flow for probability distributions with the vector probability diffusion flow.

CHAPTER SEVEN

Contrast Enhancement

Images are captured at low contrast in a number of different scenarios. The main reason for this problem is poor lighting conditions (e.g., pictures taken at night or against the Sun's rays). As a result, the image is too dark or too bright and is inappropriate for visual inspection or simple observation. The most common way to improve the contrast of an image is to modify its pixel value distribution, or histogram. A schematic example of the contrast enhancement problem and its solution by means of histogram modification is given in Fig. 7.1. On the left, we see a low-contrast image with two different squares, one inside the other, and its corresponding histogram. We can observe that the image has low contrast and the different objects cannot be identified, as the two regions have almost identical gray values. On the right we see what happens when we modify the histogram in such a way that the gray values corresponding to the two regions are separated. The contrast is improved immediately. An additional example, this time for a real image, is given in Fig. 7.2.

In this chapter, we first follow Ref. [339, 340] and show show how to obtain any gray-level distribution as the steady state of an ODE and present examples for different pixel value distributions. Uniform distributions are usually used in most contrast enhancement applications [160]. On the other hand, for specific tasks, the exact desirable distribution can be dictated by the application, and the technique presented here applies as well. After this basic equation is presented and analyzed, we combine it with the smoothing operators proposed in Chap. 4, obtaining contrast normalization and denoising at the same time. We also extend the flow to local contrast enhancement in both the image plane and in the gray-value space. Local contrast enhancement in the gray-value space is performed for example to improve the visual appearance of the image (the reason for this will be explained below). The straightforward extension of this formulation to local contrast enhancement

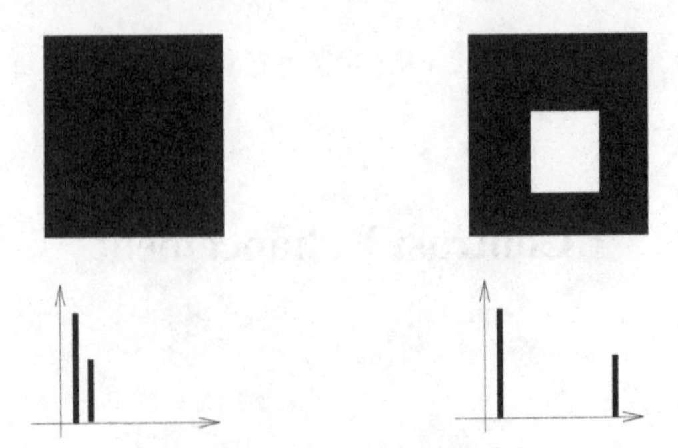

Fig. 7.1. Schematic explanation of the use of histogram modification to improve image contrast.

in the image domain, that is, in the neighborhood of each pixel, does not achieve good results. Indeed, in this case, fronts parallel to the edges are created (this is common to most contrast enhancement techniques). At the experimental level, we can avoid this by combining local and global techniques in the framework presented here or by using the approach described in the second half of this chapter, which follows Ref. [69].

After the basic image flows for histogram modification are given, a variational interpretation of the histogram modification flow and theoretical results regarding existence of solutions to the proposed equations are presented.

Basically, the approach described in the first part of this chapter has the following characteristics:

1. It presents contrast enhancement as an image deformation algorithm and not just as a distribution modification one. This progressively improves the contrast, also allowing us to choose intermediate steps.

Fig. 7.2. Example of contrast enhancement. Note how objects that are not visible on the original image on the left (e.g., the second chair and the objects through the window) are now detectable in the processed one (right).

2. It enriches the area of PDEs in image processing, showing how to solve one of the most important problems in image processing by means of image flows.
3. It allows us to perform contrast enhancement and denoising simultaneously.
4. It presents a variational framework for contrast enhancement, formulated in the image domain. This allows us to improve classical techniques, including for example image and perceptual models, as well as to better understand current approaches.
5. It holds formal existence (and uniqueness) results.

This work on differential equations for histogram modification provides then the basis for the second part of this chapter. In this second part, we design local histogram modification operations that preserve the family of level sets of the image, that is, following the morphology school, they preserve shape. Local contrast enhancement is mainly used to further improve the image contrast and facilitate the visual inspection of the data. As we will later see in Sections 7.1.3 and 7.2.4, global histogram modification does not always produce good contrast, and especially small regions are hardly visible after such a global operation. On the other hand, local histogram modification improves the contrast of small regions as well, but because the level sets are not preserved, artificial objects are created. The theory developed will enjoy the best of both words: The shape-preservation property of global techniques and the contrast improvement quality of local ones.

Before proceeding, it should be pointed out that in Ref. [313] Perona and Tartagni presented a diffusion network for image normalization. In their work, the image $I(x, y)$ is normalized by means of $\frac{I-I_a}{I_M-I_m}$, where I_a, I_M, and I_m are the average, maximum, and minimum of I over local areas, respectively. These values are computed with a diffusion flow, which minimizes a cost functional. The method was generalized by computing a full local frame of reference for the gray-level values of the image. This is achieved by changing the variables in the flow. A number of properties, including the existence of the solution of the diffusion flow, were presented as well. In contrast with their work, in the algorithms described in this chapter we have full control of the distribution of the gray levels, which means that although their work is on contrast normalization, the work now described is on histogram modification. Also, the modified image is obtained in this work as the steady-state solution of a flow. This allows straightforward combination with other PDE-based algorithms, as explained above. No energy-type interpretation or local shape-preserving enhancement is given in Ref. [313].

7.1. Global PDE-Based Approach

7.1.1. Histogram Modification

As explained in the Introduction, the most common way to improve the contrast of an image is to modify the pixel value distribution, i.e., the histogram. We do this by means of an evolution equation. We start with the equation for histogram equalization, and then we extend it for any given distribution. In histogram equalization, the goal is to achieve a uniform distribution of the image values [318]. That means, given the gray-level distribution p of the original image, the image values are mapped into new ones such that the new distribution $\hat{p}$ is uniform. In the case of digital images, $p(i), 0 \leq i \leq M$, is computed as

$$p(i) = \frac{\text{number of pixels with value } i}{\text{total number of pixels in the image}},$$

and uniformity can be obtained only approximately.

We proceed to show an image evolution equation that achieves this uniform distribution when it arrives at steady state. Assume that the continuous image $I(x, y, t) : [0, N]^2 \times [0, T) \rightarrow [0, M]$ evolves according to

$$\frac{\partial I(x, y, t)}{\partial t} = [N^2 - N^2/MI(x, y, t)] - \mathcal{A}[(v, w) : I(v, w, t) \geq I(x, y, t)], \tag{7.1}$$

where $\mathcal{A}[\cdot]$ represents area (or number of pixels in the discrete case). For the steady-state solution ($I_t = 0$) we have

$$\mathcal{A}[(v, w) : I(v, w) \geq I(x, y)] = [N^2 - N^2/M \;\; I(x, y)].$$

Then, for $a, b \in [0, M]$, $b > a$, we have

$$\mathcal{A}[(v, w) : b \geq I(v, w) \geq a] = (N^2/M) * (b - a),$$

which means that the histogram is constant. Therefore the steady-state solution of Eq. (7.1), if it exists (see below) gives the image after normalization by means of histogram equalization.

From Eq. (7.1) we can extend the algorithm to obtain any given gray-value distribution $h : [0, M] \rightarrow \mathbb{R}^+$. Let $H(s) := \int_0^s h(\xi) d\xi$, that is, $H(s)$ gives the density of points between 0 and s. Then, if the image evolves according to

$$\frac{\partial I(x, y, t)}{\partial t} = \{N^2 - H[I(x, y, t)]\} - \mathcal{A}[(v, w) : I(v, w, t) \geq I(x, y, t)], \tag{7.2}$$

the steady-state solution is given by

$$\mathcal{A}[(v, w) : I(v, w) \geq I(x, y)] = \{N^2 - H[I(x, y)]\}.$$

Therefore,

$$\mathcal{A}[(v, w) : I(x, y) \leq I(v, w) \leq I(x, y) + \delta] = H[I(x, y) + \delta] - H[I(x, y)],$$

and taking Taylor expansion when $\delta \to 0$, we obtain the desired result. Note that of course Eq. (7.1) is a particular case of Eq. (7.2), with h = constant.

Existence and Uniqueness of the Flow. Results are presented now that are related to the existence and uniqueness of the proposed flow for histogram equalization. We will see that the flow for histogram modification has an explicit solution, and its steady state is straightforward to compute. This is because, as we will prove below, the value of $\mathcal{A}$ is constant in the evolution. This is not unexpected, as it is well known that histogram modification can be performed with look-up tables. In spite of this, it is important, not only from the theoretical point of view, to first present the basic flow for histogram modification, in order to arrive to the energy-based interpretation and to derive the extensions later presented in Section 7.2. These extensions do not have explicit solutions.

Let I_0 be an image, i.e., a bounded measurable function, defined in $[0, N]^2$ with values in the range $[a, b]$, $0 \leq a < b \leq M$. We assume that the distribution function of I_0 is continuous, that is

$$\mathcal{A}[X : I_0(X) = \lambda] = 0 \tag{7.3}$$

for all $X \in [0, N]^2$ and all $\lambda \in [a, b]$. To equalize the histogram of I_0 we look for solutions of

$$I_t(t, X) = \mathcal{A}[Z : I(t, Z) < I(t, X)] - \frac{N^2}{b - a}[I(t, X) - a], \tag{7.4}$$

which also satisfy

$$\mathcal{A}[X : I(t, X) = \lambda] = 0. \tag{7.5}$$

Hence the distribution function of $I(t, X)$ is also continuous. This requirement, mainly technical, avoids the possible ambiguity of changing the sign "$<$" by "$\leq$" in the computation of $\mathcal{A}$ (see also the remarks at the end of this section).

Let us recall the definition of $\text{sign}^-(\cdot)$:

$$\text{sign}^-(r) = \begin{cases} 1, & \text{if } r < 0 \\ [0, 1], & \text{if } r = 0. \\ 0, & \text{if } r > 0 \end{cases}$$

With this notation, I that satisfies Eqs. (7.4) and (7.5) may be written as

$$I_t(t, X) = \int_{[0,N]^2} \text{sign}^-[I(t, Z) - I(t, X)]\text{d}Z - \frac{N^2}{b-a}[I(t, X) - a]. \tag{7.6}$$

Observe that, as a consequence of Eq. (7.5), the real value of sign^- at zero is unimportant, avoiding possible ambiguities. To simplify the notation, let us normalize I such that it is defined on $[0, 1]^2$ and takes values in the range $[0, 1]$. This is done just by the change of variables given by $I(t, X) \leftarrow \frac{I(\mu t, NX) - a}{b-a}$, where $\mu = \frac{b-a}{N^2}$. Then I satisfies the equation

$$I_t(t, X) = \int_{[0,1]^2} \text{sign}^-[I(t, Z) - I(t, X)]\text{d}Z - I(t, X). \tag{7.7}$$

Therefore, without loss of generality we can assume that $N = 1$, $a = 0$, and $b = 1$, and analyze Eq. (7.7). Let us make precise our notion of solution for Eq. (7.7).

Definition 7.1. *A bounded measurable function* $I : [0, \infty) \times [0, 1]^2 \to [0, 1]$ *will be called a solution of Eq. (7.7) if, for almost all* $X \in [0, 1]^2$, $I(., X)$ *is continuous in* $[0, \infty)$, $I_t(., X)$ *exists a.e. with respect to t and Eq. (7.7) holds a.e. in* $[0, \infty) \times [0, 1]^2$.

Now we may state the following result:

Theorem 7.1. *For any bounded measurable function* $I_0 : [0, 1]^2 \to [0, 1]$ *such that* $\mathcal{A}[Z : I_0(Z) = \lambda] = 0$ *for all* $\lambda \in [0, 1]$, *there exists a unique solution* $I(t, X)$ *in* $[0, \infty) \times [0, 1]^2$ *with range in* $[0, 1]$ *satisfying the flow (7.7) with initial condition given by* I_0, *and such that* $\mathcal{A}[Z : I(t, Z) = \lambda] = 0$ *for all* $\lambda \in [0, 1]$. *Moreover, as* $t \to \infty$, $I(t, X)$ *converges to the histogram equalization of* $I_0(X)$.

Proof: We start with the existence. We look for a solution $I(t, X)$ such that

$$\mathcal{A}[Z : I(t, Z) < I(t, X)] = \mathcal{A}[Z : I(0, Z) < I(0, X)]. \tag{7.8}$$

This assumption is enough to prove existence. This will also mean that a closed solution exists for Eq. (7.7), further supporting the validity of the proposed image flow. Note that having $\mathcal{A}$ constant in time transforms the equation into a family of ODEs, one for each space coordinate (x, y) (this is generally true for histogram modification techniques, as they are based on pixel value distributions). In spite of this, which makes the original equation as trivial as classical histogram equalization techniques, the formulation is crucial for later sections, in which local and simultaneous contrast enhancement and denoising models are presented. For those models, which do not have a closed solution, it is important to identify the basic histogram modification algorithm. Moreover, we will later prove that condition (7.8) actually holds for any solution of Eq. (7.7).

Thus, let $\mathcal{F}_0(X) := \mathcal{A}[Z : I(t, Z) < I(t, X)]$, which is independent of t. Then Eq. (7.7) can be rewritten as

$$I_t = \mathcal{F}_0(X) - I(t, X), \tag{7.9}$$

whose explicit solution is

$$I(t, X) = \exp\{-t\}I_0 + (1 - \exp\{-t\})\mathcal{F}_0(X), \tag{7.10}$$

which satisfies

$$\mathcal{A}[Z : I(t, Z) = \lambda] = 0 \ , \ t \geq 0 \ , \ \lambda \in [0, 1]. \tag{7.11}$$

Observe that our solution is a continuous function of t for all X and satisfies

$$I(t, X) < I(t, X') \quad \text{if and only if} \quad I_0(X) < I_0(X') \tag{7.12}$$

for all $t > 0$ and all $X, X' \in [0, 1]^2$, that is, the gray-value order is preserved. Observe also that $[0, 1]$ is the smallest interval that contains the range of the solution. As $t \to +\infty$, $I(t, X) \to \mathcal{F}_0(X)$. It is easy to see that the distribution function of $\mathcal{F}_0(X)$ is uniform.

Note that Eq. (7.10) progressively improves the image contrast until steady state is achieved. This gives an advantage over classical histogram equalization, in which intermediate contrast states are not available and, if the solution has excessive contrast, there is no possible compromise.

To prove uniqueness, we need the following Lemma:

Lemma 7.1. *Let $I(t, X)$ be a solution of Eq. (7.7). Let $X, X' \in [0, 1]^2$ be such that $I(., X)$, $I(., X')$ are continuous in $[0, \infty)$, $I_t(., X)$, $I_t(., X')$ exist a.e. with respect to t, and Eq. (7.7) holds a.e. in t. Suppose that $I(0, X) < I(0, X')$. Then*

$$I(t, X') - I(t, X) \geq \exp\{-t\}[I(0, X') - I(0, X)] \quad \text{for all} \quad t \geq 0. \tag{7.13}$$

Proof: Let $\delta > 0$. Then, using our assumptions, we find that

$$\frac{\mathrm{d}}{\mathrm{d}t}\left(-\frac{1}{2}\ln\{\delta^2 + [I(t, X') - I(t, X)]^2\}\right)$$
$$= \frac{I(t, X) - I(t, X')}{\delta^2 + [I(t, X') - I(t, X)]^2}\left\{\int \mathrm{sign}^-[I(t, Z) - I(t, X')]\mathrm{d}Z\right.$$
$$\left. - \int \mathrm{sign}^-[I(t, Z) - I(t, X)]\mathrm{d}Z\right\} + \frac{[I(t, X') - I(t, X)]^2}{\delta^2 + [I(t, X') - I(t, X)]^2}.$$

We easily verify that the first term above is negative. Therefore

$$\frac{\mathrm{d}}{\mathrm{d}t}\left(-\frac{1}{2}\ln\{\delta^2 + [I(t, X') - I(t, X)]^2\}\right) \leq \frac{[I(t, X') - I(t, X)]^2}{\delta^2 + [I(t, X') - I(t, X)]^2} \leq 1.$$

After integration we observe that

$$\sqrt{\delta^2 + [I(t, X') - I(t, X)]^2} \geq \exp\{-t\}\sqrt{\delta^2 + [I(0, X') - I(0, X)]^2}.$$

Letting $\delta \to 0$,

$$|I(t, X') - I(t, X)| \geq \exp\{-t\}|I(0, X') - I(0, X)| > 0.$$

Because $I(t, X)$, $I(t, X)$ are continuous, this implies relation (7.13). □

Let $I(t, X)$ be a solution of Eq. (7.7). As a consequence of Lemma 7.1 we have

$$[Z : I(0, Z) < I(0, X)] \subseteq [Z : I(t, Z) < I(t, X)] \subseteq [Z : I(0, Z) \leq I(0, X)],$$
$$[Z : I(t, Z) = I(t, X)] \subseteq [Z : I(0, Z) = I(0, X)]$$

for almost all $X \in [0, 1]^2$. Because $\mathcal{A}[Z : I(0, Z) = \lambda] = 0$, we also have $\mathcal{A}[Z : I(t, Z) = \lambda] = 0$, for all $\lambda \in [0, 1]$, and it follows that

$$\int \mathrm{sign}^-[I(t, Z) - I(t, X)]\mathrm{d}Z = \int \mathrm{sign}^-[I(0, Z) - I(0, X)]\mathrm{d}Z = \mathcal{F}_0(X).$$

The flow can be rewritten as Eq. (7.9), and Eq. (7.10) gives the solution. Uniqueness follows. Letting $t \to \infty$, $I(t, X)$ tends to $\mathcal{F}_0(X)$, which corresponds to the equalized histogram for $I_0(X)$.

Remarks:

1. Even if the initial image I_0 is not continuous it may happen that $\mathcal{F}_0(X)$ is continuous. Hence an edge may disappear. This is a limitation of both this and the classical methods of histogram equalization. In spite of possible ad hoc solutions, the problem requires a sound

formulation and solution. We will see in Section 7.2 below how to use the framework developed so far to investigate shape-preserving contrast enhancement techniques.

2. According to our assumption (7.5), the initial image cannot take a constant value in a region of positive area. [Note that this assumption is only in order to present the theoretical results. The image flow and its basic contrast modification characteristics remain valid also when assumption (7.5) is violated.] If we expect as final result an image whose gray-level distribution is perfectly uniform, we must assume Eq. (7.5) unless we admit that the same gray level can be mapped to different gray levels. In practice, perfect uniform is not necessary, and assumption (7.5) does not present a major difficulty.
3. The above proof can be adapted to any required gray-value distribution h as the above result can be extended for the flow

$$I_t(t, X) = \int \text{sign}^-[I(t, Z) - I(t, X)]\text{d}Z - \Psi[I(t, X)], \tag{7.14}$$

where Ψ is any strictly increasing Lipschitz continuous function. This allows us to obtain any desired distribution. As pointed out before, the specific distribution depends on the application. Uniform distributions are the most common choice. If it is known in advance for example that the most important image information is within certain gray-value region, h can be such that it allows this region to expand further, increasing the detail there. Another possible way of finding h is to equalize between local minima of the original histogram, preserving certain types of structure in the image.

Variational Interpretation of the Histogram Flow. The formulation given by Eqs. (7.6) and (7.7) not only helps to prove Theorem 7.1, but it also gives a variational interpretation of the histogram modification flow. Variational approaches are frequently used in image processing. They give explicit solutions for a number of problems and very often help to give an intuitive interpretation of this solution, interpretation that is often not so easy to achieve from the corresponding Euler–Lagrange or PDE. Variational formulations help us to derive new approaches to solve the problem as well.

Let us consider the following functional

$$\mathcal{U}(I) = \frac{1}{2}\int \left[I(X) - \frac{1}{2}\right]^2 \text{d}X - \frac{1}{4}\iint |I(X) - I(Z)|\text{d}X\text{d}Z, \tag{7.15}$$

where $I \in L^2[0, 1]^2$, $0 \leq I(X) \leq 1$. $\mathcal{U}$ is a Lyapunov functional for Eq. (7.7):

Lemma 7.2. *Let I be the solution of Eq. (7.7) with initial data I_0 as in Theorem 1. Then*

$$\frac{\mathrm{d}\mathcal{U}(I)}{\mathrm{d}t} \leq 0.$$

Proof:

$$\frac{\mathrm{d}\mathcal{U}(I)}{\mathrm{d}t} = \int \left[I(X) - \frac{1}{2} \right] I_t(X) \mathrm{d}X$$
$$- \frac{1}{4} \iint \mathrm{sign}[I(Z) - I(X)][I_t(Z) - I_t(X)] \mathrm{d}X \mathrm{d}Z.$$

Let us denote the first integrand in the equation above by A and the second by B. Observe that, because of Eq. (7.11), part B of the integral above is well defined. Let us rewrite B as

$$B = \frac{1}{4} \iint \mathrm{sign}[I(Z) - I(X)] I_t(Z) \mathrm{d}X \mathrm{d}Z$$
$$- \frac{1}{4} \iint \mathrm{sign}[I(Z) - I(X)] I_t(X) \mathrm{d}X \mathrm{d}Z.$$

Interchanging the variables X and Z in the first part of the expression above, we obtain

$$B = -\frac{1}{2} \iint \mathrm{sign}[I(Z) - I(X)] I_t(X) \mathrm{d}X \mathrm{d}Z.$$

Fixing X, we have

$$\int \mathrm{sign}[I(t,Z) - I(t,X)] \mathrm{d}Z = 1 - 2 \int \mathrm{sign}^{-}[I(t,Z) - I(t,X)] \mathrm{d}Z,$$

and we may write

$$B = -\frac{1}{2} \int I_t(X) \mathrm{d}X + \iint \mathrm{sign}^{-}[I(Z) - I(X)] I_t(X) \mathrm{d}Z \mathrm{d}X.$$

Hence

$$\frac{\mathrm{d}\mathcal{U}(I)}{\mathrm{d}t} = \int I(t,X) I_t(t,X) \mathrm{d}X - \iint \mathrm{sign}^{-}[I(t,Z) - I(t,X)] I_t(t,X) \mathrm{d}X \mathrm{d}Z$$
$$= \int \left\{ I(t,X) - \int \mathrm{sign}^{-}[I(t,Z) - I(t,X)] \mathrm{d}Z \right\} I_t(t,X) \mathrm{d}X$$
$$= - \int I_t(t,X)^2 \mathrm{d}X \leq 0.$$

This concludes the proof. □

Therefore, when solving Eq. (7.7) we are indeed minimizing the functional $\mathcal{U}$ given by Eq. (7.15), restricted to the condition that the minimizer satisfy Eq. (7.11).

This variational formulation gives a new interpretation to histogram modification and contrast enhancement in general. It is important to note that in contrast with classical techniques of histogram modification, it is completely formulated in the image domain and not in the probability one (although the spatial relation of the image values is still not important, until the formulations below). The first term in $\mathcal{U}$ stands for the variance of the signal, and the second one gives the contrast between values at different positions. To the best of my knowledge, this is the first time a formal image-based interpretation to histogram equalization is given, showing the effect of the operation to the image contrast.

From this formulation, other functionals can be proposed to achieve contrast modification while including image and perception models. One possibility is to change the metric that measures contrast (second term in the equation above) by metrics that better model for example visual perception. It is well known that the total absolute difference is not a good perceptual measurement of contrast in natural images. At least, this absolute difference should be normalized by the local average. This also explains why ad hoc techniques that segment the gray-value domain and perform (independent) local histogram modification in each one of the segments perform better than global modifications. This is because the normalization term is less important when only pixels of the same range value are considered. Note that doing this is straightforward in our framework; $\mathcal{A}$ should consider pixels only in between a certain range in relation to the value of the current pixel being updated.

From the variational formulation it is also straightforward to extend the model to local (in image space) enhancement by changing the limits in the integral from global to local neighborhoods. In the differential form, the corresponding equation is

$$\begin{aligned}\frac{\partial I}{\partial t} &= \{N^2 - H[I(x, y, t)]\}\\ &\quad - \mathcal{A}[(v, w) \in \mathrm{B}(v, w, \delta) : I(v, w, t) \geq I(x, y, t)],\end{aligned}$$

where $\mathrm{B}(v, w, \delta)$ is a ball of center (v, w) and radius δ ($\mathrm{B}(v, w)$ can also be any other surrounding neighborhood, obtained from example from previously performed segmentation). The main goal of this type of local contrast enhancement is to enhance the image for object detection. This formulation does not work perfectly in practice when the goal of the contrast enhancement algorithm is to obtain a visually pleasent image. Fronts parallel to the edges are created, as we can see in the examples below. (Further experiments with this model are presented in Subsection 7.1.3.) We can moderate the effects of the local model by combining it with the global model.

From the same approach, it is straightforward to derive models for contrast enhancement of movies, integrating over corresponding zones in different frames, when correspondence could be given, for example, by optical flow. Another advantage is the possibility of combining it with other operations. As an example, in Subsection 7.1.2, an additional smoothing term will be added to the model of Eq. (7.7).

7.1.2. Simultaneous Anisotropic Diffusion and Histogram Modification

A flow is now presented for simultaneous denoising and histogram modification. This is just an example of the possibility of combining different algorithms in the same PDE.

In Chap. 4, geometric flows were presented for edge-preserving anisotropic diffusion. The smoothing and contrast enhancement flows can be combined to obtain a new flow that performs anisotropic diffusion (denoising) while simultaneously modifying the histogram.

As we have seen in the discussion on anisotropic diffusion, in Ref. [330] (see also Ref. [243] for theoretical results), Rudin et al. proposed to minimize the total variation of the image, given by

$$\int \|\nabla I(X)\| \, \mathrm{d}X.$$

Once again, the Euler–Lagrange of this functional is given by the curvature κ of the level sets, that is

$$\operatorname{div}\left(\frac{\nabla I}{\|\nabla I\|}\right) = \kappa,$$

which leads to the gradient descent flow

$$I_t = \kappa.$$

Using this smoothing operator, together with the histogram modification part, gives results very similar to those obtained with the affine-based flow. If this smoothing operator is combined with the histogram flow, the total flow

$$\frac{\partial I}{\partial t} = \alpha\kappa + \{N^2 - H[I(x, y, t)]\} - A[(v, w) : I(v, w, t) \geq I(x, y, t)] \tag{7.16}$$

will therefore be such that it minimizes

$$\alpha \int \|\nabla I(X)\| \, \mathrm{d}X + \mathcal{U}, \tag{7.17}$$

where $\mathcal{U}$ is given by Eq. (7.15), yielding a complete variational formulation of the combined histogram equalization/smoothing approach. This is precisely the formulation we analyze below.

Existence of the Flow. A theoretical result is now presented that is related to the simultaneous smoothing and contrast modification flow of Eq. (7.16).

Before proceeding with the existence proof of variational problem (7.17), let us recall the following standard notation:

1. $\mathcal{C}([0, T], \mathcal{H}) := \{I : [0, T] \to \mathcal{H} \text{ continuous}\}$, where $T > 0$ and $\mathcal{H}$ is a Banach space (and in particular for a Hilbert space).
2. $L^p([0, T], \mathcal{H}) := \{I : [0, T] \to \mathcal{H} \text{ such that } \int_0^T \|I(t)\|^p < \infty\}$, with $1 \leq p < \infty$.
3. $L^\infty([0, T], \mathcal{H}) := \{I : [0, T] \to \mathcal{H} \text{ such that } \operatorname{ess\,sup}_{t \in [0,T]} \|I(t)\| < \infty\}$.
4. $I \in L^p_{\text{loc}}([0, \infty), \mathcal{H})$ means that $I \in L^p([0, T], \mathcal{H})$ for all $T > 0$.
5. $W^{1,2}([0, T], \mathcal{H}) := \{I : [0, T] \to \mathcal{H} \text{ such that } I, I_t \in L^2([0, T], \mathcal{H})\}$.

To simplify notation, later we will assume that $\Omega = (0, 1)^2$ and $\mathcal{H} = L^2(\Omega)$.

We proceed now to prove the existence of the solution to the Euler–Lagrange equation corresponding to variational problem (7.17), given by ($\alpha = 1$)

$$I_t = \operatorname{div}\left(\frac{\nabla I}{\|\nabla I\|}\right) + \int_{[0,1]^2} \operatorname{sign}^-[I(t, Z) - I(t, X)]\mathrm{d}Z - I(t, X), \tag{7.18}$$

together with the initial and the boundary conditions

$$I(0, X) = I_0(X), \quad X \in [0, 1]^2, \qquad \frac{\partial I}{\partial \vec{n}}(t, X) = 0, \quad t > 0, \quad X \in \partial[0, 1]^2,$$

where $\vec{n}$ stands for the normal direction. We shall use results from the theory on nonlinear semigroups on Hilbert space [47]. Before proceeding, we need a number of additional definitions. A function $I \in L^1(\Omega)$ whose derivatives in the sense of distributions are measures with finite total variation in Ω is called a function of bounded variation. The class of such functions will be denoted by $\mathrm{BV}(\Omega)$. Thus, $I \in \mathrm{BV}(\Omega)$ if there are Radon measures $\mu_1, \ldots, \mu_n$

defined in $\Omega \subset \mathbb{R}^n$ such that its total mass $|D\mu_i|(\Omega)$ is finite and

$$\int_\Omega I(X) D_i\phi(X) \mathrm{d}X = -\int_\Omega \phi(X) D\mu_i(X)$$

for all $\phi \in C_0^\infty(\Omega)$. The gradient of I will therefore be a vector-valued measure with finite total variation

$$\begin{aligned}\|\nabla I\| = \sup \Big\{ \int_\Omega I \operatorname{div} v \mathrm{d}X \ : \ v \\ = (v_1, \ldots, v_n) \in C_0^\infty(\Omega, \mathbb{R}^n), \ |v(X)| \leq 1, \ X \in \Omega \Big\}.\end{aligned}$$

The space $\mathrm{BV}(\Omega)$ will have the norm

$$\|I\|_{\mathrm{BV}} = \|I\|_1 + \|\nabla I\|.$$

The space $\mathrm{BV}(\Omega)$ is continuously embedded in $L^p(\Omega)$ for all $p \leq \frac{n}{n-1}$. The immersion is compact if $p < \frac{n}{n-1}$ ([424], Theorem 2.5.1). If I_i is a sequence of functions in $\mathrm{BV}(\Omega)$ converging to the function I in $L^1(\Omega)$, then $\|\nabla I\| \leq \lim_i \inf \|\nabla I_i\|$ ([424], Theorem 5.2.1). Moreover, given a function $I \in \mathrm{BV}(\Omega)$, there exists a sequence of functions $I \in \mathrm{BV}(\Omega)$ such that $I_i \to I$ in $L^1(\Omega)$ and such that $\|\nabla I\| = \lim_i \|\nabla I_i\|$ ([424], Theorem 5.2.3).

Let $\mathcal{H}$ be a Hilbert space and let $\phi : \mathcal{H} \to (-\infty, +\infty]$ be convex and proper. Given $X \in \mathcal{H}$, the subdifferential of ϕ at X, $\partial\phi(X)$, is given by

$$\partial\phi(X) = \{Y \in \mathcal{H} : \ \forall \xi \in \mathcal{H} \ \ \phi(\xi) - \phi(X) \geq \langle Y, \xi - X \rangle\}.$$

We write $\mathrm{dom}(\phi) := \{X \in \mathcal{H} : \ \phi(X) < +\infty\}$, $\mathrm{dom}(\partial\phi) := \{X \in \mathcal{H} : \partial\phi(X) \neq \emptyset\}$. From now on we shall write, as mentioned above, $\Omega = (0, 1)^2$ and $\mathcal{H} = L^2(\Omega)$. We also define the functionals $\phi, \psi : \mathcal{H} \to (-\infty, +\infty]$ by

$$\phi(I) := \begin{cases} \|\nabla \Phi\| + \frac{1}{2} \int_\Omega \left[I(X) - \frac{1}{2}\right]^2 \mathrm{d}X, & I \in \mathrm{BV}(\Omega) \\ +\infty, & \text{otherwise} \end{cases},$$

$$\psi(I) := \frac{1}{4} \int_\Omega \int_\Omega |I(Z) - I(X)| \mathrm{d}X \mathrm{d}Z.$$

Note that both functionals are convex, lower semicontinuous, and proper on $\mathcal{H}$. We introduced both functionals because formally Eq. (7.18) is associated with the following abstract problem:

$$I_t + \partial\phi(I) \ni \partial\psi(I). \tag{7.19}$$

To make such a formulation precise, let us recall the following ([47], Definition 3.1): Let $T > 0$, $f \in L^1([0, T], \mathcal{H})$. We call $I \in \mathcal{C}([0, T], \mathcal{H})$ a strong solution of

$$I_t + \partial\phi(I) \ni f, \tag{7.20}$$

if I is differentiable almost everywhere on $(0, T)$, $I \in \mathrm{dom}(\partial\phi)$ a.e. in t, and

$$-I_t(t) + f(t) \in \partial\phi[I(t)] \tag{7.21}$$

a.e. on $(0, T)$. In particular, if $I \in W^{1,2}([0, T], \mathcal{H})$, $I(t) \in \mathrm{dom}(\partial\phi)$ a.e. and expression (7.21) holds a.e. on $(0, T)$, then I is a strong solution of problem (7.20). We say that $I \in \mathcal{C}([0, T], \mathcal{H})$ is a strong solution of problem (7.19) if there exists $\omega \in L^1([0, T], \mathcal{H})$, $\omega(t) \in \partial\psi[I(t)]$ a.e. in $(0, T)$, such that I is a strong solution of

$$I_t + \partial\phi(I) \ni \omega. \tag{7.22}$$

With these preliminaries, we reformulate Eq. (7.18) as an abstract evolution problem of the form of problem (7.19) and use the machinery of nonlinear semigroups on Hilbert spaces to prove existence of solutions of problem (7.19).

Theorem 7.2. *For any $I_0 \in \mathrm{BV}(\Omega)$, $0 \le I_0 \le 1$, there exists a strong solution $I \in W^{1,2}([0, T], \mathcal{H})$, $\forall T > 0$, of problem (7.19) with initial condition $I(0) = I_0$, and such that $0 \le I \le 1$, $\forall t > 0$. Moreover, the functional $\mathcal{V}(I) = \phi(I) - \psi(I)$ is a Lyapunov functional for problem (7.19).*

Note that Theorem 7.2 proves the existence of a solution. There is no result so far related to uniqueness.

Before concluding this section, let us make some remarks on the asymptotic behavior of I as $t \to \infty$. Integrating Eq. (7.10) we have

$$\int_0^T \int_\Omega |I_{\epsilon t}|^2 \mathrm{d}X \mathrm{d}t = \phi_\epsilon[I_\epsilon(t)] - \phi_\epsilon[I_\epsilon(0)] + \psi_\epsilon[I_\epsilon(t)]$$
$$- \psi_\epsilon[I_\epsilon(0)] \le \phi_\epsilon[I_\epsilon(0)] + \frac{1}{4}.$$

Letting $\epsilon \to 0$, we get

$$\int_0^T \int_\Omega |I_t|^2 \mathrm{d}X \mathrm{d}t + \phi[I(t)] \le \psi[I(0)] + \frac{1}{4}.$$

Now, letting $T \to \infty$

$$\int_0^\infty \int_\Omega |I_t|^2 \mathrm{d}X \mathrm{d}t \le \psi[I(0)] + \frac{1}{4}.$$

Therefore, for subsequent $t = t_n$ we have $I_t(t_n) \to 0$ in $\mathcal{H}$ as $n \to \infty$. Because, on the other hand, $\phi(I)$ is bounded, we may assume that $I(t_n) \to \bar{I}$ in L^1. Because $0 \le I \le 1$, also $I(t_n) \to \bar{I}$ in L^2. Now, because $-I_t + \omega \in \partial\phi(I)$ a.e., we may assume that $-I_t(t_n) + \omega(t_n) \in \partial\phi[I(t_n)]$ for all n. Hence

$$\phi(\hat{I}) - \phi[I(t_n)] \ge \langle -I_t(t_n) + \omega(t_n), I(t_n) - \hat{I} \rangle, \quad \forall \hat{I} \in \mathcal{H}. \quad (7.23)$$

Moreover, we may assume that $\omega(t_n) \to \hat{\omega} \in \partial\psi(\hat{I})$ weakly in $\mathcal{H}$. Letting $n \to \infty$ in expression (7.23), we get

$$\phi(\hat{I}) - \phi(\bar{I}) \ge \langle \bar{\omega}, I - \hat{I} \rangle \quad \forall \hat{I} \in \mathcal{H}.$$

In other words, $\hat{\omega} \in \partial\phi(\hat{I})$, where $\hat{\omega} \in \partial\psi(\hat{I})$. We may say that essentially all limit points of $I(t)$ as $t \to \infty$ are critical points of $\phi(I) - \psi(I)$.

7.1.3. Experimental Results

Before experimental results are presented, let us make some remarks on the complexity of the algorithm. Each iteration of the flow requires $O(N^2)$ operations. In our examples we observed that no more than five iterations are usually required for converging. Therefore the complexity of the proposed algorithm is $O(N^2)$, which is the minimal expected for any image processing procedure that operates on the whole image.

The first example is given in Fig. 7.3. The original image is presented on the top left. On the right the image is shown after histogram equalization performed with the popular software *xv* (copyright 1993 by John Bradley), and on the bottom left the one obtained from the steady-state solution of Eq. (7.1). On the bottom right an example is given of Eq. (7.2) for $h(I)$ that is a piecewise linear function of the form $-\alpha(|I - M/2| - M/2)$, where α is a normalization constant, I is the image value, and M is the maximal image value.

Figure 7.4 shows the progress of the histogram equalization flow. The original image is shown on the left, an intermediate step in the middle, and the steady-state solution on the right.

An example of the simultaneous denoising and histogram equalization is given in Fig. 7.5 for a fingerprint image (from the National Institute of Standards and Technology, Gaithersburg, MD).

Fig. 7.3. Original image (top left) and results of the histogram equalization process with the software package *xv* (top right), the proposed image flow for histogram equalization (bottom left), and the histogram modification flow for a piecewise linear distribution (bottom right).

Figure 7.6 presents an example of combining local and global histogram modification. The original image is given on the top left. The result of global histogram equalization is on the top right, and the one for local contrast enhancement (16×16 neighborhood) is on the bottom left. We see fronts appearing parallel to the edges. Finally, on the bottom right the combination of local and global contrast modification is shown; we apply one (or several) global steps after k successive local steps. Note that the algorithm described

Fig. 7.4. Progress of the histogram equalization flow. The original image is shown on the left, an intermediate step on the middle, and the steady-state solution on the right.

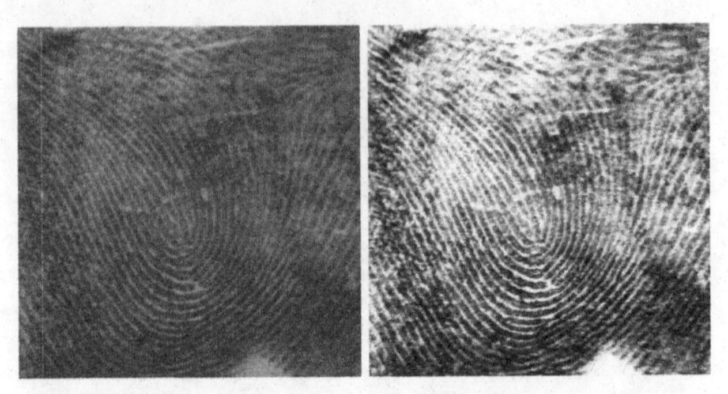

Fig. 7.5. Result (right) of simultaneous histogram equalization and anisotropic diffusion for a fingerprint image (left).

Fig. 7.6. Result of the combination of local and global contrast enhancement. The original image is given on the top left. The result of global histogram equalization is on the top right, and the one for local contrast enhancement (16×16 neighborhood) is on the bottom left. Finally, on the bottom right is shown the combination of local and global contrast modification: The image on the left is further processed by the global histogram modification flow.

is natural for this kind of combination, as all that is needed is for the area $\mathcal{A}$ to be computed in a time-dependent neighborhood.

7.2. Shape-Preserving Contrast Enhancement

As we have seen in Section 7.1, histogram modification, and in particular histogram equalization (uniform distributions), is one of the basic and most useful operations in image processing, and its description can be found in any book on image processing. This operation is a particular case of homomorphic transformations: Let $\Omega \subseteq \mathbb{R}^2$ be the image domain and $I : \Omega \to [a, b]$ be the given (low-contrast) image. Let $h : [a, b] \to [c, d]$ be a given function that we assume to be increasing. The image $\hat{I} := h(I)$ is called a homomorphic transformation of u. The particular case of histogram equalization corresponds to selecting h to be the distribution function H of u:

$$H(\lambda) := \frac{\text{area}\{x \in \Omega : I(x) \leq \lambda\}}{\text{area}(\Omega)}. \tag{7.24}$$

If we assume that H is strictly increasing, then the change of variables

$$\hat{I}(x) = (b - a)H[I(x)] + a \tag{7.25}$$

gives a new image whose distribution function is uniform in the interval $[a, b]$, $a, b \in \mathbb{R}$, $a < b$. This useful and basic operation has an important property that, in spite of being obvious, we should acknowledge: It neither creates nor destroys image information.

As argued by the *Mathematical Morphology* school [5, 254, 355, 356], the basic operations on images should be invariant with respect to contrast changes, i.e., homomorphic transformations. As a consequence, it follows that the basic information of an image is contained in the family of its binary shadows or level sets, that is, in the family of sets

$$X_\lambda I := \{x \in \Omega : I(x) \geq \lambda\} \tag{7.26}$$

for all values of λ in the range of I. Observe that, under fairly general conditions, an image can be reconstructed from its level sets by the formula $I(x) = \sup\{\lambda : x \in X_\lambda I\}$. If h is a strictly increasing function, the transformation $\hat{I} = h(I)$ does not modify the family of level sets of I; it changes its index only in the sense that

$$X_{h(\lambda)}\hat{I} = X_\lambda I \quad \text{for all } \lambda. \tag{7.27}$$

Although we can argue that if all operations in image processing must hold this principle, for the purposes of the present discussion, and following Ref. [69], we shall stick here to this basic principle. There are a number of reasons for this. First, a considerably large amount of the research in image processing is based on assuming that regions with (almost) equal gray values, which are topologically connected (see below), belong to the same physical object in the 3D world. Following this, it is natural to assume then that the shapes in a given image are represented by its level sets (we will below see how we deal with noise that produces deviations from the level sets). Furthermore, this commonly assumed image processing principle will permit us to develop a theoretical and practical framework for shape-preserving contrast enhancement. This can be extended to other definitions of shape, different from the level-set morphological approach assumed here. As we have previously observed, the level-set theory is also applicable to a large number of problems beyond image processing [294, 361].

In this section, we want to design local histogram modification operations that preserve the family of level sets of the image, that is, following the morphology school, preserve shape [68, 69]. Local contrast enhancement is mainly used to further improve the image contrast and facilitate the visual inspection of the data. We have already seen, and it will be further exemplified later in this chapter, that global histogram modification not always produces good contrast, and especially small regions are hardly visible after such a global operation. On the other hand, local histogram modification improves the contrast of small regions as well, but because the level sets are not preserved, artificial objects are created. The theory developed now will enjoy the best of both words: The shape-preservation property of global techniques and the contrast improvement quality of local ones.

The recent formalization of multiscale analysis given in Ref. [5] and discussed in Chap. 4 leads to a formulation of recursive, causal, local, morphological and geometric invariant filters in terms of solutions of certain PDEs of the geometric type, providing a new view of many of the basic mathematical morphology operations. One of their basic assumptions was the locality assumption that aimed to translate into a mathematical language the fact that we considered basic operations that were a kind of local average around each pixel or, in other words, only a few pixels around a given sample influence the output value of the operations. Obviously this excluded the case of algorithms as histogram modification. This is why operations like those in this chapter are not modeled by these equations, and a novel framework must be developed.

It is not our goal now to review the extensive research performed in contrast enhancement. We should only note that, basically, contrast enhancement techniques are divided into the two groups mentioned above, local and global, and their most popular representatives can be found in any basic book in image processing and computer vision. An early attempt to introduce shape criteria in contrast enhancement was done in Ref. [106].

7.2.1. Global Histogram Modification: A Variational Formulation

We call representatives of I all images of the form $\hat{I} = h(I)$, where h is a strictly increasing function. The question is which representative of u is the best for our purposes. That will depend, of course, on what our purposes are. We have seen above which is the function h we have to select if we want to normalize the contrast making the distribution function of u uniform. In addition, it was shown in Section 7.1 that when equalizing an image $I : \Omega \to [a, b]$ in the range $[a, b]$ we are minimizing the functional

$$E(\hat{I}) = \frac{|\Omega|}{2(b-a)} \int_{\Omega} \left[\hat{I}(x) - \frac{b-a}{2}\right]^2 \mathrm{d}x - \frac{1}{4} \int_{\Omega} \int_{\Omega} |\hat{I}(x) - \hat{I}(z)| \mathrm{d}x \mathrm{d}z.$$

The second term of the integral can be understood as a measure of the contrast of the whole image. Thus when minimizing $E(\hat{I})$ we are distributing the values of u so that we maximize the contrast. The first term tries to keep the values of u as near as possible to the mean $(b-a)/2$. When minimizing E on the class of functions with the same family of binary shadows as u, we get the equalization of u. We will see below how to modify this energy to obtain shape-preserving local contrast enhancement.

7.2.2. Connected Components

To be able to extend the global approach to a local setting we have to insist on our main constraint: We have to keep the same topographic map, that is, we have to keep the same family of level sets of u but we have the freedom to assign them a convenient gray level. To make this statement more precise, let us give some definitions (see Ref. [354]).

Definition 7.2. *Let X be a topological space. We say that X is connected if it cannot be written as the union of two nonempty closed (open) disjoint sets. A subset C of X is called a connected component if C is a maximal connected subset of X, i.e., C is connected and for any connected subset C_1 of X such that $C \subseteq C_1$, then $C_1 = C$.*

This definition will be applied to subsets X of $\mathbb{R}^2$ that are topological spaces with the topology induced from $\mathbb{R}^2$, i.e., an open set of X is the intersection of an open set of $\mathbb{R}^2$ with X. We shall need the following observation, which follows from the definition above: Two connected components of a topological space are either disjoint or they coincide; thus the topological space can be considered as the disjoint union of its connected components.

Remark: There are several notions of connectivity for a topological space. One of the most intuitive ones is the notion of arcwise connected (also called connected by arcs). A topological space X is said to be connected by arcs if any two points x, y of X can be joined by an arc, i.e., there exists a continuous function $\gamma : [0, 1] \to X$ such that $\gamma(0) = x, \gamma(1) = y$. In a similar way as above we define the connected components (with respect to this notion of connectivity) as the maximal connected sets. These notions could be used below instead of the one given in Definition 7.2.

Definition 7.3. *Let $I : \Omega \to [a, b]$ be a given image and $\lambda_1, \lambda_2 \in [a, b]$, $\lambda_1 \leq \lambda_2$. A section of the topographic map of u is a set of the form*

$$X_{\lambda_1,\lambda_2} = \cup_{\lambda \in [\lambda_1,\lambda_2]} C_\lambda, \tag{7.28}$$

where C_λ is a connected component of $[I = \lambda]$ such that for each $\lambda', \lambda'' \in [\lambda_1, \lambda_2]$, $\lambda' < \lambda''$, the set

$$X_{\lambda',\lambda''} = \cup_{\lambda \in [\lambda',\lambda'']} C_\lambda \tag{7.29}$$

is also connected.

Definition 7.4. *Let $I : \Omega \to [a, b]$ be a given image and let $\{X_\lambda : \lambda \in [a, b]\}$ be the family of its level sets. We shall say that the mapping $h : \Omega \times \mathbb{R} \to \mathbb{R}$ is a local contrast change if the following properties hold:*

P1: *h is continuous in the following sense:*

$$h(z, \lambda') \to h(x, \lambda) \quad \text{when} \quad z \to x, \lambda' \to \lambda, z \in X_{\lambda'}, x \in C_\lambda,$$

where C_λ is a connected component of $[I = \lambda]$.

P2: *$h(x, \cdot)$ is an increasing function of λ for all $x \in \Omega$.*

P3: *$h(x, \lambda) = h(y, \lambda)$ for all x, y are in the same connected component of $[I = \lambda]$, $\lambda \in \mathbb{R}$.*

P4: *Let Γ be a connected set with $u(\Gamma)$ not reduced to a point. Let $\hat{I}(x) = h[x, I(x)]$. Then $v(\Gamma)$ is not reduced to a point.*

P5: *Let $X_{\lambda_1,\lambda_2} = \cup_{\lambda \in [\lambda_1,\lambda_2]} C_\lambda$ be a section of the topographic map of I, $\lambda_1 < \lambda_2$, and let $x \in C_{\lambda_1}$, $y \in C_{\lambda_2}$. Then $h(x, \lambda_1) < h(y, \lambda_2)$.*

Definition 7.5. *Let $I : \Omega \to [a, b]$ be a given image. We shall say that $\hat{I}$ is a local representative of u if there exists some local contrast change h such that $\hat{I}(x) = h[x, I(x)]$, $x \in \Omega$.*

We collect in the next proposition some properties that follow immediately from the definitions above.

Proposition 7.1. *Let $I : \Omega \to [a, b]$ and let $\hat{I}(x) = h[x, I(x)]$, $x \in \Omega$, be a local representative of u. Then*

1. *$\hat{I}(x) = \sup\{h(x, \lambda) : x \in X_\lambda I,\ x \in \Omega\}$. We have that $x \in X_\lambda u$ if and only if $x \in X_{h(x,\lambda)}\hat{I},\ x \in \Omega, \lambda \in \mathbb{R}$.*
2. *$\hat{I}$ is a continuous function.*
3. *Let Γ (Γ') be a connected component of $[\hat{I} = \mu]$ (resp. $[I = \lambda]$) containing x, $\mu = h(x, \lambda)$. Then $\Gamma = \Gamma'$.*
4. *Let X_{λ_1,λ_2} be a section of the topographic map of I. Then X_{λ_1,λ_2} is also a section of the topographic map of $\hat{I}$.*

Remarks:

1. The previous proposition can be phrased as saying that the set of objects contained in I is the same as the set of objects contained in $\hat{I}$, if we understand the objects of u as the connected connected components of the level sets $[\lambda \le I < \mu]$, $\lambda < \mu$, and respectively for $\hat{I}$.
2. Our definition of local representative is contained in the notion of dilation as given in [355, 356, Theorem 9.3]. Let $\mathcal{U}_n$ be a lattice of functions $f : \mathbb{R}^n \to \mathbb{R}^n$. A mapping $\Gamma : \mathcal{U}_n \to \mathcal{U}_n$ is called a dilation of $\mathcal{U}_n$ if and only if it can be written as

$$\Gamma(f)(x) = \sup\{g(x; y, t) : y \in \mathbb{R}^n, t \le f(y)\}, \quad x \in \mathbb{R}^n,$$

   where $g(x; y, t)$ is a function assigned to each point $(y, t) \in \mathbb{R}^n \times \mathbb{R}$ and is possibly different from point to point. Thus, let h be a local contrast change and let $\hat{I}(x) = h[x, I(x)]$. Let us denote by $X_t(f, x)$ the connected component of $X_t f$ that contains x if $x \in X_t f$, otherwise, let $X_t(f, x) = \emptyset$. Let $g(x; y, t) := h(x, t)$ if $X_t(f, x) \cap X_t(f, y) \ne \emptyset$; and $:= 0$ if $X_t(f, x) \cap X_t(f, y) = \emptyset$. Then $v = \Gamma(u)$.
3. Extending the definition of local contrast change to include more general functions than continuous ones, i.e., to include measurable functions, we can state and prove a converse of Proposition 7.1, saying that the topographic map contains all the information of the image that is invariant by local contrast changes [70].

7.2.3. Shape-Preserving Contrast Enhancement: The Algorithm

We can now state precisely the main question we want to address: What is the best local representative v of u, when the goal is to perform local contrast enhancement while preserving the connected components (and level sets)? For that we shall use the energy formulation given in Subsection 7.2.1. Let A be a connected component of the set $[\lambda \leq I < \mu]$, $\lambda, \mu \in \mathbb{R}$, $\lambda < \mu$. We write

$$E(\hat{I}, A) := \frac{|A|}{2(\mu - \lambda)} \int_A \left[\hat{I}(x) - \frac{\mu - \lambda}{2}\right]^2 \mathrm{d}x - \frac{1}{4} \int_A \int_A |\hat{I}(x) - \hat{I}(z)| \mathrm{d}x \mathrm{d}z.$$

We then look for a local representative v of u that minimizes $E(\hat{I}, A)$ for all connected components A of all sets of the form $[\lambda \leq I < \mu]$, $\lambda, \mu \in \mathbb{R}$, $\lambda < \mu$, or, in other words, the distribution function of $\hat{I}$ in all connected components of $[\lambda \leq \hat{I} < \mu]$ is uniform in the range $[\lambda, \mu]$, for all $\lambda, \mu \in \mathbb{R}$, $\lambda < \mu$. We now show how to solve this problem.

Let us introduce some notation that will make our discussion easier. Without loss of generality we assume that $I : \Omega \to [0, 1]$. Let $\lambda_{k,j} := j/2^k$, $k = 0, 1, 2, \ldots, j = 0, \ldots, 2^k$. We need to assume that H, the distribution function of u, is continuous and strictly increasing. For that we assume that u is continuous and

$$\text{area}\{x \in \Omega : I(x) = \lambda\} = 0, \quad \text{for all } \lambda \in \mathbb{R}. \tag{7.30}$$

We shall construct a sequence of functions converging to the solution of the problem. Let $\tilde{I}_0 = H(I)$ be the histogram equalization of I. Suppose that we already constructed $\tilde{I}_0, \ldots, \tilde{I}_{i-1}$. Let us construct $\tilde{I}_i$. For each $j = 0, 1, \ldots, 2^i - 1$, let

$$O_{i,j} := [\lambda_{i,j} \leq \tilde{I}_{i-1} < \lambda_{i,j+1}], \tag{7.31}$$

and let $O_{i,j;r}$ be the connected components of $O_{i,j}$, $r = 1, \ldots, n_{i,j}$ ($n_{i,j}$ can be eventually ∞). Define

$$h_{i,j;r}(\lambda) := \frac{|[\tilde{I}_{i-1} \leq \lambda] \cap O_{i,j;r}|}{|O_{i,j;r}|} (\lambda_{i,j+1} - \lambda_{i,j}) + \lambda_{i,j}, \quad \lambda \in [\lambda_{i,j}, \lambda_{i,j+1}).$$

By assumption (7.30), $h_{i,j;r}$ is a continuous strictly increasing function in $[\lambda_{i,j}, \lambda_{i,j+1})$ and we can equalize the histogram of $\tilde{I}_{i-1}$ in $O_{i,j;r}$. Thus we define

$$\tilde{I}_{i,j;r} := h_{i,j;r}(\tilde{I}_{i-1}) \chi_{O_{i,j;r}}, \tag{7.32}$$

$j = 0, 1, \ldots, 2^i - 1, r = 1, \ldots, n_{i,j}$, and

$$\tilde{I}_i := \sum_{j=1}^{2^i-1} \sum_{r=1}^{n_{i,j}} \tilde{I}_{i,j;r} \chi_{O_{i,j;r}}. \tag{7.33}$$

We can then prove Theorem 7.3.

Theorem 7.3. *Under assumption (7.30) the functions $\tilde{I}_i$ have a uniform histogram for all connected components of all dyadic sets of the form $[\lambda \leq \tilde{I}_i < \mu]$ where $\lambda, \mu \in \{\lambda_{i,j} : j = 0, \ldots, 2^i\}$, $\lambda < \mu$. Moreover, as $i \to \infty$, $\tilde{I}_i$ converges to a function $\tilde{I}$ that has a uniform histogram for all connected components of all sets $[\lambda \leq \tilde{I} < \mu]$, for all $\lambda, \mu \in [0, 1]$, $\lambda < \mu$.*

Theorem 7.4. *Let $\tilde{I}$ be the function constructed in Theorem 7.3. Then $\tilde{I}$ is a local representative of I.*

The proof of Theorem 7.3 is based in the next two simple lemmas. All the proofs can be found in Ref. [69].

Lemma 7.3. *Let $O_1, O_2 \subseteq \Omega$ such that $O_1 \cap O_2 = \emptyset$. Let $I_i : O_i \to [a, b)$, $i = 1, 2$, be two functions with uniform histogram in $[a, b)$. Let $I : O_1 \cup O_2 \to [a, b]$ be given by*

$$u(x) = \begin{cases} u_1(x), & \text{if } x \in O_1 \\ u_2(x), & \text{if } x \in O_2 \end{cases}. \tag{7.34}$$

Then I has a uniform histogram in $[a, b]$.

Lemma 7.4. *Let $O_1, O_2 \subseteq \Omega$ such that $O_1 \cap O_2 = \emptyset$. Let $I_1 : O_1 \to [a, b)$, $I_2 : O_2 \to [b, c)$ be two functions with uniform histogram in $[a, b)$, $[b, c)$, respectively. Assume that*

$$\frac{|O_1|}{|O_1| + |O_2|} = \frac{b-a}{c-a}, \quad \frac{|O_2|}{|O_1| + |O_2|} = \frac{c-b}{c-a}. \tag{7.35}$$

Let $I : O_1 \cup O_2 \to [a, c)$ be given by

$$I(x) = \begin{cases} I_1(x), & \text{if } x \in O_1 \\ I_2(x), & \text{if } x \in O_2 \end{cases}. \tag{7.36}$$

Then I has a uniform histogram in $[a, c)$.

7.2.4. The Numerical Algorithm and Experiments

The algorithm has been described in Subsection 7.2.3. Let us summarize it here. Let $I : \Omega \to [0, M]$ be an image whose values have been normalized in $[0, M]$. Let $\lambda_{k,j} := jM/2^k$, $k = 0, 1, 2, \ldots, N$, $j = 0, \ldots, 2^k$.

Step 1: Construct $\tilde{I}_0 = H(I)$ to be the histogram equalization of u.

Step 2: Construct $\tilde{I}_i$, $i = 1, \ldots, N$. Suppose that we have already constructed $\tilde{I}_0, \ldots, \tilde{I}_{i-1}$. Let us construct $\tilde{I}_i$. For each $j = 0, 1, \ldots, 2^i - 1$, let

$$O_{i,j} := [\lambda_{i,j} \leq \tilde{I}_{i-1} < \lambda_{i,j+1}], \tag{7.37}$$

and let $O_{i,j;r}$ be the connected components of $O_{i,j}$, $r = 1, \ldots, n_{i,j}$. Let $h_{i,j;r}$ be the distribution function of $\tilde{I}_{i-1}\chi_{O_{i,j;r}}$ with values in the range $[\lambda_{i,j}, \lambda_{i,j+1}]$.

Then we define

$$\tilde{I}_i := \sum_{j=1}^{2^i-1} \sum_{r=1}^{n_{i,j}} h_{i,j;r}(\tilde{I}_{i-1})\chi_{O_{i,j;r}}. \tag{7.38}$$

Remark: An interesting variant practice consists of using the mean of $\tilde{I}_0$, denoted by $m_{0,1}$, as the value to subdivide the range of $\tilde{I}_0$:

$$O_{1,0} = [0 \leq \tilde{I}_0 < m_{0,1}], \qquad O_{1,1} = [m_{0,1} \leq \tilde{I}_0 \leq M]. \tag{7.39}$$

Then we equalize $\tilde{I}_0$ in all connected components of $O_{1,0}$ in the range $[0, m_{0,1} - 1]$, respectively, in all connected components of $O_{1,1}$ in the range $[m_{0,1}, M]$. In this way we construct $\tilde{I}_1$. Then we compute the mean values of $\tilde{I}_1$ in $O_{1,0}$, $O_{1,1}$. We denote them by $m_{1,1}$, $m_{1,3}$ ($m_{1,2} = m_{0,1}$). Now we use these values to subdivide again $\tilde{I}_1$ into four pieces and proceed to equalize the histogram of $\tilde{I}_1$ in all connected components of all these pieces. We may continue iteratively in this way until desired.

Before proceeding we should note that a number of algorithms have been proposed in the literature to efficiently compute connected components, making the algorithm here described very fast.

In Fig. 7.7 we compare the classical local technique just described. In the classical algorithm the procedure is to define an $n \times m$ neighborhood and move the center of this area from pixel to pixel. At each location we compute the histogram of the $n \times m$ points in the neighborhood and obtain a histogram equalization (or histogram specification) transformation function. This function is used to map the level of the pixel centered in the neighborhood. The center of the $n \times m$ region is then moved to an adjacent pixel location and the procedure is repeated. In practice we update the histogram

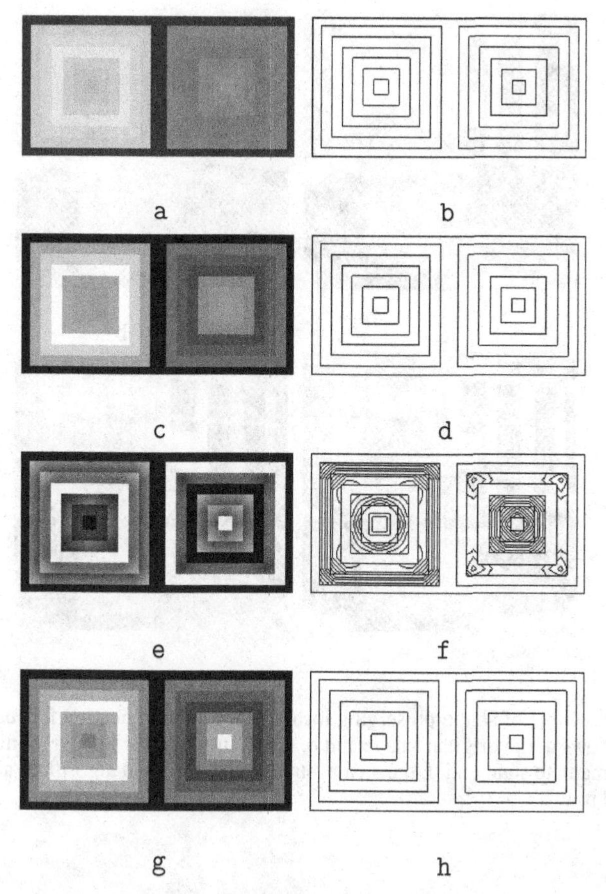

Fig. 7.7. Example of the level-set preservation. The top row shows the original image and its level sets. The second row shows the result of global histogram modification and the corresponding level sets. Results of classical local contrast enhancement and its corresponding level sets are shown in the third row. The last row shows the result of the algorithm. Note how the level sets are preserved, in contrast with the result on the third row, whereas the contrast is much better than the global modification.

obtained in the previous location with the new data introduced at each motion step. Figure 7.7a shows the original image whose level lines are displayed in Fig. 7.7b. Figure 7.7c shows the result of the global histogram equalization of Fig. 7.7a. Its level lines are displayed in Fig. 7.7d. Note how the level-set lines are preserved, whereas the contrast of small objects is reduced. Figure 7.7e shows the result of the classical local histogram

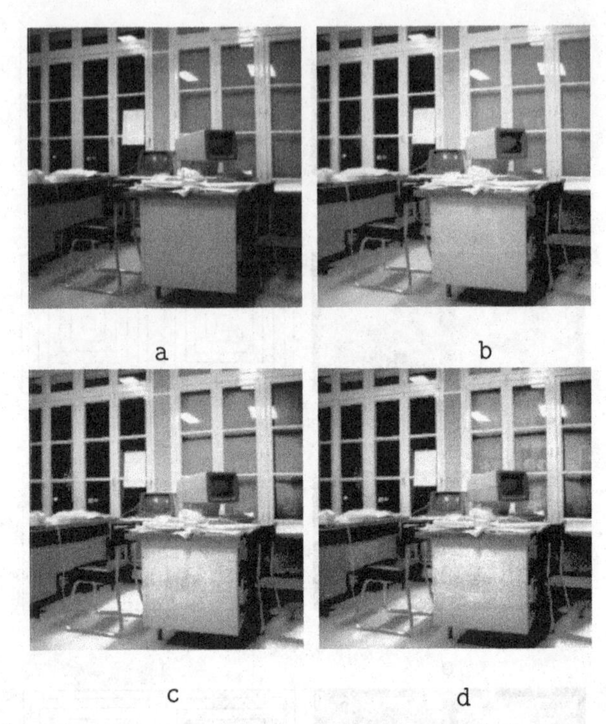

Fig. 7.8. Example of shape-preserving local histogram modification for real data. The first row shows a, the original image and b, the result of global histogram modification. c, an intermediate state and d, the steady state of the proposed algorithm, are shown in the second row.

equalization described above (31×31 neighborhood), with level lines displayed in Fig. 7.7f. All the level sets for gray-level images are displayed at intervals of 20 gray values. We see that new level lines appear, thus modifying the topographic map (the set of level lines) of the original image, introducing new objects. Figure 7.7g shows the result of the algorithm for local histogram equalization. Its corresponding level lines are displayed in Fig. 7.7h. We see that they coincide with the level lines of the original image, Fig. 7.7b.

Results for a real image are presented in Fig. 7.8. Figure 7.8a is the typical "Bureau de l'INRIA image." Figure 7.8b is the global histogram equalization of Fig. 7.8a. Figure 7.8c shows an intermediate step of the proposed algorithm, and Fig. 7.8d is the steady-state solution. Note how objects that are not visible in the global modification, such as those through the window, are now visible with the new local scheme.

Fig. 7.9. Additional example of shape-preserving local histogram modification for real data. Figure a is the original image. Figures b–d are the results of global histogram equalization, classical local scheme (61 × 61 neighborhood), and shape-preserving algorithm, respectively.

An additional example is given in Fig. 7.9. Figure 7.9a is the original image. Figures 7.9b–7.9d are the results of global histogram equalization, classical local scheme (61 × 61 neighborhood), and shape-preserving algorithm, respectively.

Experiments with a color image are given in Fig. 7.10, working on the YIQ (luminance and chrominance) color space. Figure 7.10a is the original image. In Fig. 7.10b, the shape-preserving algorithm was applied to the luminance image Y (maintaining IQ) and then the RGB color system was recomposed. In Fig. 7.10c, again, the proposed local histogram modification was applied to the color Y channel only, but the chrominance vector was rescaled to maintain the same color point on the Maxwell triangle.

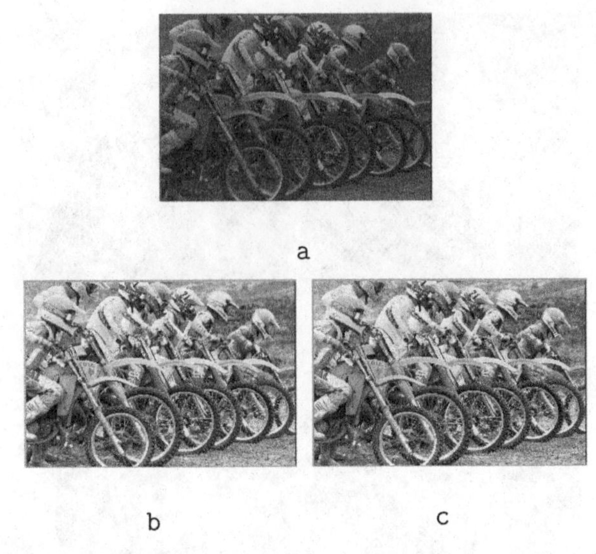

Fig. 7.10. Example of local histogram modification of a color image. The original image is shown in a. Image b is the result of applying the shape-preserving algorithm to the Y channel in the YIQ color space. In image c the algorithm is applied again only to the Y channel, the chrominance vector is rescaled to maintain the same color point on the Maxwell triangle.

In the last example, Fig. 7.11, the classical local histogram modification scheme is compared with the new one described here for a color image, following the same procedure as in Fig. 7.10. Figure 7.11a shows the original image, Fig. 7.11b is the one obtained with the classical technique, and Fig. 7.11c is the result of applying the shape preserving scheme. Note the spurious objects introduced by the classical local scheme.

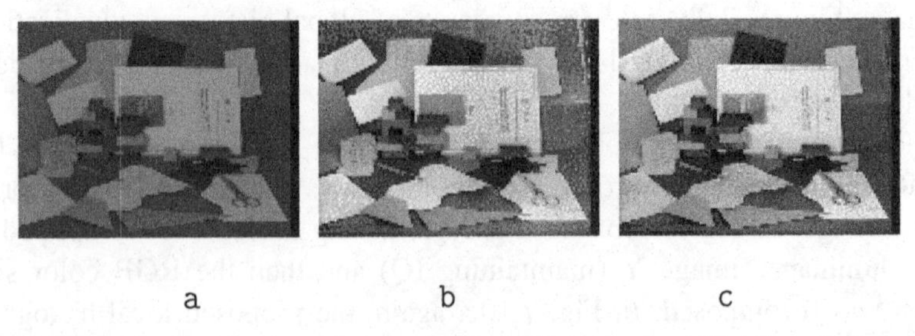

Fig. 7.11. Comparison between the classical local histogram modification scheme with the one described here for a color image. Image a is the original image, image b is the one obtained with the classical technique, and image c is the result of applying the shape-preserving scheme. Note the spurious objects introduced by the classical local scheme.

Exercises

1. Implement the differential equation for global contrast enhancement and combine it with PDEs for anisotropic diffusion. Investigate the differences between simultaneous contrast and noise enhancement and the direct algorithm obtained from first removing noise and then enhancing the contrast.
2. Compute the connected components of an image.
3. How would you extend the definition of connected components to vector-valued images? Propose a number of definitions and show examples.
4. Implement the local contrast enhancement algorithm and investigate the importance of the number of subdivision steps.

CHAPTER EIGHT

Additional Theories and Applications

8.1. Interpolation

We have seen in Chap. 4 how a series of basic axioms leads to the formulation of a number of important image processing operations such as PDEs. Examples included both isotropic and anisotropic diffusion. Of course, not all image processing algorithms hold those axioms, one example being the contrast enhancement technique from Chap. 7 (it does not hold, for example, the locality principle). We now show how PDE-based image interpolation schemes can also be obtained from a number of intuitive axioms [72]. We deal with the interpolation of images from a series of points and/or lines on the plane, following Ref. [72]. In Section 8.2 we will discuss an algorithm that fills up entire regions that are missing from the image.

Assume that the image $I : \mathbb{R}^2 \to \mathbb{R}$ is known at all pixels except one, x_0. Let us also assume that the image I is an interpolant of itself, meaning that it satisfies $I(x_0) = [\text{mean value} I(x)]$, mean value standing for all possible interpolants we have selected. Let us work out three examples:

1. $I(x_0)$ is the mean of the neighboring pixels:

$$I(x_0) := \frac{1}{4}\{I[x_0 + (h, 0)] + I[x_0 - (h, 0)] + I[x_0 + (0, h)] + I[x_0 - (0, h)]\},$$

$h > 0$. If $h \to 0$, then, by Taylor expansion,

$$\Delta I(x_0) = 0.$$

This result is independent of the specific linear combination of neighboring pixels selected.

2. $I(x_0)$ is the median value of the neighboring pixels:

$$I(x_0) := \text{median}\{I(y),\, y \in D(x_0, h)\},$$

where $D(x_0, h)$ is the disk centered at x_0 with radius h. In this case, as we have seen when connecting anisotropic diffusion with median filtering, if $h \to 0$, then

$$\kappa[I(x_0)] = 0.$$

This can also be written as

$$\kappa[I(x_0)] = \frac{1}{\|\nabla I\|^3}\nabla^2 I(\nabla I^{\perp}, \nabla I^{\perp}) = 0,$$

where $\nabla^2 I$ is the Hessian of I, $\nabla I^{\perp}$ is the unit vector perpendicular to the gradient ∇I, and we have used the notation $A(x, y) := \sum_{i,j=1}^{2} a_{ij} x_i y_j$, where $A = (a_{ij})_{i,j=1}^{2}$ is a 2×2 matrix and x, y are vectors in $\mathbb{R}^2$.

3. $I(x_0)$ is a directional interpolation (between level lines):

$$I(x_0) := \frac{1}{2}[I(x + h\nabla u) + I(x - h\nabla u)].$$

Then, if $h \to 0$,

$$\nabla^2 I(\nabla I, \nabla I) = 0.$$

We have then obtained the following three possibilities:

$$\Delta I = 0. \tag{8.1}$$

$$\nabla^2 I(\nabla I^{\perp}, \nabla I^{\perp}) = 0 \tag{8.2}$$

$$\nabla^2 I(\nabla I, \nabla I) = 0. \tag{8.3}$$

Note that the last two equations, when added, give the first one. We will later discuss the interesting fact that, based on a few intuitive axioms, these three are the only possible interpolants. The first one does not permit us to interpolate isolated points, as the problem $\Delta I = 0$, $I[\partial D(0, r)] = 0$, $I(0) = 1$, has no solution. The second equation also does not have a solution for this problem. The third equation yields a cone function for this case ($I(x) = |x| - 1$).

Let us now proceed to present the axiomatic approach for image interpolation. Let Γ be a set of continuous simple Jordan curves in $\mathbb{R}^2$. For each $\mathcal{C} \in \Gamma$, let $\mathcal{F}(\mathcal{C})$ be the set of continuous functions defined on $\mathcal{C}$. We shall consider an interpolation operator as a transformation E that associates with each $\mathcal{C} \in \Gamma$ and each $\phi \in \mathcal{F}(\mathcal{C})$ a unique function $E(\phi, \mathcal{C})$ defined on the region $D(\mathcal{C})$ enclosed by $\mathcal{C}$. This interpolant must hold the following properties:

1. Comparison principle:

$$E(\phi, \mathcal{C}) \leq E(\psi, \mathcal{C}),\ \forall \mathcal{C} \in \Gamma,\ \phi, \psi \in \mathcal{F}(\mathcal{C}),\ \phi \leq \psi.$$

2. Stability principle:

$$E(E(\phi, \mathcal{C})|_{\mathcal{C}'}, \mathcal{C}') = E(\phi, \mathcal{C})|_{D(\mathcal{C}')},$$

   for any $\mathcal{C} \in \Gamma$, any $\phi \in \mathcal{F}(\mathcal{C})$, and $\mathcal{C}' \in \Gamma$ such that $D(\mathcal{C}') \subseteq D(\mathcal{C})$. This means that no new application of the interpolation can improve a given interpolant (otherwise, the interpolation is iterated until the axiom holds).
3. Regularity principle: This is an adaptation of the regularity principle for the axiomatic approach presented in Chap. 4 [72].
4. Translation invariant: The interpolant of a translated image is the translation of the interpolant.
5. Rotation invariant: Same as for the translation property.
6. Gray-scale shift invariant:

$$E(\phi + c, \mathcal{C}) = E(\phi, \mathcal{C}) + c,$$

   for any constant $c \in \mathbb{R}$.
7. Linear gray-scale invariant:

$$E(\lambda\phi, \mathcal{C}) = \lambda E(\phi, \mathcal{C}),$$

   for any constant $\lambda \in \mathbb{R}$.
8. Zoom invariance: Same as for the translation property.

The basic result then says that any function that is an interpolant satisfying all the eight properties above is a viscosity solution of

$$G\left[\nabla^2 u\left(\frac{\nabla u}{\|\nabla u\|}, \frac{\nabla u}{\|\nabla u\|}\right), \nabla^2 u\left(\frac{\nabla u}{\|\nabla u\|}, \frac{\nabla u^{\perp}}{\|\nabla u\|}\right),\right.$$
$$\left.\nabla^2 u\left(\frac{\nabla u^{\perp}}{\|\nabla u\|}, \frac{\nabla u^{\perp}}{\|\nabla u\|}\right)\right] = 0,$$

where $G(\cdot)$ is a nondecreasing scale-invariant function.

A number of additional conditions can help to further simplify the form of G. For example, if it does not depend on its first argument (last), then it depends on only its last (first). If G is differentiable at (0, 0, 0), then it can be written as

$$a\nabla^2 I\left(\frac{\nabla I}{\|\nabla I\|}, \frac{\nabla I}{\|\nabla I\|}\right) + 2b\nabla^2 I\left(\frac{\nabla I}{\|\nabla I\|}, \frac{\nabla I^{\perp}}{\|\nabla I\|}\right) + c\nabla^2 I\left(\frac{\nabla I^{\perp}}{\|\nabla I\|}, \frac{\nabla I^{\perp}}{\|\nabla I\|}\right), \tag{8.4}$$

where $a, c \geq 0$ and $ac - b^2 \geq 0$. Let us further explore which selection of a, b, c can lead to useful interpolators. Let us consider once again a ball of center in the origin and radius 1, such that I is equal to 1 at the center and 0 at the boundary. Assume we that have both existence and uniqueness of the solution of Eq. (8.4). Because both the equation and the data are rotational invariant, we search for interpolants of the form $I = f(r)$, with $r := (x_1^2 + x_2^2)^{1/2}$. Because I satisfies Eq. (8.4), then f satisfies

$$arf'' + cf' = 0,$$

for $0 < r < 1$, and such that $f(0) = 1$ and $f(1) = 0$. We then have that if $a = 0$, then $b = 0$. If also $c = 0$, we have no equation. If $c > 0$, $f' = 0$, and the only solution is $f =$ constant, which does not satisfy the initial conditions. There is no interpolation operator in this case. Consider now the case of $a > 0$. The solutions are then of form $1, r^z$, or $\log r$. If $0 \leq c < a$, then $z = 1 - c/a$ and $f(r) = 1 - r^z$. The gradient ∇I is bounded if and only if $z = 1$, and then $c = 0$, obtaining

$$\nabla^2 I\left(\frac{\nabla I}{\|\nabla I\|}, \frac{\nabla I}{\|\nabla I\|}\right) = 0. \tag{8.5}$$

If $c > 0$, then the gradient is not bounded at the origin, and if $c = a$, the conditions at the origin and boundary cannot be satisfied. The interpolant of Eq. (8.5) is the only one that satisfies the axioms and is always bounded for bounded data [72]. This is also true as the data are not just given on curves, but on regions and isolated points as well.

Moreover, it can be shown that the corresponding evolution problem, given by

$$\frac{\partial I}{\partial t} = \nabla^2 I\left(\frac{\nabla I}{\|\nabla I\|}, \frac{\nabla I}{\|\nabla I\|}\right), \tag{8.6}$$

with the corresponding initial and boundary conditions, has also a unique viscosity solution, bounded, and converging to the viscosity solution of Eq. (8.5) when $t \to 0$. This is the equation used to numerically interpolate data, with an implicit scheme of the form

$$I_{ij}^{n+1} = I_{ij}^{n} + \Delta t \nabla^2 I_{ij}^{n+1} \left(\frac{\nabla I_{ij}^{n+1}}{\|\nabla I_{ij}^{n+1}\|}, \frac{\nabla I_{ij}^{i+1}}{\|\nabla I_{ij}^{n+1}\|} \right),$$

which is solved with nonlinear over relaxation methods. An example, courtesy of Caselles et al. [72], is given in Fig. 8.1.

Fig. 8.1. Interpolation by means of PDEs. From left to right, top to bottom: Original image, level lines for $\delta = 30$, quantized image for $\delta = 20$, the interpolant for $\delta = 20$, quantized image for $\delta = 30$, the interpolant for $\delta = 30$.

8.2. Image Repair: Inpainting

The modification of images in a way that is nondetectable for an observer who does not know the original image is a practice as old as artistic creation itself. Medieval artwork started to be restored as early as the Renaissance, the motives being often as much to bring medieval pictures up to date as to fill in any gaps [122, 400]. This practice is called retouching or inpainting. The object of inpainting is to reconstitute the missing or damaged portions of the work in order to make it more legible and to restore its unity [122].

The need to retouch the image in an unobtrusive way extended naturally from paintings to photography and film. The purposes remain the same: to revert deterioration (e.g., cracks in photographs or scratches and dust spots in film) or to add or remove elements (e.g., removal of stamped date and red eye from photographs, the infamous "airbrushing" of political enemies [217]).

Digital techniques are starting to be a widespread way of performing inpainting, ranging from attempts at fully automatic detection and removal of scratches in film [222, 223] all the way to software tools that allow a sophisticated but mostly manual process [46].

In this section an algorithm is described for automatic digital inpainting, its main motivation being to replicate the basic techniques used by professional restorators [25]. The only user interaction required by the algorithm presented here is to mark the regions to be inpainted. Although a number of techniques exist for the semiautomatic detection of image defects (mainly in films), addressing this is out of the scope of this section. Moreover, because the inpainting algorithm here presented can be used not just to restore damaged photographs but also to remove undesired objects and writing on the image, the regions to be inpainted must be marked by the user, as they depend on his/her subjective selection. Here we are concerned on how to fill in the regions to be inpainted, once they have been selected. Marked regions are automatically filled with the structure of their surrounding, in a form that will be explained later in this section.

8.2.1. Related Work

We should first note that classical image denoising algorithms do not apply to image inpainting. In common image enhancement applications, the pixels contain both information about the real data and the noise (e.g., image plus noise for additive noise), whereas in image inpainting, there is no significant information in the region to be inpainted. The information is mainly in the regions surrounding the areas to be inpainted. There is then a need to develop specific techniques to address these problems.

Mainly three groups of works can be found in the literature related to digital inpainting. The first one deals with the restoration of films, the second one is related to texture synthesis, and the third one, a significantly less-studied class, although connected to the work presented here, is related to disocclusion.

Kokaram et al. [223] use motion estimation and autoregressive models to interpolate losses in films from adjacent frames. The basic idea is to copy into the gap the right pixels from neighboring frames. The technique cannot be applied to still images or to films for which the regions to be inpainted span many frames.

Hirani and Totsuka [183] combine frequency- and spatial-domain information in order to fill a given region with a selected texture. This is a very simple technique that produces incredibly good results. On the other hand, the algorithm deals mainly with texture synthesis (and not with structured background) and requires the user to select the texture to be copied into the region to be inpainted. For images for which the region to be replaced covers several different structures, the user would need to go through the tremendous work of segmenting them and searching corresponding replacements throughout the picture. Although part of this search can be done automatically, this is extremely time consuming and requires the nontrivial selection of many critical parameters; see e.g., Ref. [121]. Other texture synthesis algorithms, e.g., Refs. [121, 178, and 367], can be used as well to recreate a preselected texture to fill in a (square) region to be inpainted.

In the group of disocclusion algorithms, a pioneering work is described in Ref. [270]; Nitzberg et al. presented a technique for removing occlusions with the goal of image segmentation. Because the region to be inpainted can be considered as occluding objects, removing occlusions is analogous to image inpainting. The basic idea is to connect T-junctions at the same gray level with elastica-minimizing curves. The technique was mainly developed for simple images, with only a few objects with constant gray levels, and will not be applicable for the examples with natural images presented later in this paper. Masnou and Morel [236] extended these ideas, presenting a very inspiring general variational formulation for disocclusion and a particular practical algorithm that implements some of the ideas in this formulation. The algorithm performs inpainting by joining with geodesic curves the points of the isophotes arriving at the boundary of the region to be inpainted. As reported by the authors, the regions to be inpainted are limited to having simple topology, e.g., holes are not allowed. This is not intrinsic to the general variational formulation they propose, only to the specific discrete implementation they perform. In addition, the angle with which the level lines arrive at the boundary of the inpainted region is not (well) preserved: The algorithm mainly uses straight lines to join equal gray-value pixels. This is the closest technique to the one here described now and has motivated in part and inspired the work now reported.

8.2.2. The Digital Inpainting Algorithm

Fundamentals. Let Ω be for the region to be inpainted and $\partial\Omega$ its boundary (note once again that no assumption on the topology of Ω is made). Intuitively, the technique we propose will prolong the isophote lines arriving at $\partial\Omega$, while maintaining the angle of arrival. We proceed, drawing from $\partial\Omega$ inward in this way, while curving the prolongation lines progressively to prevent them from crossing each other.

Before the detailed description of this technique is presented, let us analyze how experts inpaint. Inpainting is a very subjective procedure, different for each work of art and for each professional. There is no such thing as "the" way to solve the problem, but the underlying methodology is as follows: (1) The global picture determines how to fill in the gap, the purpose of inpainting being to restore the unity of the work; (2) The structure of the area surrounding Ω is continued into the gap, contour lines are drawn by means of the prolongation of those arriving at $\partial\Omega$; (3) The different regions inside Ω, as defined by the contour lines, are filled with color, matching those of $\partial\Omega$; and (4) The small details are painted (e.g., little white spots on an otherwise uniformly blue sky): in other words, texture is added.

A number of lessons can immediately be learned from these basic inpainting rules used by professionals. The algorithm described simultaneously, and iteratively, performs steps (2) and (3) above. We progressively shrink the gap Ω by prolonging inward, in a smooth way, the lines arriving at the gap boundary $\partial\Omega$.

The Inpainting Algorithm. We need to translate the manual inpainting concepts expressed above into a mathematical and algorithmic language. We proceed to do this now, presenting the basic underlying concepts first. The implementation details are given in the next subsection. Let

$$I_0(i, j) : [0, M] \times [0, N] \to \mathbb{R}, \text{ with } [0, M] \times [0, N] \subset \mathbb{N} \times \mathbb{N}),$$

be a discrete 2D gray-level image. From the description of manual inpainting techniques, an iterative algorithm seems a natural choice. The digital inpainting procedure will construct a family of images $I(i, j, n) : [0, M] \times [0, N] \times \mathbb{N} \to \mathbb{R}$ such that $I(i, j, 0) = I_0(i, j)$ and $\lim_{n\to\infty} I(i, j, n) =$

$I_R(i, j)$, where $I_R(i, j)$ is the output of the algorithm (inpainted image). Any general algorithm of that form can be written as

$$I^{n+1}(i, j) = I^n(i, j) + \Delta t I_t^n(i, j), \forall (i, j) \in \Omega \tag{8.7}$$

where the superscript n denotes the inpainting time n, (i, j) are the pixel coordinates, Δt is the rate of improvement and $I_t^n(i, j)$ stands for the update of the image $I^n(i, j)$. Note that the evolution equation runs inside only Ω, the region to be inpainted.

With this equation, the image $I^{n+1}(i, j)$ is an improved version of $I^n(i, j)$, with the improvement given by $I_t^n(i, j)$. As n increases, we achieve a better image. We need now to design the update $I_t^n(i, j)$.

As suggested by manual inpainting techniques, we need to continue the lines arriving at the boundary $\partial\Omega$ of the region Ω to be inpainted [see point (2) in the Fundamentals subsection of Subsection 8.2.2]. In other words, we need to smoothly propagate information from outside Ω into Ω (points (2) and (3) in the fundamentals subsection). With $L^n(i, j)$ being the information that we want to propagate and (i, j) the propagation direction, this means that we must have

$$I_t^n(i, j) = \overrightarrow{\delta L^n}(i, j) \cdot \quad (i, j), \tag{8.8}$$

where $\overrightarrow{\delta L^n}(i, j)$ is a measure of the change in the information $L^n(i, j)$. (Borrowing notation from continuous mathematics, we could also write $\overrightarrow{\delta L^n}(i, j)$ as ∇L.) With this equation, we estimate the information $L^n(i, j)$ of our image and compute its change along the direction. Note that at steady state, that is, when the algorithm converges, $I^{n+1}(i, j) = I^n(i, j)$ and from Eqs. (8.7) and (8.8) we have that $\overrightarrow{\delta L^n}(i, j) \cdot \quad (i, j) = 0$, meaning exactly that the information L has been propagated in the direction .

What is left now is to express the information L being propagated and the direction of propagation .

Because we want the propagation to be smooth, $L^n(i, j)$ should be an image smoothness estimator. For this purpose we may use a simple discrete implementation of the Laplacian: $L^n(i, j) := I_{xx}^n(i, j) + I_{yy}^n(i, j)$ (subscripts represent derivatives in this case). Other smoothness estimators might be used, although satisfactory results were already obtained with this very simple selection.

Then we must compute the change $\overrightarrow{\delta L^n}(i, j)$ of this value along . To do this we must first define what the direction for the 2D information propagation will be. One possibility is to define as the normal to the

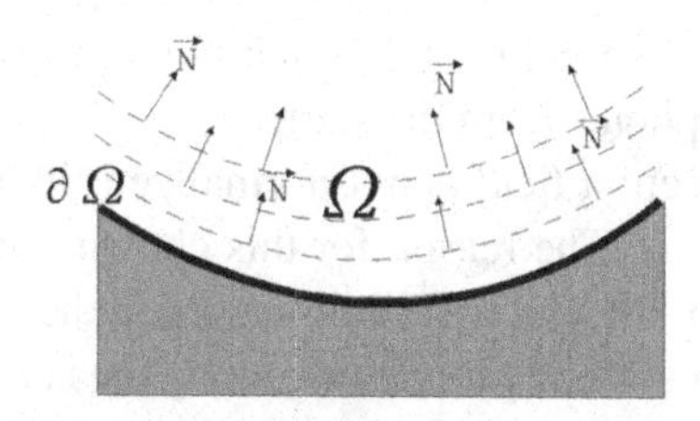

Fig. 8.2. Propagation direction as the normal to the signed distance to the boundary of the region to be inpainted.

signed distance to $\partial\Omega$, i.e., at each point (i, j) in Ω the vector (i, j) will be normal to the shrinked version of $\partial\Omega$ to which (i, j) belongs; see Fig. 8.2. This choice is motivated by the belief that a propagation normal to the boundary would lead to the continuity of the isophotes at the boundary. Instead, what happens is that the lines arriving at $\partial\Omega$ curve in order to align with ; see Fig. 8.3. This is of course not what we expect. Note that the orientation of $\partial\Omega$ is not intrinsic to the image geometry, as the region to be inpainted is arbitrary.

If isophotes tend to align with , the best choice for is then the isophotes directions. This is a bootstrapping problem: Having the isophotes directions inside Ω is equivalent to having the inpainted image itself, as we can easily recover the gray-level image from its isophote direction field (see the discussion section and Ref. [199]).

We use then a time varying estimation of the isophotes direction field: for any given point (i, j), the discretized gradient vector $\nabla I^n(i, j)$ gives the direction of largest spatial change, and its 90° rotation$\nabla^{\perp} I^n(i, j)$ is the direction of smallest spatial change, so the vector $\nabla^{\perp} I^n(i, j)$ gives the isophotes direction. Our field is then given by the time-varying $(i, j, n) = \nabla^{\perp} I^n(i, j)$. We are using a time-varying estimation that is coarse at the beginning but progressively achieves the desired continuity

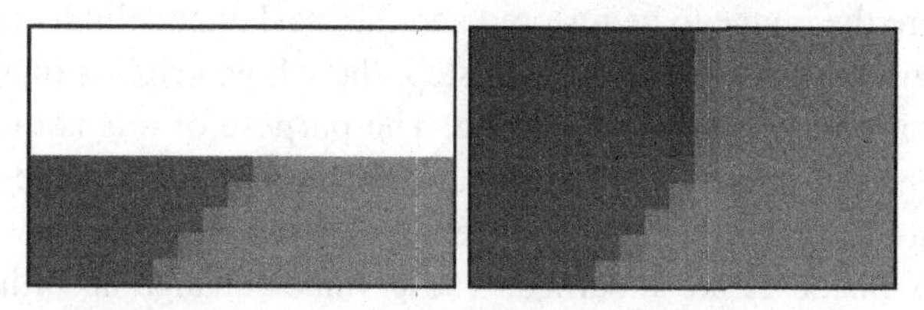

Fig. 8.3. Unsuccessful choice of the information propagation direction. Left: detail of the original image; region to be inpainted is in white. Right: restoration.

at $\partial\Omega$, instead of a fixed field (i, j) that would imply knowledge of the directions of the isophotes from the start.

Note that the direction field is not normalized; its norm is the norm of the gradient of $I^n(i, j)$. The reason for this choice relies on the numerical stability of the algorithm, and will be discussed in the following subsection.

Because we are performing inpainting along the isophotes, it is irrelevant if $\nabla^{\perp} I^n(i, j)$ is obtained as a clockwise or counterclockwise rotation of $\nabla I^n(i, j)$. In both cases, the change of $I^n(i, j)$ along those directions should be minimum.

Recapping, we estimate a variation of the smoothness, given by a discretization of the 2D Laplacian in our case, and project this variation into the isophotes direction. This projection is used to update the value of the image inside the region to be inpainted.

To ensure a correct evolution of the direction field, a diffusion process is interleaved with the image inpainting process described above. That is, every few steps (see below), we apply a few iterations of image diffusion. This diffusion corresponds to the periodical curving of lines to avoid them from crossing each other, as was mentioned in Subsection 8.2.2. We use anisotropic diffusion in order to achieve this goal without losing sharpness in the reconstruction. In particular, we apply a straightforward discretization of the following continuous-time/continuous-space anisotropic diffusion equation:

$$\frac{\partial I}{\partial t}(x, y, t) = g_{\epsilon}(x, y)\kappa(x, y, t)\left|\nabla I(x, y, t)\right|, \forall(x, y) \in \Omega^{\epsilon}, \quad (8.9)$$

where Ω^{ϵ} is a dilation of Ω with a ball of radius ϵ, κ is the Euclidean curvature of the isophotes of I, and $g_{\epsilon}(x, y)$ is a smooth function in Ω^{ϵ} such that $g_{\epsilon}(x, y) = 0$ in $\Omega^{\epsilon} \setminus \Omega$ and $g_{\epsilon}(x, y) = 1$ at the set of points of Ω whose distance to $\partial\Omega$ is larger that ϵ (this is a way to impose Dirichlet boundary conditions for the Eq. (8.9)).

Discrete Scheme and Implementation Details. The only inputs to our algorithm are the image to be restored and the mask that delimits the portion to be inpainted. As a preprocessing step, the whole original image undergoes anisotropic diffusion smoothing. The purpose of this is to minimize the influence of noise on the estimation of the direction of the isophotes arriving at $\partial\Omega$. After this, the image enters the inpainting loop, where only the values inside Ω are modified. These values change according to the discrete implementation of the inpainting procedure, which we proceed to describe. Every few iterations, a step of anisotropic diffusion is applied (a straightforward, central-difference implementation of Eq. (8.9) is used). This process is repeated until a steady state is achieved.

Let $I^n(i, j)$ stand for each one of the image pixels inside the region Ω at the inpainting time n. Then the discrete inpainting equation borrows from the numerical analysis literature and is given by

$$I^{n+1}(i, j) = I^n(i, j) + \Delta t I_t^n(i, j), \forall (i, j) \in \Omega \tag{8.10}$$

where

$$I_t^n(i, j) = \left[\overrightarrow{\delta L^n}(i, j) \cdot \frac{(i, j, n)}{|\quad (i, j, n)|} \right] |\nabla I^n(i, j)|, \tag{8.11}$$

$$\overrightarrow{\delta L^n}(i, j) := [L^n(i + 1, j) - L^n(i - 1, j), L^n(i, j + 1) - L^n(i, j - 1)], \tag{8.12}$$

$$L^n(i, j) = I_{xx}^n(i, j) + I_{yy}^n(i, j), \tag{8.13}$$

$$\frac{(i, j, n)}{|\quad (i, j, n)|} := \frac{\left[-I_y^n(i, j), I_x^n(i, j)\right]}{\sqrt{\left[I_x^n(i, j)\right]^2 + \left[I_y^n(i, j)\right]^2}}, \tag{8.14}$$

$$\beta^n(i, j) = \overrightarrow{\delta L^n}(i, j) \cdot \frac{(i, j, n)}{|\quad (i, j, n)|}, \tag{8.15}$$

$$|\nabla I^n(i, j)| = \begin{cases} \sqrt{\left(I_{xbm}^n\right)^2 + \left(I_{xfM}^n\right)^2 + \left(I_{ybm}^n\right)^2 + \left(I_{yfM}^n\right)^2}, \\ \text{when } \beta^n > 0 \\ \sqrt{\left(I_{xbM}^n\right)^2 + \left(I_{xfm}^n\right)^2 + \left(I_{ybM}^n\right)^2 + \left(I_{yfm}^n\right)^2}, \\ \text{when } \beta^n < 0 \end{cases}. \tag{8.16}$$

We first compute the 2D smoothness estimation L in Eq. (8.13) and the isophote direction $/|\quad|$ in Eq. (8.14). Then in Eq. (8.15) we compute β^n, the projection of $\overrightarrow{\delta L}$ onto the (normalized) vector , that is, we compute the change of L along the direction of . Finally, we multiply β^n by a slope-limited version of the norm of the gradient of the image, $|\nabla I|$, in Eq. (8.16). A central-difference realization would turn the scheme unstable, and that is the reason for using slope limiters. The subscripts b and f denote backward and forward differences, respectively, and the subscripts m and M denote the minimum or the maximum, respectively, between the derivative and zero (we have omitted the space coordinates (i, j) for simplicity); see, for example, Refs. [294 and 330] for details. Finally, let us note that the choice of a nonnormalized field instead of a normalized version of it allows for a simpler and more stable numerical scheme; see Refs. [235 and 330].

Note once again that when the inpainting algorithm arrives at steady state, that is, $I_t = 0$, we have geometrically solved $\nabla(\text{smoothness}) \cdot \nabla^{\perp} I = 0$, meaning that the smoothness is constant along the isophotes. This type of information propagation is related to the work on velocity-field extension in level-set techniques described earlier in this book, e.g., Refs. [291 and 426].

When applying Eqs. (8.10)–(8.16) to the pixels in the border $\partial\Omega$ of the region Ω to be inpainted, known pixels from outside this region are used. That is, conceptually, we compute Eqs. (8.10)–(8.16) in the region Ω^{ϵ} (an ϵ dilation of Ω), although we update the values only inside Ω (that is, Eq. (8.10) is applied only inside Ω). The information in the narrow band $\Omega^{\epsilon} - \Omega$ is propagated inside Ω. Propagation of this information, both gray values and isophotes directions, is fundamental for the success of the algorithm.

In the restoration loop we perform A steps of inpainting with Eq. (8.10), then B steps of diffusion with Eq. (8.9), again A steps of Eq. (8.10), and so on. The total number of steps is T. This number may be preestablished or the algorithm may stop when changes in the image are below a given threshold. The values we use are $A = 15$, $B = 2$, at speed $\Delta t = 0.1$. The value of T depends on the size of Ω. If Ω is of considerable size, a multiresolution approach is used to speed up the process.

Color images are considered as a set of three images, and the above-described technique is applied independently to each one. To avoid the appearance of spurious colors, we use a color model which is very similar to the Luv model, with one luminance and two chroma components; see Fig. 8.4.

8.2.3. Results

The CPU time required for inpainting depends on the size of Ω. In all the color examples presented here, the inpainting process was completed in less than 5 min (for the three color planes), with the nonoptimized C++ code running on a PentiumII PC (128-Mbytes RAM, 300 MHz) under Linux. All the examples use images available from public databases over the Internet.

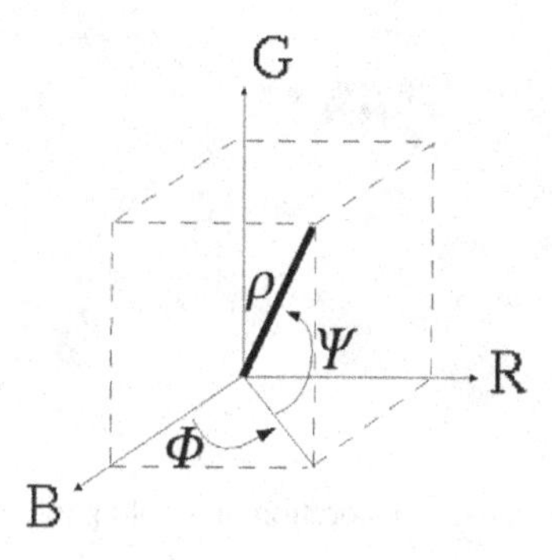

Fig. 8.4. Relation between the (R,G,B) color model and the one used in this section, $(\rho, \sin\phi, \sin\psi)$.

Figure 8.5 shows, on the left, a synthetic image with the region to inpaint in white. Here Ω is large (30 pixels in diameter) and contains a hole. The inpainted reconstruction is shown on the right. Note that contours are recovered, joining points from the inner and outer boundaries. Also, these reconstructed contours follow smoothly the direction of the isophotes arriving at $\partial\Omega$.

Figure 8.6 shows a deteriorated black-and-white image and its reconstruction. As in all the examples in this article, the user supplied only the "mask" image, shown in Fig. 8.7. This image was drawn manually with a paintbrushlike program. The variables were set to the values specified in Subsection 8.2.2 and the number of iterations T was set to 3000. When multiresolution is not used, the CPU time required by the inpainting procedure is approximately 7 min. With a two-level multiresolution scheme, only 2 min were needed. Observe that details in the nose and right eye of the middle girl could not be completely restored. This is in part due to the fact that the mask covers most of the relevant information, and there is not

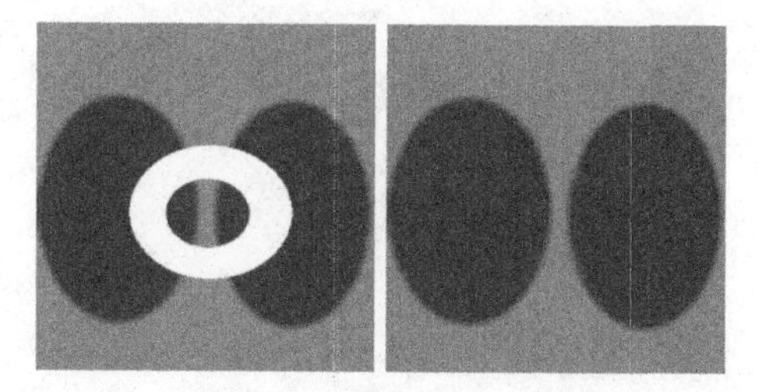

Fig. 8.5. Synthetic example: Ω is shown in white. Topology is not an issue, and the recovered contours smoothly continue the isophotes.

Fig. 8.6. Restoration of an old photograph.

much to be done without the use of high-level prior information (e.g., the fact that it is an eye). These minor errors can be corrected by the manual procedures mentioned in the introduction, and still the overall inpainting time would be reduced by orders of magnitude.

Figure 8.8 shows a vandalized image and its restoration, followed by an example in which overimposed text is removed from the image. These are typical examples in which texture synthesis algorithms such as those described in the introduction cannot be used, since the number of different regions to be filled is very large.

Figure 8.9 shows the progressive nature of the algorithm: Several intermediate steps of the inpainting procedure are shown for a detail of Fig. 8.8.

Finally, Fig. 8.10 shows an entertainment application. The bungee cord and the knot tying the man's legs have been removed. Given the size of Ω, a two-level multiresolution scheme was used. Here it becomes apparent that it is the user who has to supply the algorithm with the masking image, as the choice of the region to inpaint is completely subjective.

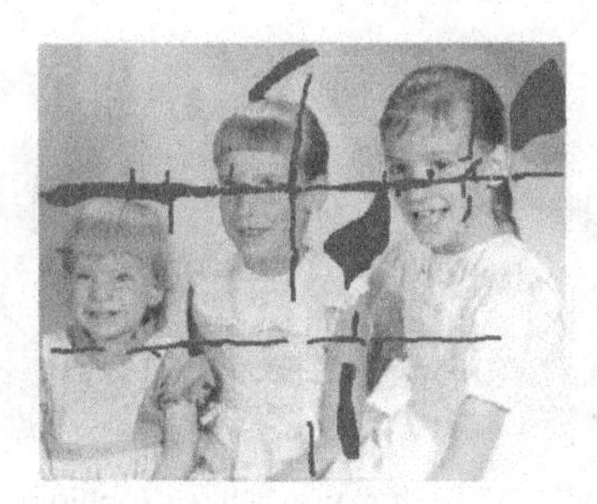

Fig. 8.7. The user defines the region to inpaint (here shown in red).

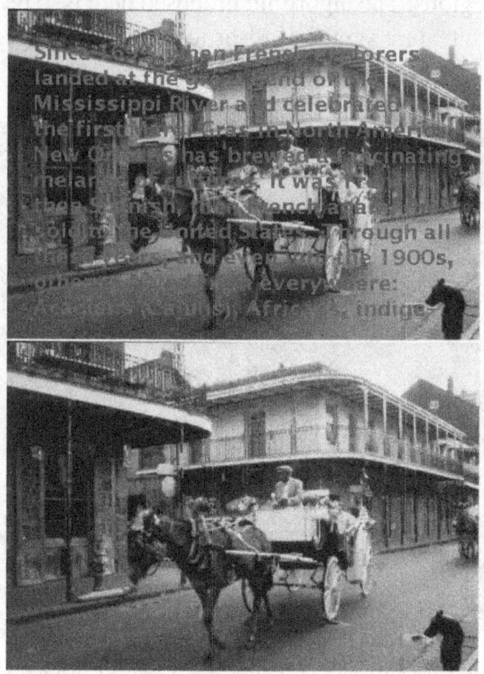

Fig. 8.8. Restoration of a color image and removal of superimposed text.

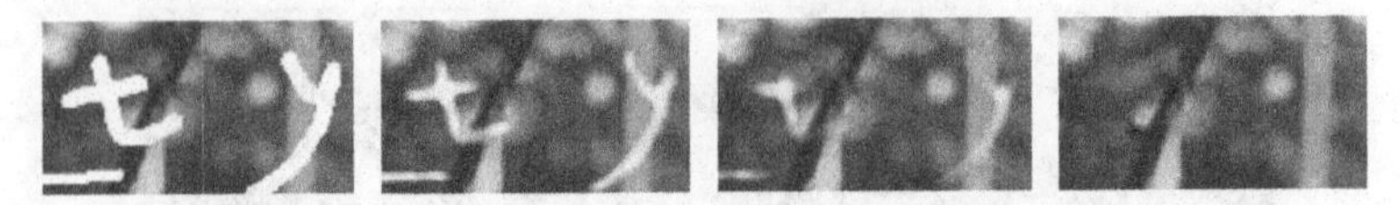

Fig. 8.9. Progressive nature of the algorithm: several intermediate steps of the reconstruction of Fig. 8.8.

8.2.4. Comments

In this section we have introduced a novel algorithm for image inpainting that attempts to replicate the basic techniques used by professional restorers. The basic idea is to smoothly propagate information from the surrounding areas in the isophotes direction. The user needs only to provide the region to be inpainted; the rest is automatically performed by the algorithm in a few minutes. The examples shown suggest a wide range of applications such as the restoration of old photographs and damaged film, removal of superimposed text, and removal of objects. The results can either be adopted as a final restoration or be used to provide an initial point for manual restoration, thereby reducing the total restoration time by orders of magnitude.

One of the main problems with the technique is the reproduction of large textured regions. The algorithm proposed here is to be investigated in conjunction with texture synthesis ideas to address this issue.

Fig. 8.10. The bungee cord and the knot tying the man's feet have been removed.

The inpainting algorithm presented here has been clearly motivated by and has borrowed from the intensive work on the use of PDE reported in the rest of this book. When "blindly" letting the grid go to zero, the inpainting technique in Eqs. (8.10)–(8.16) naively resembles a third-order equation, for which too many boundary conditions are imposed (all of them being essential) and for which a complete understanding is beyond the current state of mathematical knowledge (although results for other high-order equations, which might be relevant for image processing as well, are available, e.g., Ref. [28].) Nevertheless, this suggests the investigation of the use of lower, second-order PDEs to address the inpainting problem. We can split the inpainting problem into two coupled variational formulations, one for the isophotes direction (point (2) in the fundamentals subsection of Subsection 8.2.2) and one for the gray-values, consistent with the estimated directions (point (3) in the fundamentals subsection: this is like the multi-valued to single-valued images map previously reported in this book). The corresponding gradient descent flows will give two coupled second-order PDEs for which formal results regarding existence and uniqueness of the solutions can be shown.

8.3. Shape from Shading

According to the so-called Lambertian shading rule, the 2D array of pixel gray levels, corresponding to the shading of a 3D object, is proportional to the cosine of the angle between the light-source direction and the surface normal. The shape from shading problem is the inverse problem of reconstructing the 3D surface from this shading data. The history of this problem is extensive. A basic technique, developed by Kimmel and Bruckstein [210] to address this problem, is described here. See this reference for details and an extensive literature. See also [276] for a different formal approach.

Consider a smooth surface, actually a graph, given by $z(x, y)$. According to the Lambertian shading rule, the shading image $I(x, y)$ is equal (or proportional) to the inner product between the light direction $\hat{l} = (0, 0, 1)$ and the normal $\vec{\mathcal{N}}(x, y)$ to the parameterized surface. This gives the so-called irradiance equation:

$$I(x, y) = \hat{l} \cdot \vec{\mathcal{N}} = \frac{1}{\sqrt{1 + p^2 + q^2}},$$

where $p := \partial z/\partial x$ and $q := \partial z/\partial y$. Starting from a small circle around a singular point, Bruckstein [50] observed that equal-height contours $\mathcal{C}(p,t) : S \to \mathbb{R}^2$ of the surface z (t stands for the height) hold

$$\frac{\partial \mathcal{C}}{\partial t} = \frac{I}{\sqrt{1-I^2}}\vec{n},$$

where now $\vec{n}$ is the 2D unit normal to the equal height contour (or level set of z). This means that the classical shape from the shading problem is simply a curve evolution problem, and, as such, we can use all the curve evolution machinery to solve it. In particular, we can use both the level-set and the fast-marching numerical techniques (the weight for the distance is always positive and given by $\sqrt{1/I^2 - 1}$). An example, courtesy of Kimmel and Bruckstein [210], is presented in Fig. 8.11.

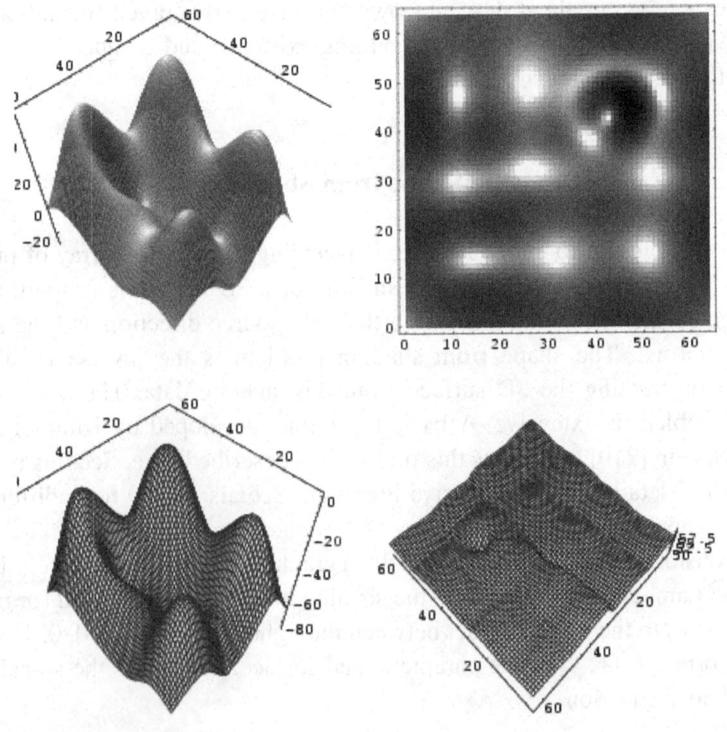

Fig. 8.11. Example of shape from shading by means of curve evolution. The figure shows the original surface, the simulated shading, the reconstructed surface, and the reconstruction error.

8.4. Blind Deconvolution

Assume that the image $I(x, y) : \mathbb{R}^2 \to \mathbb{R}$ has been blurred with the unknown filter $h(x, y) : \mathbb{R}^2 \to \mathbb{R}$, obtaining $\hat{I}(x, y) : \mathbb{R}^2 \to \mathbb{R}$ ($\hat{I} = I * h + \text{noise}$), and we want to recover I. It is well known that this is an ill-posed problem, and regularization conditions must be added in order to obtain any reasonable solution. Here we present the technique described in Refs. [81, 419 and 420], which is based on the simultaneous estimation of the image and the blurring kernel by use of an edge-preserving regularization function. We have seen in Chap. 4 a number of edge-preserving regularization forms, like Tukey's function and the L_1 norm. Here, we follow the developments by Chan and Wong, and use the L_1 norm, although more robust norms can be used as well.

The basic idea is to find the minimum (with respect to I, h) of the energy $\mathcal{F}(I, v)$ given by

$$\frac{1}{2}\|h * I - \hat{I}\|_{L_2}^2 + \alpha_1 \iint \|\nabla I\| \, dxdy + \alpha_2 \iint \|\nabla h\| \, dxdy.$$

The corresponding Euler–Lagrange or first variation is

$$\frac{\partial \mathcal{F}}{\partial h} = I(-x, -y) * (u * h - \hat{I}) - \alpha_2 \operatorname{div}\left(\frac{\nabla h}{\|\nabla h\|}\right) = 0,$$

$$\frac{\partial \mathcal{F}}{\partial I} = h(-x, -y) * (I * h - \hat{I}) - \alpha_1 \operatorname{div}\left(\frac{\nabla I}{\|\nabla I\|}\right) = 0.$$

For I given (respectively h), the functional $\mathcal{F}$ is convex, although it is not jointly convex. Therefore, with an initial guess (I^0, h^0), the idea is to to perform an alternating minimization [81]. Assuming we have (I^n, h^n), then we proceed as follows:

- Solve for h^{n+1}:

$$I^n(-x, -y) * (I^n * h^{n+1} - z) - \alpha_2 \operatorname{div}\left(\frac{\nabla h^{n+1}}{\|\nabla h^{n+1}\|}\right) = 0,$$

- Solve for I^{n+1}:

$$h^{n+1}(-x, -y) * (I^{n+1} * h^{n+1} - z) - \alpha_2 \operatorname{div}\left(\frac{\nabla I^{n+1}}{\|\nabla I^{n+1}\|}\right) = 0.$$

This system of equations can be solved in different forms. The most straightforward one is to use the time-marching schemes we discussed in

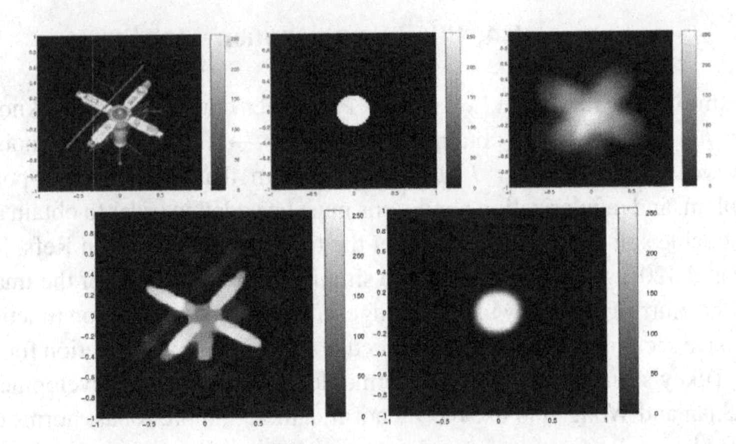

Fig. 8.12. Edge-preserving blind deconvolution. The first row shows, from left to right, the original image, the out-of-focus blur, and the blurred image. The recovered image and blurring function are shown on the second row.

Chap. 4, which means that we look for steady-state solutions of the corresponding PDE. Other more sophisticated methods can be used as well [79, 399]. The example in Fig. 8.12, courtesy of Chan and Wong [81], was obtained with these schemes.

Exercises

1. Take a picture of a face and run the curve evolution approach to the shape from shading problem. Plot the isolevels of the reconstructed 3D function and analyze their shape.
2. Extend the curve evolution approach to shape from shading for light coming from the side.
3. Write the corresponding blind deconvolution equations when Tukey's biweight functions are used instead of the L_1 norm.

Bibliography

[1] U. Abresch and J. Langer, "The normalized curve shortening flow and homothetic solutions," *J. Diff. Geom.* **23**, 175–196 (1986).

[2] D. Adalsteinsson and J. A. Sethian, "A fast level set method for propagating interfaces," *J. Comput. Phys.* **118**, 269 (1995).

[3] D. Adalsteinsson and J. A. Sethian, "The fast construction of extension velocities in level-sets methods," *J. Comput. Phys.* **148**, 2 (1999).

[4] F. Alouges, "An energy decreasing algorithm for harmonic maps," in *Nematics*, J. M. Coron et al., eds., Nato ASI Series, Kluwer Academic, Dordrecht, The Netherlands, 1991, pp. 1–13.

[5] L. Alvarez, F. Guichard, P. L. Lions, and J. M. Morel, "Axioms and fundamental equations of image processing," *Arch. Ration. Mech. and Anal.* **123**, 199–257 (1993).

[6] L. Alvarez, P. L. Lions, and J. M. Morel, "Image selective smoothing and edge detection by nonlinear diffusion," *SIAM J. Numer. Anal.* **29**, 845–866 (1992).

[7] L. Alvarez and L. Mazorra, "Signal and image restoration by using shock filters and anisotropic diffusion," *SIAM J. Numer. Anal.* **31**, 590–605 (1994).

[8] L. Alvarez and J. M. Morel, "Morphological approach to multiscale analysis: From principles to equations," in Ref. [324], pp. 229–254.

[9] L. Ambrosio and M. Soner, "Level set approach to mean curvature flow in arbitrary codimension," *J. Diff. Geom.* **43**, 693–737 (1996).

[10] F. Andreu, C. Ballester, V. Caselles, and J. M. Mazon, "Minimizing total variation," preprint, 1988.

[11] B. Andrews, "Contraction of convex hypersurfaces by their affine normal," submitted for publication, *J. Diff. Geom.* **43**, 207–230 (1996).

[12] S. Angenent, "Parabolic equations for curves on surfaces, Part II. Intersections, blow-up, and generalized solutions," *Ann. Math.* **133**, 171–215 (1991).

[13] S. Angenent, "On the formation of singularities in the curve shortening flow," *J. Diff. Geom.* **33**, 601–633 (1991).

[14] S. Angenent, S. Haker, A. Tannenbaum, and R. Kikinis, "Laplace-Beltrami operator and brain flattening," ECE Department Report, University of Minnesota, Summer 1998.

[15] S. Angenent, G. Sapiro, and A. Tannenbaum, "On the affine invariant heat equation for nonconvex curves," *J. Am. Math. Soc.* **11**, 601–634 (1998).

[16] G. Anzellotti, "The Euler equation for functionals with linear growth," *Trans. Am. Math. Soc.* **290**, 483–501 (1985).
[17] G. Aubert and L. Vese, "A variational method for image recovery," *SIAM J. Appl. Math.* **34**, 1948–1979 (1997).
[18] J. Babaud, A. P. Witkin, B. Baudin, and R. O Duda, "Uniqueness of the Gaussian kernel for scale-space filtering," *IEEE Trans. Pattern Anal. Mach. Intell.* **8**, 26–33 (1986).
[19] C. Ballester, M. Bertalmio, V. Caselles, and G. Sapiro, "Interpolation of planar vector fields and its applications," ECE Department Technical Report, University of Minnesota, April 2000.
[20] G. Barles, "Remarks on flame propagation," in *Reports de Recherche*, INRIA, Sophia Antipolis, 1985, Vol 464.
[21] G. Barles and C. Georgelin, "A simple proof of convergence for an approximation scheme for computing motions by mean curvature," *SIAM J. Num. Anal.* **32**, 484–500 (1995).
[22] R. H. Bartles, J. C. Beatty, and B. A. Barsky, *An Introduction to Splines for Use in Computer Graphics and Geometric Modeling*, Morgan Kaufmann, Los Altos, CA, 1987.
[23] M. Bertalmio, "Morphing Active Contours," M.Sc. Thesis, I. I. E., Universidad de la Republica, Uruguay, June 1998.
[24] M. Bertalmio, L. T. Cheng, S. Osher, and G. Sapiro, "PDE's on implicit (level-sets) manifolds: The framework and applications in image processing and computer graphics," University of Minnesota IMA Report, June 2000.
[25] M. Bertalmio, G. Sapiro, V. Caselles, and C. Ballester, "Image inpainting," *IMA Rep.* **1655**, December 1999. Also in *SIGGRAPH 2000*.
[26] M. Bertalmio, G. Sapiro, and G. Randall, "Morphing active contours: a geometric, topology-free, technique for image segmentation and tracking," in *Proceedings of the IEEE International Conference on Image Processing*, IEEE, New York, October 1998.
[27] M. Bertalmio, G. Sapiro, and G. Randall, "Region tracking on level-sets methods," *IEEE Trans. Med. Imaging* **18**, 448–451 (1999).
[28] A. Bertozzi "The mathematics of moving contact lines in thin liquid films," *Not. Am. Math. Soc.* **45**, 689–697 (1998).
[29] F. Bethuel and X. Zheng, "Density of smooth functions between two manifolds in Sobolev spaces," *J. Funct. Anal.* **80**, 60–75 (1988).
[30] J. Besag, "On the statistical analysis of dirty pictures," *J. R. Stat. Soc.* **48**, 259–302 (1986).
[31] P. J. Besl, J. B. Birch, and L. T. Watson, "Robust window operators," in *Proceedings of the International Conference on Computer Vision, ICCV-88*, IEEE Publications, Los Alamitos, CA, 1988, pp. 591–600.
[32] M. J. Black and P. Anandan, "Robust dynamic motion estimation over time," in *Proceedings of the Conference on Computer Vision and Pattern Recognition, CVPR-91*, IEEE Publications, Los Alamitos, CA, 1991, pp. 296–302.
[33] M. Black and P. Anandan, "A framework for the robust estimation of optical flow," in *Proceedings of the Fourth International Conference on Computer Vision*, IEEE Publications, Los Alamitos, CA, 1993, pp. 231–236.
[34] M. Black and A. Rangarajan, "On the unification of line processes, outlier

rejection, and robust statistics with applications in early vision," *Int. J. Comput. Vis.* **19**, 57–92 (July 1996).

[35] M. Black and G. Sapiro, "Edges as outliers: Anisotropic smoothing using local image statistics," in *Proceedings of the Scale-Space Conference*, Springer-Verlag, Berlin 1999.

[36] M. Black, G. Sapiro, D. Marimont, and D. Heeger, "Robust anisotropic diffusion," *IEEE Trans. Image Process.* **7**, 421–432 (1998).

[37] A. Blake and M. Isard, *Active Contours*, Springer-Verlag, New York, 1998.

[38] A. Blake and A. Zisserman, *Visual Reconstruction*, MIT Press, Cambridge, MA, 1987.

[39] W. Blaschke, *Vorlesungen uber Differentialgeometrie II: Affine Differentialgeometrie*, Springer, Berlin, 1923.

[40] P. Blomgren and T. Chan, "Color TV: total variation methods for restoration of vector valued images," *IEEE Trans. Image Process.* **7**, 304–309 (1998).

[41] H. Blum, "Biological shape and visual science," *J. Theor. Biol.* **38**, 205–287 (1973).

[42] G. Borgefors, "Distance transformations in digital images," *Comput. Graph. Image Process.* **34**, 344–371 (1986).

[43] C. de Boor, *A Practical Guide to Splines*, Vol. 27 of Springer-Verlag Applied Mathematical Sciences Series, Springer-Verlag, New York, 1978.

[44] C. de Boor, *Spline Toolbox for use with MATLAB*™, The MathWorks, Inc., Natick, MA,1990.

[45] M. Born and W. Wolf, *Principles of Optics*, 6th ed., Pergamon, New York, 1986.

[46] C. Braverman, *Photoshop Retouching Handbook*, IDG Books Worldwide, Foster City, CA, 1998.

[47] H. Brézis, *Operateurs Maximaux Monotones*, Vol. 50 of Notes de Matematice Series, North-Holland Mathematical Studies, Amsterdam, 1973.

[48] H. Brezis, J. M. Coron, and E. H. Lieb, "Harmonic maps with defects," *Commun. Math. Phys.* **107**, 649–705 (1986).

[49] R. W. Brockett and P. Maragos, "Evolution equations for continuous-scale morphology," *IEEE Int. Conf. Acoustics and Signal Process.* IEEE, New York, 1992.

[50] A. M. Bruckstein, "On shape from shading," *Comp. Vision Graph. Image Process.* **44**, 139–154 (1988).

[51] A. M. Bruckstein and A. N. Netravali, "On differential invariants of planar curves and recognizing partially occluded planar shapes," *AT&T Tech. Rep.*, July 1990; also in *Proceedings of the Visual Form Workshop*, Plenum, New York, 1991.

[52] A. M. Bruckstein, G. Sapiro, and D. Shaked, "Evolutions of planar polygons," *Int. J. Pattern Recog. Artif. Intell.* **9**, 991–1014 (1995).

[53] S. Buchin, *Affine Differential Geometry*, Gordon & Breach, New York, 1983.

[54] P. Buchard, L. T. Cheng, B. Merriman, and S. Osher, "Motion of curves in three spatial dimensions using a level set approach," preprint, 1999.

[55] B. Cabral and C. Leedom. "Imaging vector fields using line integral convolution," in *ACM Computer Graphics (SIGGRAPH '93 Proceedings)* ACM Inc., New York 1993, pp. 263–272.

[56] M. P. Do Carmo, *Differential Geometry of Curves and Surfaces*, Prentice-Hall, Englewood Cliffs, NJ, 1976.

[57] E. Calabi, P. J. Olver, and A. Tannenbaum, "Affine geometry, curve flows, and invariant numerical approximations," *Adv. Math.* **124**, 154–196 (1996).

[58] E. Calabi, P. J. Olver, C. Shakiban, A. Tannenbaum, and S. Haker, "Differential and numerical invariant signature curves applied to object recognition," *Int. J. Comput. Vis.* **26**, 107–135 (1988).

[59] J. Canny, "A computational approach to edge detection," *IEEE Trans. Pattern Anal. Mach. Intell.* **8**, 679–698 (1986).

[60] I. Carlbom, D. Terzopoulos, and K. Harris, "Computer-assisted registration, segmentation, and 3D reconstruction from images of neuronal tissue sections," *IEEE Trans. Med. Imag.* **13**, 351–362 (1994).

[61] E. Cartan, *La Methode du Repere Mobile, la Theorie des Groupes Continus et les Espaces Generalises*, Hermann et cie, Paris, 1935.

[62] E. Cartan, *La Théorie des Groupes Finis et Continus et la Géometrie Différentielle traitée par le Méthode du Repère Mobile*, Gauthier-Villars, Paris, 1937.

[63] V. Caselles, F. Catte, T. Coll, and F. Dibos, "A geometric model for active contours," *Num. Math.* **66**, 1–31 (1993).

[64] V. Caselles, R. Kimmel, and G. Sapiro, "Geodesic active contours," in *Proceedings of the International Conference on Computer Vision '95*, 1995.

[65] V. Caselles, R. Kimmel, and G. Sapiro, "Geodesic active contours," *Int. J. Comput. Vis.* **22**, 61–79 (1997).

[66] V. Caselles, R. Kimmel, G. Sapiro, and C. Sbert, "Minimal surfaces: A geometric three-dimensional segmentation approach," *Num. Math.* **77**, 423–451 (1997).

[67] V. Caselles, R. Kimmel, G. Sapiro, and C. Sbert, "Minimal surfaces based object segmentation," *IEEE Trans. Pattern Anal. Mach. Intell.* **19**, 394–398 (1997).

[68] V. Caselles, J.-L. Lisani, J.-M. Morel, and G. Sapiro, "Shape-preserving local contrast enhancement," *Proceedings of the IEEE International Conference on Image Processing*, IEEE, New York, 1997.

[69] V. Caselles, J.-L. Lisani, J.-M. Morel, and G. Sapiro, "Shape-preserving local contrast enhancement," *IEEE Trans. Image Process.* **8**, 220–230 (1999).

[70] V. Caselles, J.-L. Lisani, J.-M. Morel and G. Sapiro, "The information of an image invariant by local contrast changes," preprint, 1998.

[71] V. Caselles, J.-M. Morel, G. Sapiro, and A. Tannenbaum, "Introduction to the special issue on PDE's and geometry driven diffusion in image processing and analysis," *IEEE Trans. Image Process.* **7**, 269–273 (1998).

[72] V. Caselles, J.-M. Morel, and C. Sbert, "An axiomatic approach to image interpolation," *IEEE Trans. Image Process.* **7**, 376–386 (1998).

[73] V. Caselles and G. Sapiro, "Vector median filters, morphology, and PDE's: theoretical connections," ECE Technical Report, University of Minnesota, September 1998. Also in V. Caselles, G. Sapiro, and D. H. Chung, "Vector median filters, inf-sup operations, and coupled PDE's: theoretical connections," *J. Math. Imag. Vis.* **12**, 109–120 (April 2000).

[74] V. Caselles and C. Sbert, "What is the best causal scale-space for 3D images?," Technical Report, Department of Mathematics and Computer Sciences, University of Illes Balears, 07071 Palma de Mallorca, Spain, March 1994.

[75] F. Catte, P.-L. Lions, J.-M. Morel, and T. Coll, "Image selective smoothing and edge detection by nonlinear diffusion," *SIAM J. Num. Anal.* **29**, 182–193 (1992).

[76] A. Chambolle, "Partial differential equations and image processing," *Proceedings of the IEEE International Conference on Image Processing*, IEEE, New York, 1994.

[77] A. Chambolle, R. A. DeVore, N. Lee, and B. J. Lucier, "Nonlinear wavelet image processing: variational problems, compression, and noise removal through wavelet shrinkage," *IEEE Trans. Image Process.* **7**, 319–335 (1998).

[78] A. Chambolle and P.-L. Lions, "Image recovery via total variation minimization and related problems," preprint, CEREMADE, University of Paris IX-Dauphine, 1995.

[79] T. F. Chan, G. H. Golub, and P. Mulet, "A nonlinear primal-dual method for total variation-based image restoration," Technical Report, UCLA, 1995.

[80] T. Chan and J. Shen, "Variational restoration of non-flat image features: models and algorithms," CAM-Technical Report, 99-20, UCLA, June 1999.

[81] T. F. Chan and C.-K. Wong, "Total variation blind deconvolution," *IEEE Trans. Image Process.* **7**, 370–375 (1998).

[82] T. F. Chan and L. A. Vese, "Active contours without edges," CAM Report 98-53, UCLA, December 1998.

[83] T. F. Chan, B. Y. Sandberg, and L. A. Vese, "Active contours without edges for vector-valued images," CAM Report 99-35, UCLA, October 1999.

[84] K. C. Chang, W. Y. Ding, and R. Ye, "Finite-time blow-up of the heat flow of harmonic maps from surfaces," *J. Diff. Geom.* **36**, 507–515 (1992).

[85] P. Charbonnier, L. Blanc-Feraud, G. Aubert, and M. Barlaud, "Deterministic edge-preserving regularization in computed imaging," *IEEE Trans. Image Process.* **6**, 298–311 (1997).

[86] Y. Chen, "The weak solutions of the evolution problems of harmonic maps," *Math. Z.* **201**, 69–74 (1989).

[87] Y. Chen, M. C. Hong and N. Hungerbuhler, "Heat flow of p-harmonic maps with values into spheres," *Math. Z.* **205**, 25–35 (1994).

[88] Y. Chen, J. Li, and F. H. Lin, "Partial regularity for weak heat flows into spheres," *Commun. Pure Appl. Math.* **XLVIII**, 429–448 (1995).

[89] S. Chen, B. Merriman, S. Osher, and P. Smereka, "A simple level-set method for solving Stefan problems," *J. Comput. Phys.* **135**, 8–29 (1997).

[90] Y. G. Chen, Y. Giga, and S. Goto, "Uniqueness and existence of viscosity solutions of generalized mean curvature flow equations," *J. Diff. Geom.* **33**, 749–786 (1991).

[91] D. S. Chen and B. G. Schunck, "Robust statistical methods for building classification procedures," in *Proceedings of the International Workshop on Robust Computer Vision*, IEEE Publications, Los Alamitos, CA, 1990, pp. 72–85.

[92] P. B. Chou and C. M. Brown, "The theory and practice of Bayesian image labeling," *Int. J. Comput. Vis.* **4**, 185–210 (1990).

[93] D. Chopp, "Computing minimal surfaces via level set curvature flows," Lawrence Berkeley Laboratory Technical Report, University of Berkeley, CA, 1991.

[94] B. Chow, "Deforming convex hypersurfaces by the nth root of the Gaussian curvature," *J. Diff. Geom.* **22**, 117–138 (1985).

[95] D. H. Chung and G. Sapiro, "On the geometry of multi-valued images," ECE Department Technical Report, University of Minnesota, December 1999.

[96] L. D. Cohen, "On active contour models and balloons," *Comput. Vis. Graph. Image Process.* **53**, 211–218 (1991).

[97] L. D. Cohen and I. Cohen, "Finite element methods for active contour models and balloons for 2D and 3D images," *IEEE Trans. Pattern Anal. Mach. Intell.* **15**, 1131–1147 (1993).

[98] I. Cohen, L. D. Cohen, and N. Ayache, "Using deformable surfaces to segment 3D images and infer differential structure," *Comput. Vis. Graph. Image Process.* **56**, 242–263 (1992).

[99] R. Cohen, R. M. Hardt, D. Kinderlehrer, S. Y. Lin, and M. Luskin, "Minimum energy configurations for liquid crystals: computational results," in *Theory and Applications of Liquid Crystals*, J. L. Ericksen and D. Kinderlehrer, eds., IMA Volumes in Mathematics and its Applications, Springer-Verlag, New York, 1987, pp. 99–121.

[100] L. D. Cohen, and R. Kimmel, "Global minimum for active contours models: a minimal path approach," *Int. J. Comput. Vis.* (to be published). (A short version appeared in *Proceedings of Conference on Computer Vision and Pattern Recognition*, IEEE Publications, Los Alamitos, CA, 1996).

[101] J. M. Coron, "Nonuniqueness for the heat flow of harmonic maps," *Ann. Inst. H. Poincaré, Anal. Non Linéaire* **7**, 335–344 (1990).

[102] J. M. Coron and R. Gulliver, "Minimizing p-harmonic maps into spheres," *J. Reine Angew. Math.* **401**, 82–100 (1989).

[103] R. Courant, K. O. Friedrichs, and H. Lewy, "On the partial differential equations of mathematical physics," *IBM J.* **11**, 215–234 (1967).

[104] R. Courant and D. Hilbert, *Methods of Mathematical Physics*, IBM Research, Yorktown Heights, NY Interscience, New York, 1962.

[105] M. G. Crandall, H. Ishii, and P. L. Lions, "User's guide to viscosity solutions of second order partial linear differential equations," *Bull. Am. Math. Soc.* **27**, 1–67 (1992).

[106] R. Cromartie and S. M. Pizer, "Edge-affected context for adaptive contrast enhancement," in *Proceedings of the conference on Information Processing in Medical Imaging,* Vol. 511 of Springer-Verlag Lecture Notes in Computer Science Series, Springer-Verlag, New York, 1991.

[107] A. Cumani, "Edge detection in multispectral images," *Comput. Vis. Graph. Image Process.* **53**, 40–51 (1991).

[108] P. E. Danielsson, "Euclidean distance mapping," *Comput. Graph. Image Process.* **14**, 227–248 (1980).

[109] R. Deriche, C. Bouvin, and O. Faugeras, "A level-set approach for stereo," INRIA Technical Report, Sophia-Antipolis, 1996.

[110] S. Di Zenzo, "A note on the gradient of a multi-image," *Comput. Vis. Graph. Image Process.* **33**, 116–125 (1986).

[111] J. Dieudonné and J. Carrell, *Invariant Theory: Old and New*, Academic, London, 1970.

[112] D. Donoho and I. Johnstone, "Ideal spatial adaptation by wavelet shrinkage," *Biometrika* **81**, 425–455 (1994).

[113] B. A. Dubrovin, A. T. Fomenko, and S. P. Novikov, *Modern Geometry – Methods and Applications I*, Springer-Verlag, New York, 1984.

[114] M. Eck, T. DeRose, T. Duchamp, H. Hoppe, M. Lounsbery, and W. Stuetzle. "Multiresolution analysis of arbitrary meshes," in *Computer Graphics (SIGGRAPH '95 Proceedings)*, ACM Inc., New York. 1995, pp. 173–182.

[115] M. Eck and H. Hoppe, "Automatic reconstruction of B-spline surfaces of arbitrary topological type," in *Computer Graphics (SIGGRAPH '96 Proceedings)*, ACM Inc., New York. 1996, pp. 325–334.
[116] K. Ecker and G. Huisken, "Mean curvature evolution of entire graphs," *Ann. Math.* **130**, 453–471 (1989).
[117] K. Ecker and G. Huisken, "Interior estimates for hypersurfaces moving by mean curvature," *Invent. Math.* **105**, 547–569 (1991).
[118] J. Eells and L. Lemarie, "A report on harmonic maps," *Bull. London Math. Soc.* **10**, 1–68 (1978).
[119] J. Eells and L. Lemarie, "Another report on harmonic maps," *Bull. London Math. Soc.* **20**, 385–524 (1988).
[120] J. Eells and J. H. Sampson, "Harmonic mappings of Riemannian manifolds," *Am. J. Math.* **86**, 109–160 (1964).
[121] A. Efros and T. Leung, "Texture synthesis by non-parametric sampling," in *Proceedings of the IEEE International Conference on Computer Vision*, IEEE, New York, 1999, pp. 1033–1038.
[122] G. Emile-Male, *The Restorer's Handbook of Easel Painting*, Van Nostrand Reinhold, New York, 1976.
[123] C. L. Epstein and M. Gage, "The curve shortening flow," in *Wave Motion: Theory, Modeling, and Computation*, A. Chorin and A. Majda, eds., Springer-Verlag, New York, 1987.
[124] L. C. Evans, "Convergence of an algorithm for mean curvature motion," *Indiana Univ. Math. J.* **42**, 553–557 (1993).
[125] L. C. Evans, *Partial Differential Equations*, American Mathematical Society, Providence, RI, 1998.
[126] L. C. Evans and J. Spruck, "Motion of level sets by mean curvature, I," *J. Diff. Geom.* **33**, 635–681 (1991).
[127] L. C. Evans and J. Spruck, "Motion of level sets by mean curvature, II," *Trans. Am. Math. Soc.* **330**, 321–332 (1992).
[128] M. Falcone, "The minimum time problem and its applications to front propagation," in *Motion by Mean Curvature and Related Topics*, Walter de Gruyter, New York, 1994.
[129] M. Falcone, T. Giorgi, and P. Loretti, "Level-sets of viscosity solution: some applications to fronts and rendez-vous problems," *SIAM J. Appl. Math.* **54**, 1335–1354 (1994).
[130] O. D. Faugeras, *Three-Dimensional Computer Vision: A Geometric Viewpoint*, MIT Press, Cambridge, MA, 1993.
[131] O. D. Faugeras, "On the evolution of simple curves of the real projective plane," *CR Acad. Sci. Paris* **317**, 565–570 (1993).
[132] O. D. Faugeras, "Cartan's moving frame method and its application to the geometry and evolution of curves in the Euclidean, affine and projective planes," in *Applications of Invariance in Computer Vision*, J. L. Mundy, A. Zisserman, and D. Forsyth, eds., Vol. 825 of Springer-Verlag Lecture Notes in Computer Science Series, Springer-Verlag, New York, 1994, pp. 11–46.
[133] O. D. Faugeras and M. Berthod, "Improving consistency and reducing ambiguity in stochastic labeling: an optimization approach," IEEE *Trans. Pattern Anal. Mach. Intell.* **3**, 412–423 (1981).

[134] O. D. Faugeras and R. Keriven, "Some recent results on the projective evolution of 2-D curves," in *Proceedings of the International Conference on Image Processing*, IEEE Publications, Los Alamitos, CA, 1995, Vol. III, pp. 13–16.

[135] O. D. Faugeras and R. Keriven, "On projective plane curve evolution," in *Proceedings of the 12th International Conference on Analysis and Optimization of Systems, Images, Wavelets and PDE's*, Springer-Verlag, Berlin, 1996.

[136] O. D. Faugeras and R. Keriven, "Scale-spaces and affine curvature," in *Proceedings of the Europe-China Workshop on Geometrical Modeling and Invariants for Computer Vision*, R. Mohr and C. Wu, eds., 1995, pp. 17–24.

[137] O. D. Faugeras and R. Keriven, "Variational principles, surface evolution, PDE's, level-set methods, and the stereo problem," *IEEE Trans. Image Process.* **7**, 336–344 (1998).

[138] J. Favard, *Cours de Géométrie Différentielle Locale*, Gauthier-Villars, Paris, 1957.

[139] M. Feldman, "Partial regularity for harmonic maps of evolutions into spheres," *Commun. Partial Diff. Equat.* **19**, 761–790 (1994).

[140] M. Fels and P. J. Olver, "Moving coframes. I. A practical algorithm," *Acta Appl. Math.* **51**, 161–213 (1998).

[141] M. Fels and P. J. Olver, "Moving coframes II. Regularization and theoretical foundations," *Acta Appl. Math.* **55**, 127–208 (1999).

[142] L. Florack, B. Romeny, J. J. Koenderink, and M. Viergever, "Scale and the differential structure of images," *Image Vis. Comput.* **10**, 376–388 (1992).

[143] W. T. Freeman and E. H. Adelson, "The design and use of steerable filters," *IEEE Trans. Pattern Anal. Mach. Intell.* **9**, 891–906 (1991).

[144] A. Freire, "Uniqueness for the harmonic map flow in two dimensions," *Calc. Var.* **3**, 95–105 (1995).

[145] P. Fua and Y. G. Leclerc, "Model driven edge detection," *Mach. Vis. Appl.* **3**, 45–56 (1990).

[146] D. Gabor, "Information theory in electron microscopy," *Lab. Invest.* **14**, 801–807 (1965).

[147] M. Gage, "An isoperimetric inequality with applications to curve shortening," *Duke Math. J.* **50**, 1225–1229 (1983).

[148] M. Gage, "Curve shortening makes convex curves circular," *Invent. Math.* **76**, 357–364 (1984).

[149] M. Gage, "On an area-preserving evolution equation for plane curves," *Contemp. Math.* **51**, 51–62 (1986).

[150] M. Gage and R. S. Hamilton, "The heat equation shrinking convex plane curves," *J. Diff. Geom.* **23**, 69–96 (1986).

[151] D. Geiger, A. Gupta, L. A. Costa, and J. Vlontzos, "Dynamic programming for detecting, tracking, and matching deformable contours," *IEEE Trans. Pattern Anal. Mach. Intell.* **17**, 294–302 (1995).

[152] D. Geiger and A. Yuille, "A common framework for image segmentation," *Int. J. Comput. Vis.* **6**, 227–243 (1991).

[153] S. Geman and D. Geman. "Stochastic relaxation, Gibbs distributions, and the Bayesian restoration of images," *IEEE Trans. Pattern Anal. Mach. Intell.* **6**, 721–742 (1984).

[154] D. Geman and G. Reynolds, "Constrained restoration and the recovery of discontinuities," *IEEE Trans. Pattern Anal. Mach. Intell.* **14**, 367–383 (1992).

[155] D. Geman and C. Yang, "Nonlinear image recovery with half-quadratic regularization," *IEEE Trans. Image Process.* **4**, 932–946 (1995).

[156] C. Gerhardt, "Flow of nonconvex hypersurfaces into spheres," *J. Diff. Geom.* **32**, 299–314 (1990).

[157] G. Gerig, O. Kubler, R. Kikinis, and F. A. Jolesz, "Nonlinear anisotropic filtering of MRI data," *IEEE Trans. Med. Imag.* **11**, 221–232 (1992).

[158] M. Giaquinta, G. Modica, and J. Soucek, "Variational problems for maps of bounded variation with values in S^1," *Cal. Var.* **1**, 87–121 (1993).

[159] J. Gomez and O. Faugeras, "Reconciling distance functions and level-sets," in *Proceedings of the Scale-Space Workshop*, Springer-Verlag, Berlin, 1999.

[160] R. C. Gonzalez and P. Wintz, *Digital Image Processing*, Addison-Wesley, Reading, MA, 1987.

[161] G. H. Granlund and H. Knutsson, *Signal Processing for Computer Vision*, Kluwer, Boston, 1995.

[162] M. Grayson, "The heat equation shrinks embedded plane curves to round points," *J. Diff. Geom.* **26**, 285–314 (1987).

[163] M. Grayson, "Shortening embedded curves," *Ann. Math.* **129**, 285–314 (1989).

[164] M. Grayson, "A short note on the evolution of a surface by its mean curvature," *Duke Math. J.* **58**, 555–558 (1989).

[165] W. E. L. Grimson, *From Images to Surfaces*, MIT Press, Cambridge, MA, 1981.

[166] H. W. Guggenheimer, *Differential Geometry*, McGraw-Hill, New York, 1963.

[167] F. Guichard, "Multiscale analysis of movies: theory and algorithms," Ph.D. Dissertation, CEREMADE-Paris, 1993.

[168] F. Guichard and J. M. Morel, "Introduction to partial differential equations in image processing," Tutorial Notes, *IEEE International Conference on Image Processing*, IEEE, New York, 1995.

[169] S. Haker, S. Angenent, A. Tannenbaum, R. Kikinis, G. Sapiro, and M. Halle, "Conformal surface parametrization for texture mapping," IMA Preprint Series 1611, University of Minnesota, April 1999.

[170] S. Haker, G. Sapiro, and A. Tannenbaum, "Knowledge based segmentation of SAR data," in *Proceedings of the IEEE International Conference on Image Processing*, IEEE, New York, 1998.

[171] S. Haker, G. Sapiro, and A. Tannenbaum, "Knowledge-based segmentation of SAR data with learned priors," *IEEE Trans. Image Process.* **9**, 299–301 (2000).

[172] F. R. Hampel, E. M. Ronchetti, P. J. Rousseeuw, and W. A. Stahel, *Robust Statistics: The Approach Based on Influence Functions*, Wiley, New York, 1986.

[173] R. M. Haralick, S. R. Stenberg, and X. Zhuang, "Image analysis using mathematical morphology," *IEEE Trans. Pattern Anal. Mach. Intell.* **9**, 523–550 (1987).

[174] G. Harikumar and Y. Bresler, "Feature extraction techniques for exploratory visualization of vector-valued imagery," *IEEE Trans. Image Process.* **5**, 1324–1334 (1996).

[175] R. M. Hardt, "Singularities of harmonic maps," *Bull. Am. Math. Soc.* **34**, 15–34 (1997).

[176] R. M. Hardt and F. H. Lin, "Mappings minimizing the L^p norm of the gradient," *Commun. Pure Appl. Math.* **XL**, 555–588 (1987).

[177] J. G. Harris, C. Koch, E. Staats, and J. Luo, "Analog hardware for detecting discontinuities in early vision," *Int. J. Comput. Vis.* **4**, 211–223 (1990).

[178] D. Heeger and J. Bergen, "Pyramid based texture analysis/synthesis," *Computer Graphics (SIGGRAPH '95 Proceedings)*, ACM Inc., New York, 1995.

[179] H. J. A. M. Heijmans and J. B. T. M. Roerdink, eds., *Mathematical Morphology and its Applications to Image and Signal Processing*, Kluwer, Dordrecht, The Netherlands, 1998.

[180] J. Helmsen, E. G. Puckett, P. Collela, and M. Dorr, "Two new methods for simulating photolithography development in 3D," in *Optical Microlithography IX*, G. E. Fuller, ed., *Proc. SPIE* **2726**, 253–261 (1996).

[181] D. Henry, *Geometric Theory of Semilinear Parabolic Equations*, Springer-Verlag, New York, 1981.

[182] G. Hermosillo, O. Faugeras, and J. Gomes, "Cortex unfolding using level set methods," INRIA Sophia Antipolis Technical Report, April 1999.

[183] A. Hirani and T. Totsuka, "Combining frequency and spatial domain information for fast interactive image noise removal," in *Computer Graphics (SIGGRAPH '96 Proceedings)*, ACM Inc., New York, 1996, pp. 269–276.

[184] B. K. P. Horn, *Robot Vision*, MIT Press, Cambridge, MA, 1986.

[185] B. K. P. Horn and E. J. Weldon, Jr., "Filtering closed curves," *IEEE Trans. Pattern Anal. Mach. Intell.* **8**, 665–668 (1986).

[186] P. J. Huber, *Robust Statistics*, Wiley, New York, 1981.

[187] T. Hughes, *The Finite Element Method*, Prentice-Hall, Englewood Cliffs, NJ, 1987.

[188] G. Huisken, "Flow by mean curvature of convex surfaces into spheres," *J. Diff. Geom.* **20**, 237–266 (1984).

[189] G. Huisken, "Non-parametric mean curvature evolution with boundary conditions," *J. Diff. Equat.* **77**, 369–378 (1989).

[190] G. Huisken, "Local and global behavior of hypersurfaces moving by mean curvature," in *Proceedings of Symposia in Pure Mathematics* **54**, 175–191 (1993).

[191] R. A. Hummel, "Representations based on zero-crossings in scale-space," in *Proceedings of the IEEE Conference on Computer Vision and Pattern Recognition*, IEEE, New York, 1986, pp. 204–209.

[192] R. A. Hummel and S. W. Zucker, "On the foundations of relaxation labeling processes," *IEEE Trans. Pattern Anal. Mach. Intell.* **5**, 267–286 (1983).

[193] H. Ishii, "A generalization of Bence, Merriman, and Osher algorithm for motion by mean curvature," in *Curvature Flows and Related Topics*, A. Damlamian, J. Spruck, and A. Visintin, eds., Gakkôtosho, Tokyo, 1995, pp. 111–127.

[194] H. Ishii, G. E. Pires, and P. E. Souganidis, "Threshold dynamics type schemes for propagating fronts," preprint, 1997.

[195] S. Izumiya and T. Sano, "Generic affine differential geometry of plane curves," *Proc. Edinburgh Math. Soc.* **41**, 315–324 (1998).

[196] A. K. Jain, "Partial differential equations and finite-difference methods in image processing, part 1: image representation," *J. Optimiz. Theory Appl.* **23**, 65–91 (1977).

[197] D. G. Karakos and P. E. Trahanias, "Generalized multichannel image-filtering structures," *IEEE Trans. Image Process.* **6**, 1038–1045 (1997).

[198] M. Kass, A. Witkin, and D. Terzopoulos, "Snakes: active contour models," *Int. J. Comput. Vis.* **1**, 321–331 (1988).

[199] C. Kenney and J. Langan, "A new image processing primitive: reconstructing

images from modified flow fields," preprint, University of California Santa Barbara, 1999.

[200] S. Kichenassamy, "Edge localization via backward parabolic and hyperbolic PDE," preprint, University of Minnesota, 1996.

[201] S. Kichenassamy, A. Kumar, P. Olver, A. Tannenbaum, and A. Yezzi, "Gradient flows and geometric active contour models," in *Proceedings of the International Conference on Computer Vision '95*, IEEE Publications, Los Alamitos, CA, 1995, pp. 810–815.

[202] S. Kichenassamy, A. Kumar, P. Olver, A. Tannenbaum, and A. Yezzi, "Conformal curvature flows: from phase transitions to active vision," *Arch. Ration. Mech. Anal.* **134**, 275–301 (1996).

[203] B. B. Kimia, "Toward a computational theory of shape," Ph.D. dissertation, Department of Electrical Engineering, McGill University, Montreal, Canada, August 1990.

[204] B. B. Kimia, A. Tannenbaum, and S. W. Zucker, "Toward a computational theory of shape: an overview," in Vol. 427 of Springer-Verlag Lecture Notes in Computer Science Series, Springer-Verlag, New York, 1990, pp. 402–407.

[205] B. B. Kimia, A. Tannenbaum, and S. W. Zucker, "On the evolution of curves via a function of curvature, I: the classical case," *J. Math. Anal. Appl.* **163**, 438–458 (1992).

[206] B. B. Kimia, A. Tannenbaum, and S. W. Zucker, "Shapes, shocks, and deformations, I," *Int. J. Comput. Vis.* **15**, 189–224 (1995).

[207] R. Kimmel, "Numerical geometry of images: theory, algorithms, and applications," Technion CIS Report 9910, October 1999.

[208] R. Kimmel, A. Amir, A. M. Bruckstein, "Finding shortest paths on surfaces using level sets propagation," *IEEE Trans. Pattern Anal. Mach. Intell.* **17**, 635–640 (1995).

[209] R. Kimmel A. M. Bruckstein, "Shape offsets via level sets," *CAD* **25**(5), 154–162 (1993).

[210] R. Kimmel and A. M. Bruckstein, "Tracking level sets by level sets: a method for solving the shape from shading problem," *Comput. Vis. Image Underst.* **62**, 47–58 (1995).

[211] R. Kimmel, R. Malladi, and N. Sochen, "Image processings via the Beltrami operator," in *Proceedings of the third Asian Conference on Computer Vision*, 1998.

[212] R. Kimmel, personal communication.

[213] R. Kimmel and G. Sapiro, "Shortening three dimensional curves via two dimensional flows," *Int. J. Comput. Math. Appl.* **29**, 49–62 (1995).

[214] R. Kimmel and J. A. Sethian, "Fast marching method for computation of distance maps," Lawrence Berkeley National Laboratory, Report 38451, University of California Berkeley, February, 1996

[215] R. Kimmel and J. A. Sethian, "Computing geodesic paths on manifolds," *Proc. Natl. Acad. Sci.* **95**, 8431–8435 (1998).

[216] R. Kimmel, D. Shaked, N. Kiryati, and A. Bruckstein, "Skeletonization via distance maps and level-sets," *Comput. Vis. Image Underst.* **62** (1995).

[217] D. King, *The Commissar Vanishes*. Holt, New York, 1997.

[218] J. J. Koenderink, "The structure of images," *Biol. Cyber.* **50**, 363–370 (1984).

[219] J. J. Koenderink, *Solid Shape*, MIT Press, Cambridge, MA, 1990.

[220] J. J. Koenderink and A. J. van Doorn, "Dynamic shape," *Biol. Cyber.* **53**, 383–396 (1986).
[221] J. J. Koenderink and A. J. van Doorn, "Representations of local geometry in the visual system," *Biol. Cyber.* **55**, 367–375 (1987).
[222] A. C. Kokaram, R. D. Morris, W. J. Fitzgerald, P. J. W. Rayner, "Detection of missing data in image sequences," *IEEE Trans. Image Process.* **11**, 1496–1508 (1995).
[223] A. C. Kokaram, R. D. Morris, W. J. Fitzgerald, P. J. W. Rayner, "Interpolation of missing data in image sequences," *IEEE Trans. Image Process.* **11**, 1509–1519 (1995).
[224] E. Kreyszig, *Differential Geometry*, University of Toronto Press, Toronto, 1959.
[225] R. Kumar and A. R. Hanson, "Analysis of different robust methods for pose refinement," in *Proceedings of the International Workshop on Robust Computer Vision*, IEEE Publications, Los Alamitos, CA, 1990, pp. 167–182.
[226] E. P. Lane, *A Treatise on Projective Differential Geometry*, University of Chicago Press, Chicago, 1941.
[227] S. R. Lay, *Convex Sets and Their Applications*, Wiley, New York, 1982.
[228] P. D. Lax, "Weak solutions of nonlinear hyperbolic equations and their numerical computations," *Commun. Pure Appl. Math.* **7**, 159 (1954).
[229] P. D. Lax, *Hyperbolic Systems of Conservation Laws and the Mathematical Theory of Shock Waves,* Vol. 11 of SIAM Regional Conference Series in Applied Mathematics, Society for Industrial and Applied Mathematics, Philadelphia, 1972.
[230] Y. G. Leclerc, "Constructing simple stable descriptions for image partitioning," *Int. J. Comput. Vis.* **3**, 73–102 (1989).
[231] H-C. Lee and D. R. Cok, "Detecting boundaries in a vector field," *IEEE Trans. Signal Process.* **39**, 1181–1194 (1991).
[232] T. S. Lee, D. Mumford, and A. L. Yuille, "Texture segmentation by minimizing vector-valued energy functionals: the coupled-membrane model," in *ECCV'92*, Vol. 588 of Springer-Verlag Lecture Notes in Computer Science Series, 165–173, Springer-Verlag, New York, 1992.
[233] S. Lie, "Klassifikation und Integration von gewöhnlichen Differentialgleichungen zwischen x, y, die eine Gruppe von Transformationen gestatten I, II" *Math. Ann.* **32**, 213–281 (1888). See also *Gesammelte Abhandlungen*, Teubner, Leipzig, 1924, Vol. 5, pp. 240–310.
[234] L. M. Lorigo, O. Faugeras, W. E. L. Grimson, R. Keriven, and R. Kikinis, "Segmentation of bone in clinical knee MRI using texture-based geodesic active contours," in *Proceedings on Medical Image Computing and Computer-Assisted Intervention, MICCAI '98*, Springer, New York, 1998, pp. 1195–1204.
[235] A. Marquina and S. Osher, "Explicit algorithms for a new time dependent model based on level set motion for nonlinear debluring and noise removal," UCLA CAM Report 99-5, UCLA, January 1999.
[236] S. Masnou and J. M. Morel. "Level-lines based disocclusion," in *Proceedings of the fifth IEEE International Conference on Image Processing*, IEEE, New York, 1998.
[237] F. Memoli, G. Sapiro, and S. Osher, "Harmonic maps onto implicit manifolds," University of Minnesota IMA Report, August 2000.
[238] F. Leitner and P. Cinquin, "Dynamic segmentation: detecting complex topology

3D objects," in *Proceedings of Engineering in Medicine and Biology Society*, IEEE Publications, Los Alamitos, CA, 1991.

[239] M. Leyton, *Symmetry, Causality, Mind*, MIT Press, Cambridge, MA, 1992.

[240] R. J. LeVeque, *Numerical Methods for Conservation Laws*, Birkhäuser, Boston, 1992.

[241] S. Z. Li, H. Wang, and M. Petrou, "Relaxation labeling of Markov random fields," in *Proceedings of the International Conference on Pattern Recognition*, 1994.

[242] T. Lindeberg, *Scale-Space Theory in Computer Vision*, Kluwer, Dordrecht, The Netherlands, 1994, pp. 488–492.

[243] P. L. Lions, S. Osher, and L. Rudin, "Denoising and debluring algorithms with constrained nonlinear PDE's," *SIAM J. Num. Anal.* (to be published).

[244] D. G. Lowe, "Organization of smooth image curves at multiple scales," *Int. J. Comput. Vis.* **3**, 69–87 (1989).

[245] L. Lui, B. G. Schunck, and C. C. Meyer, "On robust edge detection," in *Proceedings of the International Workshop on Robust Computer Vision*, IEEE Publications, Los Alamitos, CA, 1990, pp. 261–286.

[246] E. Lutwak, "On the Blaschke–Santalo inequality," *Ann. NY Acad. Sci. Discrete Geom. Convex.* **440**, 106–112 (1985).

[247] E. Lutwak, "On some affine isoperimetric inequalities," *J. Diff. Geom.* **23**, 1–13 (1986).

[248] D. L. MacAdam, "Visual sensitivities to color differences in daylight," *J. Opt. Soc. Am.* **32**, 247 (1942).

[249] R. Malladi, R. Kimmel, D. Adalsteinsson, G. Sapiro, V. Caselles, and J. A. Sethian, "A geometric approach to segmentation and analysis of 3D medical images," in *Proceedings of the Mathematical Methods in Biomedical Image Analysis Workshop*, 1996.

[250] R. Malladi, J. A. Sethian, and B. C. Vemuri, "Evolutionary fronts for topology independent shape modeling and recovery," in *Proceedings of the Third European Conference on Computer Vision*, Springer-Verlag, New York, 1994, pp. 3–13.

[251] R. Malladi, J. A. Sethian, and B. C. Vemuri, "Shape modeling with front propagation: a level set approach," *IEEE Trans. Pattern Anal. Mach. Intell.* **17**, 158–175 (1995).

[252] R. Malladi, J. A. Sethian, and B. C. Vemuri, "A fast level set based algorithm for topology independent shape modeling," *J. Math. Imag. Vis.* special issue on Topology and Geometry, A. Rosenfeld and Y. Kong, eds. Vol. 6, 1996.

[253] S. G. Mallat, "Multiresolution approximations and wavelet orthonormal bases of $L^2(\mathbb{R})$," *Trans. Am. Math. Soc.* **315**, 69–87 (1989).

[254] P. Maragos, "A representation theory for morphological image and signal processing," *IEEE Trans. Pattern Anal. Mach. Intell.* **6**, 586–599 (1989).

[255] G. Matheron, *Random Sets and Integral Geometry*, Wiley, New York, 1975.

[256] T. McInerney and D. Terzopoulos, "Topologically adaptable snakes," in *Proceedings of the International Conference on Computer Vision*, IEEE Publications, Los Alamitos, CA, 1995.

[257] C. Mead, *Analog VLSI and Neural Systems*, Addison-Wesley, New York, 1989.

[258] P. Meer, D. Mintz, and A. Rosenfeld, "Robust recovery of piecewise polynomial image structure," in *Proceedings of the International Workshop on Robust Computer Vision*, IEEE Publications, Los Alamitos, CA, 1990, pp. 109–126.

[259] P. Meer, D. Mintz, A. Rosenfeld, and D. Y. Kim, "Robust regression methods for computer vision: a review," *Int. J. Comput. Vis.* **6**, 59–70 (1991).

[260] B. Merriman, J. Bence, and S. Osher, "Diffusion generated motion by mean curvature," in *Computational Crystal Growers Workshop*, J. E. Taylor, ed., American Mathematical Society, Providence, RI, 1992, pp. 73–83.

[261] B. Merriman, J. Bence, and S. Osher, "Motion of multiple junctions: a level-set approach," *J. Comput. Phys.* **112**, 334–363 (1994).

[262] B. Merriman, R. Caflisch, and S. Osher, "Level set methods, with an application to modeling the growth of thin films, UCLA CAM Report 98-10, ULCA, CA, February 1998.

[263] F. Mokhatarian and A. Mackworth, "Scale-based description of planar curves and two dimensional shapes," *IEEE Trans. Pattern Anal. Mach. Intell.* **8**, 34–43 (1986).

[264] F. Mokhatarian and A. Mackworth, "A theory of multiscale, curvature-based shape representation for planar curves," *IEEE Trans. Pattern Anal. Mach. Intell.* **14**, 789–805 (1992).

[265] D. Mumford and J. Shah, "Optimal approximations by piecewise smooth functions and variational problems," *Commun. Pure Appl. Math.* **42**, 577–685 (1989).

[266] D. W. Murray and B. F. Buxton, "Scene segmentation from visual motion using global optimization," *IEEE Trans. Pattern Anal. Mach. Intell.* **PAMI 9**, 220–228 (1987).

[267] R. Nevatia, "A color edge detector and its use in scene segmentation," *IEEE Trans. Syst. Man, Cybern.* **7**, 820–826 (1977).

[268] R. Nevatia and K. R. Babu, "Linear feature extraction and description," *Comput. Graph. Image Process.* **13**, 257–269 (1980).

[269] W. J. Niessen, B. M. ter Haar Romeny, L. M. J. Florack, and A. H. Salden, "Nonlinear diffusion of scalar images using well-posed differential operators," Technical Report, Utrecht University, The Netherlands, October 1993.

[270] M. Nitzberg, D. Mumford, and T. Shiota, *Filtering, Segmentation, and Depth*, Springer-Verlag, Berlin, 1993.

[271] M. Nitzberg and T. Shiota, "Nonlinear image filtering with edge and corner enhancement," *IEEE Trans. Pattern Anal. Mach. Intell.* **14**, 826–833 (1992).

[272] K. Nomizu and T. Sasaki, *Affine Differential Geometry*, Cambridge University Press, Cambridge, UK, 1993.

[273] N. Nordström, "Biased anisotropic diffusion: a unified regularization and diffusion approach to edge detection," *Image Vis. Comput.* **8**, 318–327 (1990).

[274] T. Ohta, D. Jasnow, and K. Kawasaki, "Universal scaling in the motion of random interfaces," *Phys. Rev. Lett.* **47**, 1223–1226 (1982).

[275] J. Oliensis, "Local reproducible smoothing without shrinkage," *IEEE Trans. Pattern Anal. Mach. Intell.* **15**, 307–312 (1993).

[276] J. Oliensis and P. Dupuis, "Direct method for reconstructing shape from shading," in *Geometric Methods in Computer Vision*, B. C. Vemuri, ed., Proc. SPIE **1570**, 116–128 (1991).

[277] V. I. Oliker, "Evolution of nonparametric surfaces with speed depending on curvature I. The Gauss curvature case," *Indiana Univ. Math. J.* **40**, 237–258 (1991).

[278] V. I. Oliker, "Self-similar solutions and asymptotic behavior of flows of nonparametric surfaces driven by Gauss or mean curvature," *Proc. Symposia Pure Math.* **54**, 389–402 (1993).

[279] V. I. Oliker and N. N. Uraltseva, "Evolution of nonparametric surfaces with speed depending on curvature II. The mean curvature case," *Commun. Pure Appl. Math.* **46**, 97–135 (1993).

[280] V. I. Oliker and N. N. Uraltseva, "Evolution of nonparametric surfaces with speed depending on curvature, III. Some remarks on mean curvature and anisotropic flows," IMA Volumes 53, in Mathematics, Springer-Verlag, New York, 1993.

[281] P. J. Olver, *Applications of Lie Groups to Differential Equations,* 2nd ed., Springer-Verlag, New York, 1993.

[282] P. J. Olver, "Differential invariants," *Acta Appl. Math.* **41**, 271–284 (1995).

[283] P. Olver, *Equivalence, Invariants, and Symmetry*, Cambridge University Press, Cambridge, UK, 1995.

[284] P. Olver, G. Sapiro, and A. Tannenbaum, "Classification and uniqueness of invariant geometric flows," *CR Acad. Sci. Paris*, 339–344 (August 1994).

[285] P. Olver, G. Sapiro, and A. Tannenbaum, "Differential invariant signatures and flows in computer vision: a symmetry group approach," in *Geometry Driven Diffusion in Computer Vision,* B. Romeny, ed., Kluwer, Dordrecht, The Netherlands, September 1994.

[286] P. Olver, G. Sapiro, and A. Tannenbaum, "Invariant geometric evolutions of surfaces and volumetric smoothing," *SIAM J. Appl. Math.* **57**, 176–194 (1997).

[287] P. Olver, G. Sapiro, and A. Tannenbaum, "Affine invariant edge maps and active contours," Geometry Center Technical Report 90, University of Minnesota, October 1995.

[288] P. Olver, G. Sapiro, and A. Tannenbaum, "Affine invariant detection: edges, active contours, and segments," in *Proceedings of the Conference on Computer Vision and Pattern Recognition*, IEEE Publications, Los Alamitos, CA, 1996.

[289] S. Osher, "A level-set formulation for the solution of the Dirichlet problem for Hamilton–Jacobi equations," *SIAM J. Num. Anal.* **24**, 1145 (1993).

[290] S. Osher, UCLA Technical Reports, located at http:://www.math.ucla.edu/applied/cam/index.html.

[291] S. Osher, personal communication, October 1999.

[292] S. Osher and J. Helmsen, "A generalized fast algorithm with applications to ion etching," in preparation.

[293] S. Osher and L. I. Rudin, "Feature-oriented image enhancement using shock filters," *SIAM J. Num. Anal.* **27**, 919–940 (1990).

[294] S. J. Osher and J. A. Sethian, "Fronts propagation with curvature dependent speed: algorithms based on Hamilton-Jacobi formulations," *J. Comput. Phys.* **79**, 12–49 (1988).

[295] S. Osher and L. Vese, personal communication, May 1999.

[296] R. Osserman, *Survey of Minimal Surfaces*, Dover, New York, 1986.

[297] L. V. Ovsiannikov, *Group Analysis of Differential Equations*, Academic, New York, 1982.

[298] N. Paragios and R. Deriche, "A PDE-based level-set approach for detection and tracking of moving objects," INRIA Technical Report 3173, Sophia-Antipolis, May 1997.

[299] N. Paragios and R. Deriche, "A PDE-based level-set approach for detection and tracking of moving objects," in *Proceedings of the International Conference on Computer Vision '98*, IEEE Publications, Los Alamitos, CA, 1998.

[300] N. Paragios and R. Deriche, "Geodesic active regions for tracking," in *Proceedings of the European Symposium on Computer Vision and Mobile Robotics CVMR'98*, 1998.

[301] N. Paragios and R. Deriche, "Geodesic active regions for motion estimation and tracking," INRIA Technical Report 3631, Sophia-Antipolis, March 1999.

[302] N. Paragios and R. Deriche, "Geodesic active contours for supervised texture segmentation," in *Proceedings of the Conference on Computer Vision and Pattern Recognition*, IEEE Publications, Los Alamitos, CA, 1999.

[303] N. Paragios and R. Deriche, "Geodesic active regions for supervised texture segmentation," in *Proceedings of the International Conference on Computer Vision*, IEEE Publications, Los Alamitos, CA, 1999.

[304] N. Paragios and R. Deriche, "Geodesic active contours and level sets for detection and tracking of moving objects," *IEEE Trans. Pattern Anal. Mach. Intell.* **22**, 266–280, (2000).

[305] A. Pardo and G. Sapiro, "Vector probability diffusion," IMA Technical Report, University of Minnesota, October 1999.

[306] E. J. Pauwels, P. Fiddelaers, and L. J. Van Gool, "Shape-extraction for curves using geometry-driven diffusion and functional optimization," in *Proceedings of the International Conference on Computer Vision*, IEEE Publications, Los Alamitos, CA, 1995.

[307] E. J. Pauwels, P. Fiddelaers, and L. J. Van Gool, "Coupled geometry-driven diffusion equations for low-level vision," in Ref. [324].

[308] D. Peng, B. Merriman, S. Osher, H. Zhao, and M. Kang, "A PDE-based fast local level-set method," *J. Comput. Phys.* **155**, 410–438 (1999).

[309] P. Perona, "Orientation diffusion," *IEEE Trans. Image Process.* **7**, 457–467 (1998).

[310] P. Perona and J. Malik, "Scale-space and edge detection using anisotropic diffusion," *IEEE Trans. Pattern Anal. Mach. Intell.* **12**, 629–639 (1990).

[311] P. Perona and J. Malik, "Detecting and localizing edges composed of steps, peaks, and roofs," CICS Technical Report, MIT, Cambridge, MA, October 1991.

[312] P. Perona, T. Shiota, and J. Malik, "Anisotropic diffusion," in Ref. [324].

[313] P. Perona and M. Tartagni, "Diffusion network for on-chip image contrast normalization," in *Proceedings of the IEEE International Conference on Image Processing*, IEEE, New York, 1994, Vol. 1, pp. 1–5.

[314] C. M. Petty, "Affine isoperimetric problems," *Ann. NY Acad. Sci. Discrete Geom. Convex.* **440**, 113–127 (1985).

[315] L. M. Pismen and J. Rubinstein, "Dynamics of defects," in *Nematics*, J. M. Coron et al., eds., Nato ASI Series, Kluwer Academic, Dordrecht, The Netherlands, 1991, pp. 303–326.

[316] A. Polden, "Compact surfaces of least total curvature," Technical Report, University of Tübingen, Germany, 1997.

[317] L. C. Polymenakos, D. P. Bertsekas, and J. N. Tsitsiklis, "Implementation of efficient algorithms for globally optimal trajectories," *IEEE Trans. Autom. Control* **43**, 278–283 (1988).

[318] W. K. Pratt, *Digital Image Processing*, Wiley, New York, 1991.

[319] C. B. Price, P. Wambacq, and A. Oosterlink, "Image enhancement and analysis with reaction-diffusion paradigm," IEE Proc. **137**, 136–145 (1990).

[320] *Proceedings of the International Workshop on Robust Computer Vision*, IEEE Publications, Los Alamitos, CA, 1990.

[321] M. H. Protter and H. Weinberger, *Maximum Principles in Differential Equations*, Springer-Verlag, New York, 1984.

[322] M. Proesmans, E. Pauwels, and J. van Gool, "Coupled geometry-driven diffusion equations for low-level vision," in Ref. [324].

[323] J. Qing, "On singularities of the heat flow for harmonic maps from surfaces into spheres," *Commun. Anal. Geom.* **3**, 297–315 (1995).

[324] B. Romeny, ed., *Geometry Driven Diffusion in Computer Vision*, Kluwer, Dordrecht, The Netherlands, 1994.

[325] A. Rosenfeld, R. Hummel, and S. Zucker, "Scene labeling by relaxation operations," *IEEE Trans. Syst. Man Cybern.* **6**, 420–433 (1976).

[326] P. J. Rousseeuw and A. M. Leroy, *Robust Regression and Outlier Detection*, Wiley, New York, 1987.

[327] E. Rouy and A. Tourin, "A viscosity solutions approach to shape-from-shading," *SIAM. J. Num. Anal.* **29**, 867–884 (1992).

[328] J. Rubinstein, P. Sternberg, and J. B. Keller, "Fast reaction, slow diffusion, and curve shortening," *SIAM J. Appl. Math.* **49**, 116–133 (1989).

[329] L. I. Rudin and S. Osher, "Total variation based image restoration with free local constraints," in *Proceedings of the IEEE International Conference on Image Processing*, IEEE, New York, 1994, Vol. 1, pp. 31–35.

[330] L. I. Rudin, S. Osher, and E. Fatemi, "Nonlinear total variation based noise removal algorithms," *Physica D* **60**, 259–268 (1992).

[331] L. Rudin, S. Osher and E. Fatemi "Nonlinear total variation based noise removal algorithms" in *Proc. Modélisations Matématiques pour le Traitement d'Images*, INRIA, 1992, pp. 149–179.

[332] S. J. Ruuth and B. Merriman, "Convolution generated motion and generalized Huygen's principles for interface motion," UCLA Technical Report, ULCA, CA, 1998.

[333] G. Sapiro, "Color snakes," Hewlett-Packard Technical Report 113, September 1995.

[334] G. Sapiro "From active contours to anisotropic diffusion: connections between the basic PDE's in image processing," in *Proceedings of the IEEE International Conference on Image Processing*, IEEE, New York, 1996.

[335] G. Sapiro, "Color snakes," *Comput. Vis. Image Underst.* **68**, 247–253 (1997).

[336] G. Sapiro, "Vector-valued active contours," in *Proceedings of the Conference on Computer Vision and Pattern Recognition*, IEEE Publications, Los Alamitos, CA, 1996.

[337] G. Sapiro "Vector (self) snakes: a geometric framework for color, texture, and multiscale image segmentation," in *Proceedings of the IEEE International Conference on Image Processing*, IEEE, New York, 1996.

[338] G. Sapiro and A. M. Bruckstein, "The ubiquitous ellipse," *Acta Appl. Math.* **38**, 149–161 (1995).

[339] G. Sapiro and V. Caselles, "Histogram modification via differential equations," *J. Diff. Equat.* **135**, 238–268 (1997).

[340] G. Sapiro and V. Caselles, "Contrast enhancement via image evolution flows," *Graph. Models Image Process.* **59**, 407–416 (1997).

[341] G. Sapiro, A. Cohen, and A. M. Bruckstein, "A subdivision scheme for continuous scale B-splines and affine invariant progressive smoothing," *J. Math. Imag. Vis.* **7**, 23–40 (1997).

[342] G. Sapiro, R. Kimmel, and V. Caselles, "Object detection and measurements in medical images via geodesic deformable contours," in *Vision Geometry* IV, R. A. Melter, A. Y. Wu, F. L. Bookstein, and W. D. Green, eds., Proc. SPIE **2573**, xx–xx (1995).

[343] G. Sapiro, R. Kimmel, D. Shaked, B. B. Kimia, and A. M. Bruckstein, "Implementing continuous-scale morphology via curve evolution," *Pattern Recog.* **26** (1993).

[344] G. Sapiro and D. Ringach, "Anisotropic diffusion of multivalued images with applications to color filtering," *IEEE Trans. Image Process.* **5**, 1582–1586 (1996).

[345] G. Sapiro and A. Tannenbaum, "Area and length preserving geometric invariant scale-space," LIDS Technical Report 2200, MIT, Cambridge, MA, 1993.

[346] G. Sapiro and A. Tannenbaum, "Affine invariant scale-space," *Int. J. Comput. Vis.* **11**, 25–44 (1993).

[347] G. Sapiro and A. Tannenbaum, "On invariant curve evolution and image analysis," *Indiana Univ. Math. J.* **42**, 985–1009 (1993).

[348] G. Sapiro and A. Tannenbaum, "On affine plane curve evolution," *J. Function. Anal.* **119**, 79–120 (1994).

[349] G. Sapiro and A. Tannenbaum, "Area and length preserving geometric invariant scale-spaces," *IEEE Trans. Pattern Anal. Mach. Intell.* **17**, 67–72 (1995).

[350] I. J. Schoenberg, *Cardinal Spline Interpolation*, Society for Industrial and Applied Mathematics, Philadelphia, 1973.

[351] I. J. Schoenberg, *Selected Papers II*, C. de Boor, ed., Birkhauser, Boston, 1988.

[352] B. G. Schunck, "Image flow segmentation and estimation by constraint line clustering," *IEEE Trans. Pattern Anal. Mach. Intell.* **11**, 1010–1027 (1989).

[353] B. G. Schunck, "Robust computational vision," in *Proceedings of the International Workshop on Robust Computer Vision*, IEEE Publications, Los Alamitos, CA, 1990.

[354] L. Schwartz, *Analyse I. Theorie des Ensembles et Topologie*, Hermann, Paris, 1991.

[355] J. Serra, *Image Analysis and Mathematical Morphology*, Academic, New York, 1982.

[356] J. Serra, *Image Analysis and Mathematical Morphology: Theoretical Advances*, Academic, New York, 1988, Vol. 2.

[357] J. A. Sethian, "Curvature and the evolution of fronts," *Commun. Math. Phys.* **101**, 487–499 (1985).

[358] J. A. Sethian, "A review of recent numerical algorithms for hypersurfaces moving with curvature dependent flows," *J. Diff. Geom.* **31**, 131–161 (1989).

[359] J. Sethian, "Fast marching level set methods for three-dimensional photolithography development," in *Optical Microlithography IX*, G. E. Fuller, ed., Proc. SPIE **2726**, 262–272 (1996).

[360] J. A. Sethian, "A fast marching level-set method for monotonically advancing fronts," *Proc. Nat. Acad. Sci.* **93**, 1591–1595 (1996).

[361] J. A. Sethian, *Level Set Methods: Evolving Interfaces in Geometry, Fluid Mechanics, Computer Vision and Materials Sciences,* Cambridge University Press, Cambridge, U.K., 1996.

[362] J. Shah, "Segmentation by nonlinear diffusion, II" in *Proceedings of the Conference on Computer Vision and Pattern Recognition*, IEEE Publications, Los Alamitos, CA, 1992, pp. 644–647.

[363] J. Shah, "A common framework for curve evolution, segmentation, and anisotropic diffusion," in *Proceedings of the Conference on Computer Vision and Pattern Recognition*, IEEE Publications, Los Alamitos, CA, 1996.

[364] K. Siddiqi and B. Kimia, "Parts of visual form: computational aspects," *IEEE Trans. Pattern Anal. Mach. Intell.* **17**, 239–251 (1995).

[365] K. Siddiqi and B. Kimia, "A shock grammar for recognition," in *Proceedings of the Conference on Computer Vision and Pattern Recognition*, IEEE Publications, Los Alamitos, CA, 1996, pp. 507–513.

[366] K. Siddiqi, Berube, A. Tannenbaum, and S. Zucker, "Area and length minimizing flows for shape segmentation," *IEEE Trans. Image Process.* **7**, 433–443 (1998).

[367] E. Simoncelli and J. Portilla, "Texture characterization via joint statistics of wavelet coefficient magnitudes," in *Proceedings of the Fifth IEEE International Conference on Image Processing*, IEEE, New York, 1998.

[368] S. S. Sinha and B. G. Schunck, "A two-stage algorithm for discontinuity-preserving surface reconstruction," *IEEE Trans. Pattern Anal. Mach. Intell.* **14**, 36–55 (1992).

[369] J. Smoller, *Shock Waves and Reaction-Diffusion Equations*, Springer-Verlag, New York, 1983.

[370] N. Sochen, R. Kimmel, and R. Malladi, "A general framework for low-level vision," *IEEE Trans. Image Process.* **7**, 310–318 (1998).

[371] G. Sod, *Numerical Methods in Fluid Dynamics*, Cambridge University Press, New York, 1985.

[372] H. M. Soner, "Motion of a set by the curvature of its boundary," *J. Diff. Equat.* **101**, 313–372 (1993).

[373] H. M. Soner and P. E. Souganidis, "Singularities and uniqueness of cylindrically symmetric surfaces moving by mean curvature," *Comm. Partial Diff. Equat.* **18**, 859–894 (1993).

[374] M. Spivak, *A Comprehensive Introduction to Differential Geometry*, Publish or Perish, Berkeley, CA, 1979.

[375] G. Strang, *Introduction to Applied Mathematics*, Wellesley-Cambridge Press, Wellesley, MA, 1986.

[376] M. Struwe, "On the evolution of harmonic mappings of Riemannian surfaces," *Comm. Math. Helvetici* **60**, 558–581 (1985).

[377] M. Struwe, *Variational Methods*, Springer-Verlag, New York, 1990.

[378] M. Sussman, P. Smereka, and S. Osher, "A level-set method for computing solutions of incompressible two-phase flows," *J. Comput. Phys.* **114**, 146–159 (1994).

[379] R. Szeliski, D. Tonnesen, and D. Terzopoulos, "Modeling surfaces of arbitrary topology with dynamic particles," in *Proceedings of the Conference on Computer Vision and Pattern Recognition*, IEEE Publications, Los Alamitos, CA, 1993. pp. 82–87.

[380] B. Tang, G. Sapiro, and V. Caselles, "Diffusion of general data on non-flat manifolds via harmonic maps theory: the direction diffusion case," *Int. J. Comput. Vis.* **36**, 149–161 (2000).

[381] B. Tang, G. Sapiro, and V. Caselles, "Color image enhancement via chromaticity diffusion," IEEE Trans. Image Processing, to be published.

[382] G. Taubin, "Estimation of planar curves, surfaces, and nonplanar space curves defined by implicit equations with applications to edge and range image segmentation," *IEEE Trans. Pattern Anal. Mach. Intell.* **13**, 1115–1138 (1991).

[383] H. Tek and B. B. Kimia, "Image segmentation by reaction-diffusion bubbles," in *Proceedings of the International Conference on Computer Vision*, IEEE Publications, Los Alamitos, CA, 1995, pp. 156–162.

[384] P. Teo, G. Sapiro, and B. Wandell, "Creating connected representations of cortical gray matter for functional MRI visualization," *IEEE Trans. Med. Imag.* **16**, 852–863 (1997).

[385] P. Teo, G. Sapiro, and B. Wandell, "Anisotropic diffusion of posterior probabilities," in *Proceedings of the IEEE International Conference on Image Processing*, IEEE, New York, 1997.

[386] D. Terzopoulos and R. Szeliski, "Tracking with Kalman snakes," in *Active Vision*, A. Blake and A. Zisserman, eds., MIT, Cambridge, MA, 1992.

[387] D. Terzopoulos, A. Witkin, and M. Kass, "Constraints on deformable models: recovering 3D shape and nonrigid motions," *Artif. Intell.*, **36**, 91–123 (1988).

[388] A. P. Tirumalai, B. G. Schunck, and R. C. Jain, "Robust dynamic stereo for incremental disparity map refinement," in *Proceedings of the International Workshop on Robust Computer Vision*, IEEE Publications, Los Alamitos, CA, pp. 412–434.

[389] A. W. Toga, *Brain Warping*, Academic, New York, 1998.

[390] V. Torre and T. Poggio, "On edge detection," *IEEE Pattern Anal. Mach. Intell.* **8**, 147–163 (1986).

[391] P. E. Trahanias and A. N. Venetsanopoulos, "Vector directional filters – new class of multichannel image processing filters," *IEEE Trans. Image Process.* **2**, 528–534 (1993).

[392] P. E. Trahanias, D. Karakos, and A. N. Venetsanopoulos, "Directional processing of color images: theory and experimental results," *IEEE Trans. Image Process.* **5**, 868–880 (1996).

[393] J. N. Tsitsiklis, "Efficient algorithms for globally optimal trajectories," *IEEE Trans. Autom. Control* **40**, 1528–1538 (1995).

[394] G. Turk, "Generating synthetic textures using reaction-diffusion," *Comput. Graph.* **25(3)** 289–298 (July 1991).

[395] J. I. E. Urbas, "On the expansion of starshaped hypersurfaces by symmetric functions of their principal curvatures," *Math. Z.* **205**, 355–372 (1990).

[396] J. I. E. Urbas, "An expansion of convex hypersurfaces," *J. Diff. Geom.* **33**, 91–125 (1991).

[397] J. I. E. Urbas, "Correction to "an expansion of convex hypersurfaces"," *J. Diff. Geom.* **35**, 763–765 (1992).

[398] L. Vazquez, G. Sapiro, and G. Randall, "Segmenting neurons in electronic microscopy via geometric tracing," in *Proceedings of the IEEE International Conference on Image Processing*, IEEE, New York, 1998.

[399] C. Vogel and M. Oman, "Iterative methods for total variation denoising," *SIAM J. Sci. Stat. Comput.* **17**, 227–238 (1996).

[400] S. Walden, *The Ravished Image*, St. Martin's, Sunderland, MA, New York, 1985.

[401] B. Wandell, *Foundations of Vision*, Sinauer, 1995.

[402] J. Weickert, "Foundations and applications of nonlinear anisotropic diffusion filtering," *Z. Angew. Math. Mech.* **76**, 283–286 (1996).

[403] J. Weickert. "Non-linear diffusion scale-spaces: from the continuous to the discrete setting," in *Proceedings of Int. Conf. on Analysis and Optimization of Systems*, Springer, Berlin, 1996, pp. 111–118.
[404] J. Weickert, *Anisotropic Diffusion in Image Processing*, ECMI Series, Teubner-Verlag, Stuttgart, Germany, 1998.
[405] J. Weickert, "Coherence-enhancing diffusion of color images," *Image Vis. Comput.* **17**, 201–212 (1999).
[406] J. Weickert, B. Romeny, and M. Viergever, "Efficient and reliable schemes for nonlinear diffusion filtering," *IEEE Trans. Image Process.* **7**, 398–410 (1998).
[407] Y. Weiss and E. Adelson, "Perceptually organized EM: a framework for motion segmentation that combines information about form and motion," in *Proceedings of the International Conference on Computer Vision and Pattern Recognition*, IEEE Publications, Los Alamitos, CA, 1996, pp. 312–326.
[408] J. Weng and P. Cohen, "Robust motion and structure estimation using stereo vision," in *Proceedings of the International Workshop on Robust Computer Vision*, IEEE Publications, Los Alamitos, CA, 1990, pp. 367–388.
[409] R. T. Whitaker and G. Gerig, "Vector-valued diffusion," in Ref. [324].
[410] R. T. Whitaker, "Algorithms for implicit deformable models," in *Proceedings of the International Conference on Computer Vision*, IEEE Publications, Los Alamitos, CA, 1995, pp. 822–827.
[411] B. White, "Some recent developments in differential geometry," *Math. Intell.* **11**, 41–47 (1989).
[412] E. J. Wilczynski, *Projective Differential Geometry of Curves and Ruled Surfaces*, Teubner, Leipzig, 1906.
[413] A. P. Witkin, "Scale-space filtering," in *Proceedings of the International Joint Conference on Artificial Intelligence*, ACM Inc. New York, 1983, pp. 1019–1021.
[414] A. P. Witkin and M. Kass, "Reaction-diffusion textures," *Comput. Graph.* **25**(4) 299–308 (1991).
[415] G. Wyszecki and W. S. Stiles, *Color Science: Concepts and Methods, Qualitative Data and Formulae*, 2nd ed., Wiley, New York, 1982.
[416] A. Yezzi, "Modified curvature motion for image smoothing and enhancement," *IEEE Trans. Image Process.* **7**, 345–352 (1998).
[417] A. Yezzi, S. Kichenassamy, P. Olver, and A. Tannenbaum, "A gradient surface approach to 3D segmentation," in *Proceedings of 49th Information Science and Technology*, 1996.
[418] A. Yezzi, S. Kichenassamy, P. Olver, and A. Tannenbaum, "Geometric active contours for segmentation of medical imagery," *IEEE Trans. Med. Imag.* **16**, 199–210 (1997).
[419] Y. L. You and M. Kaveh, "A regularization approach to joint blur identification and image restoration," *IEEE Trans. Image Process.* **5**, 416–428 (1996).
[420] Y. L. You and M. Kaveh, "Anisotropic blind image restoration," in *Proceedings of the IEEE International Conference on Image Processing*, IEEE, New York, 1996.
[421] Y. L. You, W. Xu, A. Tannenbaum, and M. Kaveh, "Behavioral analysis of anisotropic diffusion in image processing," *IEEE Trans. Image Process.* **5**, 1539–1553 (1996).

[422] A. L. Yuille, "The creation of structure in dynamic shape," in *Proceedings of the Conference on Computer Vision and Pattern Recognition*, IEEE Publications, Los Alamitos, CA, 1988.

[423] A. L. Yuille and T. A. Poggio, "Scaling theorems for zero crossings," *IEEE Trans. Pattern Anal. Mach. Intell.* **8**, 15–25 (1986).

[424] W. P. Ziemer, *Weakly Differentiable Functions*, Springer-Verlag, New York, 1989.

[425] D. Zhang and M. Hebert, "Harmonic maps and their applications in surface matching," in *Proceedings of the Conference on Computer Vision and Pattern Recognition*, IEEE Publications, Los Alamitos, CA, 1999.

[426] H. K. Zhao, T. Chan, B. Merriman, and S. Osher, "A variational level-set approach to multiphase motion," *J. Comput. Phys.* **127**, 179–195 (1996).

[427] H. Zhao, S. Osher, B. Merriman, and M. Kang, "Implicit, nonparametric shape reconstruction from unorganized points using a variational level set method," UCLA CAM Report 98-7, UCLA, CA, February 1998.

[428] S. C. Zhu, T. S. Lee, and A. L. Yuille, "Region competition: Unifying snakes, region growing, energy/Bayes/MDL for multi-band image segmentation," in *Proceedings of the International Conference on Computer Vision*, IEEE Publications, Los Alamitos, CA, 1995, pp. 416–423.

[429] S. C. Zhu and D. Mumford, "GRADE: Gibbs reaction and diffusion equations: Sixth Int. Conf. Computer Vision, pp. 847–854, Bombay, 1998, IEEE Publications, Los Alamitos, CA.

[430] S. W. Zucker and R. A. Hummel, "A three-dimensional edge operator," *IEEE Trans. Pattern Anal. Mach. Intell.* **3**, 324–331 (1981).

Index